Remove Carbon

Politics and Po

into Acti

ements in

Innovate!

Health and Equity

Plan for Solving

Transp

Electrify

onize the Grid Fix

Trans

Speed & Scale

Speed & Scale

An Action Plan
for Solving Our
Climate Crisis Now

John Doerr
Portfolio · Penguin

Portfolio / Penguin

An imprint of
Penguin Random House LLC
penguinrandomhouse.com

Grateful acknowledgment is made for permission to reprint the following:
"Earthrise." Text copyright © 2018 by Amanda Gorman. Reprinted by permission of Writers House LLC acting as agent for the author/illustrator.

Image credits may be found on pages 411–416

ISBN: 9780593420478 (hardcover)
ISBN: 9780593420485 (ebook)
ISBN: 9780593421345 (international edition)

Printed in the United States of America
1st Printing

Book design by Order

The author's proceeds from this book will be donated to global climate justice work.

For Ann, Mary,
and Esther, and the
wonder of their
unconditional love.

By	John Doerr
With	Ryan Panchadsaram
Team	Alix Burns, Jeffrey Coplon, Justin Gillis, Anjali Grover, Quinn Marvin, and Evan Schwartz
In consultation with	Climate Champions, The Climate Reality Project, Countdown, Energy Innovation, Environmental Defense Fund, and World Resources Institute
Editor	Trish Daly
Design	Order: Emily Klaebe, Megan Nardini, and Jesse Reed
Carbon footprint	Negative (-1kg CO_2e)
Proceeds	A portion of the proceeds for the book will be donated to climate justice work globally.

In conversation with

Chris Anderson, TED

Mary Barra, General Motors

Jeff Bezos,
Amazon & Bezos Earth Fund

David Blood, Generation
Investment Management

Kate Brandt, Alphabet

Ethan Brown, Beyond Meat

Margot Brown, Environmental
Defense Fund

Amol Deshpande,
Farmers Business Network

Christiana Figueres, Global Optimism

Larry Fink, Blackrock

Taylor Francis, Watershed

Bill Gates, Breakthrough Energy

Jonah Goldman, Breakthrough Energy

Al Gore, The Climate Reality Project

Patrick Graichen,
Agora Energiewende

Steve Hamburg,
Environmental Defense Fund

Hal Harvey, Energy Innovation

Per Heggenes, IKEA Foundation

Kara Hurst, Amazon

Safeena Husain, Educate Girls

Lynn Jurich, Sunrun

Nat Keohane, Environmental
Defense Fund

John Kerry, U.S. State Department

Jennifer Kitt,
Climate Leadership Initiative

Badri Kothandaraman, Enphase

Fred Krupp,
Environmental Defense Fund

Kelly Levin, World Resources Institute

Lindsay Levin, Future Stewards

Dawn Lippert, Elemental Excelerator

Amory Lovins, RMI

Megan Mahajan, Energy Innovation

Doug McMillon, Walmart

Bruce Nilles, Climate Imperative

Robbie Orvis, Energy Innovation

Sundar Pichai, Alphabet

Ryan Popple, Proterra

Henrik Poulsen, Ørsted

Laurene Powell Jobs, Emerson
Collective

Nan Ransohoff, Stripe

Carmichael Roberts,
Breakthrough Energy

Matt Rogers, Incite

Anumita Roy Chowdhury,
Centre for Science and Environment

Jonathan Silver, Guggenheim Partners

Jagdeep Singh, Quantumscape

Andrew Steer, Bezos Earth Fund

Eric Toone, Breakthrough Energy

Eric Trusiewicz, Breakthrough Energy

Jan Van Dokkum,
Imperative Science Ventures

Brian Von Herzen, Climate Foundation

James Wakibia, The Flipflopi Project

Tensie Whelan, Rainforest Alliance

Table of Contents

Prologue

Prologue

"I'm scared, and I'm angry."

In 2006, I hosted a dinner after a screening of *An Inconvenient Truth*, former vice president Al Gore's seminal documentary on the climate crisis. We went around the table for everyone's reaction to the film's urgent message. When it came to my fifteen-year-old daughter, Mary, she declared with her typical candor: "I'm scared, and I'm angry." Then she added, "Dad, your generation created this problem. You better fix it."

The conversation stopped cold. All eyes turned to me. I didn't know what to say.

As a venture capitalist, my job is to find big opportunities, target big challenges, and invest in big solutions. I was best known for backing companies like Google and Amazon early on. But the environmental crisis dwarfed any challenge I'd ever seen. Eugene Kleiner, the late cofounder of Kleiner Perkins, the Silicon Valley firm I've been with for forty years, left behind a set of twelve laws that have stood the test of time. The first goes as follows: *No matter how groundbreaking a new technology may seem, make sure customers actually want it*. But this problem led me to invoke a lesser-known Kleiner law: *There is a time when panic is the appropriate response*.

That time had come. We could no longer afford to underestimate our climate emergency. To avert irreversible, catastrophic consequences, we needed to act urgently and decisively. For me, that evening changed everything.

My partners and I made climate a top priority. We got serious about investing in clean and sustainable technologies—or "cleantech," as they're known in Silicon Valley. We even brought in Al Gore as the firm's newest partner. But despite Al's excellent company, my journey into the world of zero-emissions investing was pretty lonely at first. After the iPhone debuted in 2007, Steve Jobs invited us to launch our iFund for mobile apps from Apple's headquarters. We were hearing great pitches from mobile app startups; I could see opportunities left and right.

So why commit a chunk of capital to the uncharted territory of solar panels, electric car batteries, and meatless proteins? Because it seemed like the right thing to do, for the firm and for the planet. I thought the cleantech market was a monster in the making. I believed we could do well by doing good.

We pursued mobile apps and climate ventures at the same time, despite doubters on both fronts. Our mobile app investments gave us a burst of quick wins. Our climate investments were slower out of the gate, and many of them failed. It's hard to build a durable company under any circumstances, and doubly hard to build one to take on the climate crisis.

Kleiner Perkins got beaten up in the press. But with patience and persistence, we stood by our founders. By 2019, our surviving cleantech investments began to hit one home run after the next. Our $1 billion in green venture investments is now worth $3 billion.

But we have no time for a victory lap. As the years roll by, the climate clock keeps ticking. Atmospheric carbon already exceeds the upper limit for climate stability. At our current pace, we will blow past 1.5 degrees Celsius (or 2.7 degrees Fahrenheit) over the Earth's preindustrial mean temperatures—the threshold, scientists say, for severe planetary damage. The effects of runaway global warming are already plain to see: devastating hurricanes, biblical flooding, uncontrollable wildfires, killer heat waves, and extreme droughts.

I must warn you up front: we're not cutting our emissions fast enough to outrun the damage on our doorstep. I said this in 2007, and I say it today: what we're doing is not nearly enough. Unless we course correct with urgent speed and at massive scale, we'll be staring at a doomsday scenario. The melting polar ice caps will drown coastal cities. Failed crops will lead to widespread famine. By midcentury, a billion souls worldwide could be climate refugees.

Fortunately, we have a powerful ally in this fight: innovation. Over the past fifteen years, prices for solar and wind power have plunged 90 percent. Clean energy sources are growing faster than anyone expected. Batteries are expanding the range of electrified vehicles at an ever lower cost. Greater energy efficiency has sharply reduced greenhouse gas emissions.

While a good many solutions are in hand, their deployment is nowhere near where it needs to be. We'll need massive investment and robust policy to make these innovations more affordable. We need to scale the ones we have—immediately—and invent the ones we still need. In short, we need both the now and the new.

So where's the plan for getting the job done? Frankly, that's what's been missing: *an actionable plan*. Sure, there are lots of ways on paper to get to net-zero carbon emissions, the point where we add no more greenhouse gas into the atmosphere than we can remove. But lists of goals are not plans. A long menu of options, however excellent they might be, is not a plan. Anger and despair aren't plans; neither are hopes and dreams.

More than anything, we need a clear course of action. That's why I've written this book. With help from some of the world's foremost experts in climate and cleantech, I created *Speed & Scale* to show precisely how we can drive greenhouse emissions to net zero by 2050. My hope is to build on the hard-earned triumphs and lessons of our climate pioneers and heroes, many of them hailed in these pages. They're the ones who blaze new trails by executing better and smarter.

A plan is only as good as its implementation. To achieve this monumental mission, we'll need to hold ourselves accountable every step of the way. That's the great lesson I learned from my mentor, Andy Grove, the legendary CEO of Intel. It's a mantra I've seen proven over and again: *Ideas are easy. Execution is everything.*

To execute a plan, we need the right tools. In my previous book, *Measure What Matters*, I outlined a simple but powerful goal-setting protocol that Andy Grove invented at Intel. Known as OKRs, or Objectives and Key Results, they guide organizations to focus on a few essential targets, to align at every level, to stretch for ambitious results, and to track their progress as they go—to measure what matters.

Now I'm proposing we apply OKRs to solve the climate crisis, the greatest challenge of our lifetimes. But before going all in (and this is an all-or-nothing proposition), we must answer three basic questions.

Do we have enough time?

We hope so, but we're fast running out of it.

Do we have much margin for error?

No, we don't. Not anymore.

Do we have enough money?

Not yet. Investors and governments are stepping up. But we'll need a lot more funding, from both public and private sectors, to develop and scale technologies for a clean economy. Most of all, we'll need to divert the trillions spent on dirty energy over to clean energy options, and use that energy more efficiently.

The data is clear. The moment is now. I am committed to using my time, my resources, and whatever knowledge I have to work with you to build a net-zero future. I invite you to join our effort at speedandscale.com. To put our plan into action, we need all hands on deck. Above all, we'll need to execute our plan with unprecedented speed and unprecedented scale. That's what matters most.

I've written this book for leaders of all kinds, for anyone anywhere who can move others to act with them. It's for entrepreneurs and business leaders who can mobilize the power of markets. For political and policy leaders willing to fight for our planet. For citizens and community leaders who can press their elected officials. And, not least, for leaders from the rising generation, like Greta Thunberg and Varshini Prakash, who will be showing the way to 2050 and beyond.

Speed & Scale is written for the leader inside you. I'm not here to prod consumers to change their behavior. Individual actions are both needed and expected, but they won't be nearly enough to reach this huge goal. Only concerted, collective, *global* action can get us past the finish line in time.

I might seem an unlikely advocate for this call to action. I'm an American, a citizen of the biggest historic polluter on Earth. I am an affluent white man, born in St. Louis, Missouri, from a generation whose negligence helped create this problem in the first place.

Yet from the home office where I wrote this book, not far from San Francisco, I've looked out over the hills and seen the bright orange skies of the wildfires, the signposts of drought and devastation. They've devoured millions of acres of forests each year in California alone, spitting back more carbon dioxide into the atmosphere than all the state's emissions from fossil fuels. It is the most vicious of circles, and I cannot stand idly by. Whatever my flaws as a messenger, I am impelled to act.

In my fifteen years on this path, I've collected my share of scar tissue. Cleantech ventures demand more money, more guts, more time, and more perseverance than just about anything else. Their horizons stretch longer than most investors can stomach. The washouts are acutely painful. But the success stories—however few and far between—are worth all the setbacks and then some. These companies are more than turning a profit. They are helping to heal the Earth.

In large part, this book is a collection of stories from my own trek through these minefields and those of dozens of other climate leaders, many of whom I've been proud to back as an investor. Their behind-the-scenes narratives illustrate the potential of our plan to reach net zero by 2050, and the hurdles we'll need to overcome. My hope is that they'll offer the reader some respite from the more technical, data-drenched sections. In making this journey, I've been inspired by both the problem and the people. I hope you will be, too.

Entrepreneurs are those hardy individuals who do more with less than anyone thinks possible—and do it faster than anyone thinks possible. Today, bold risk-takers are innovating like mad as they rewrite the rules to avert a climate apocalypse. We need to bottle their entrepreneurial energy and distribute it as widely as we can—to governments, companies, and communities worldwide.

A plan is not a guarantee. A timely transition to a net-zero future is no sure thing. But though I may be less optimistic than some, consider me hopeful—and impatient. With the right tools and technology, with precision-honed policies, and most of all with science on our side, we still have a fighting chance.

But the time is *now*.

—John Doerr
July 2021

What's the Plan?

Introduction

What's the Plan?

On a chilly day in March 1942, three months after the attack on Pearl Harbor, President Franklin D. Roosevelt met at the White House with Henry "Hap" Arnold, commanding general of the U.S. Army Air Forces. The agenda had one item: Roosevelt's plan to win World War II. It was a challenge of historic proportions– especially at that moment, when things looked especially grim. FDR might have expounded upon geopolitics or cataloged every conceivable battlefront. He might have plunged into complexity and intricate detail. Instead, the president grabbed a cocktail napkin and sketched a three-point plan stripped to its bare bones:

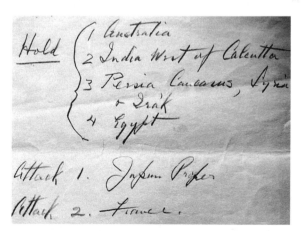

In March 1942, FDR outlined his plan for winning World War II on this cocktail napkin.

1. Hold four key territories

2. Attack Japan

3. Defeat the Nazis in occupied France

The points were focused, action-oriented, and definitive. Roosevelt's napkin provided what the nation's military leadership desperately needed: clarity.

Not coincidentally, the plan wound up succeeding. After the meeting ended, General Arnold took FDR's napkin with him back to the Pentagon. It was kept top secret through D-Day and remained classified for decades. In 2000, the entrepreneur and book collector Jay Walker bought it at auction for display in his library.

"Whenever anyone tells me a problem is too complex to solve with a clear, simple plan," Walker says, "I show them the napkin. Is the problem you are trying to solve really more complicated than World War II?"

What are greenhouse gases?

They're the gases in our atmosphere that absorb heat. The sun radiates energy; you can feel it when you step out of the shade. Some of it is absorbed by the Earth and is radiated back into the air. Nitrogen and oxygen, the predominant gases in the atmosphere, allow this thermal energy to pass freely into space. But greenhouse gases are more loosely bound, complex molecules that trap a portion of the energy and radiate it back again to the Earth's surface. Hence the "greenhouse effect," the extra warming that adds to direct heating from the sun.

We need greenhouse gases, in moderation; warmth is vital to life. But too much of them is a problem. Carbon dioxide, the most abundant, is odorless, invisible, and stubbornly enduring. Once released from a tailpipe or a chimney, it stays in the atmosphere for centuries.

Methane is a different beast. The primary ingredient in natural gas, it heats our homes and lights our stoves. Cows release it in abundance. Though methane endures in the atmosphere far more briefly than CO_2, it is many times more potent for the short-term trapping of heat.

Other gases heat the planet too. They include nitrous oxide, a by-product of fertilizers, as well as some common refrigerants. All these greenhouse gases can be calibrated by a single measure: carbon dioxide equivalents, or CO_2e. This umbrella metric accounts for the gases' uneven warming impacts and makes for more meaningful comparisons.

How much greenhouse gas is in our atmosphere?

In the preindustrial era, every million molecules of air contained around 283 molecules of CO_2e. In 2018, the Intergovernmental Panel on Climate Change warned that we needed to keep CO_2e below 485 parts per million. The problem is that we've already crossed that threshold and are now at more than 500 parts per million. (This data comes from eighty collection sites around the world and is rigorously measured by the National Oceanic and Atmospheric Administration.)

To stave off a climate catastrophe, our goal must be to prevent any additional greenhouse gas accumulation, drive CO_2e back under 430 parts per million, and keep it there.

Carbon dioxide in the atmosphere has risen dramatically over the last 200 years
Annual concentration of carbon dioxide (CO_2)
measured in parts per million (ppm).

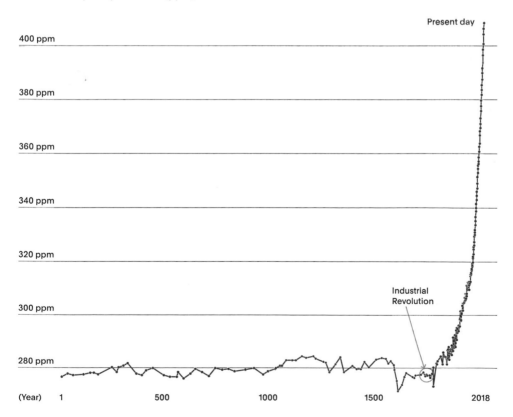

Adapted from data and visuals by NOAA/ESRL (2018) and Our World in Data

Introduction

When assessed on a planetary scale, CO_2e is typically measured by the gigaton, or one billion metric tons—the weight of 10,000 fully loaded aircraft carriers. In terms of emissions, burning 110 gallons of gasoline emits one ton of CO_2e. Powering 12,000 homes with fossil fuels for one year emits 100,000 tons of CO_2e. Driving 200,000 gasoline-fueled cars an average of 12,000 miles apiece emits 1 million tons of CO_2e. Operating 220 coal-fired power plants for one year emits a gigaton of CO_2e. The annual sum of all human-caused emissions is 59 gigatons of CO_2e.

Why do these numbers matter?

Unabated greenhouse gas emissions have created runaway warming on Earth. All told, the average global temperature has risen by about 1 degree Celsius—or almost 2 degrees Fahrenheit—since 1880. Though they may not sound like a whole lot, these small numbers have a massive impact.

Our climate crisis has been a long time coming. Since the dawn of the Industrial Revolution, the burning of fossil fuels and other human activities have emitted more than 1.6 trillion tons of greenhouse gases into the atmosphere—more than half of those emissions since 1990. Many of us are part of the problem—anyone who's ever traveled by car or by plane, had a cheeseburger for lunch, or enjoyed the comforts of a well-heated home.

Only drastic cuts to emissions—*before* they enter the atmosphere—can begin to prevent ecosystem collapse and an uninhabitable Earth. Consider these dire projections for the year 2100:

We will far exceed our 1.5°C limit.

Policy scenarios, emissions, and temperature range projections

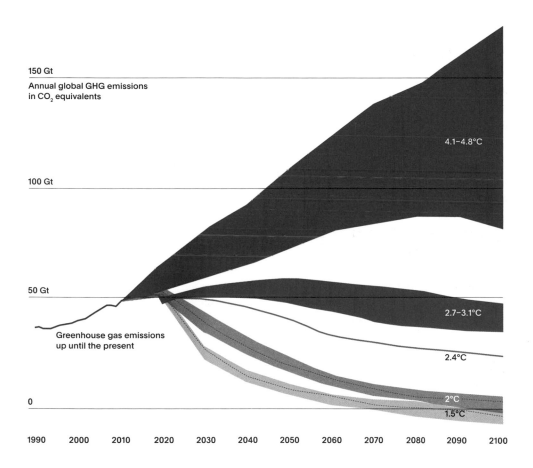

No climate policies
4.1–4.8°C

Expected warming
without current
climate reduction
policies

Current policies
2.7–3.1°C

Expected warming
with current
policies
in place

Pledges & targets
2.4°C

Expected warming
if all countries
deliver on
reduction pledges

2°C Pathways

1.5°C Pathways

150 Gt

Annual global GHG emissions
in CO_2 equivalents

4.1–4.8°C

100 Gt

50 Gt

2.7–3.1°C

Greenhouse gas emissions
up until the present

2.4°C

0

2°C

1.5°C

1990 2000 2010 2020 2030 2040 2050 2060 2070 2080 2090 2100

Adapted from data and visuals by Climate Tracker and Our World in Data

Introduction

Based on numerous studies, 4 degrees Celsius of warming would devastate the global economy, especially in the Southern Hemisphere. The scale of the disaster would far surpass the 2008 financial crisis, and it would come to stay. We would enter a permanent climate depression.

But frankly, warnings of this sort aren't likely to move us onto a planet-saving path. Eighty-year projections are too remote for the human brain. A few degrees of warming sound too innocuous to be ominous. Here's the biggest obstacle: without a road map, people are slow to commit to change. Real change requires a clear and achievable plan.

Can you show me the plan? That's the question I began asking after investing hundreds of millions of dollars of venture capital into all kinds of climate solutions. As you and I know by now, a portfolio of solutions isn't a plan. The Beatles noted the difference in "Revolution": "You say you got a real solution," they sang. "We'd all love to see the plan."

And so: Just how do we keep the climate crisis from becoming a climate catastrophe? What's the focused, actionable, measurable plan that can actually stave off this looming disaster? Where's our cocktail napkin when we need it?

I've been troubled by these questions for some time now. Over the last fifteen years, I've read everything I can on this endlessly complex subject. I've consulted with world-class authorities on the theory and practice of fighting climate change. The more I learned, the more I worried. In 2009, I shared my concern with a U.S. Senate committee. The energy tech revolution, I said, was being hamstrung by bad federal policy and insufficient funding for research and development.

The following year, to build a cleantech innovation network, my partners and I organized a workshop on the climate crisis. We brought together Nobel Prize-winner and then Secretary of Energy Steven Chu and some of the world's top climate and economic thinkers, including Al Gore, Sally Benson, Abby Cohen, Tom Friedman, Hal Harvey, and Amory Lovins.

As we began to grasp the scale of the problem, Kleiner Perkins expanded our investments in cleantech from about 10 percent to nearly half our portfolio. At the same time, I began to advocate in Sacramento for vanguard climate and energy policies in California. I gave an emotional TED talk on climate change and investing, urging others to join the crusade.

As a founding member of the American Energy Innovation Council, I worked to urge the U.S. government to increase its funding for climate research and development. With some like-minded advocates, I visited labs and factories in Brazil to see how sugarcane became biofuel. We journeyed to solar thermal farms in the Mojave Desert. We hiked through Amazon rainforests and climbed up California wind turbines. We met with President Obama in the White House. Our doggedness paid off with some initial federal funding for a new federal agency called ARPA-E, the Advanced Research Projects Agency for Energy, and a basket of loan guarantees for early-stage companies.

Internationally, the Paris Agreement of 2015 rallied the global community to declare their own emissions reduction targets—a historic advance. But as John Kerry, the United States' climate envoy, has observed, these commitments are not adequate to the task. Even if the pledges made in Paris were met in full, they would result in a much warmer world—by 3 degrees Celsius or more by 2100, well past the tipping point for global catastrophe.

In my hunt for a comprehensive plan, I've pored over more analyses of our options than I can count, from the rigorously scientific to the heartily optimistic to the deeply bleak and gloomy. It's not hard to get confused or overwhelmed. But here is what I've learned from helping generations of new companies succeed: **to execute a big plan, you need clear and measurable goals.** My first book showed how OKRs, Objectives and Key Results, can drive success for all sorts of organizations, from Google to the Bill and Melinda Gates Foundation, from modest startups to titans of the Fortune 500. I believe they can work for global emergencies too.

To learn more about OKRs, see resources at whatmatters.com.

OKRs stand for Objectives and Key Results. They address the two critical facets of any goal worth achieving: the "what" and the "how." Objectives (Os) are simply *what* you aim to accomplish. Key Results (KRs) tell us *how* we'll get the objectives done; typically they cascade down to more granular goals.

A well-formed objective is significant, action-oriented, durable, and inspirational. Each objective is supported by carefully chosen and crafted key results. Strong key results are specific, timebound, aggressive (yet realistic), and most of all, measurable and verifiable.

OKRs are not meant to be the sum of all tasks. Rather, they focus on what's most important, the handful of essential action steps for a given pursuit. They enable us to track our progress as we go. And they're designed to aim high—to stretch for ambitious but still reachable goals.

Introduction

Net zero is our goal line. The "net" signifies that there's no plausible route to zero through emissions reductions alone. We'll also need to lean on nature and technology to remove and store emissions from hard-to-abate sources. But to be clear, we cannot fall back on future atmospheric cleansing as an excuse to keep on burning fossil fuels today. The primary work ahead of us is to cut emissions.

Speed & Scale's top-line OKR is to reach net-zero emissions by 2050—and to get halfway there by 2030, a critical milestone. In the face of such an enormous challenge, Objectives and Key Results will keep us clear-eyed and practical. They'll prevent us from promising pie in the sky. They'll save us from distractions by bright and shiny objects, the seemingly brilliant innovations that can't yet compete on cost or work at scale. By holding us accountable to our own quantitative targets, they make us less tempted to rely on the slender reed of hope. We'll focus ruthlessly on the biggest, most fruitful opportunities, the ones that will get us to net zero on time.

How our greenhouse gas emissions add up

24 Gt	**12** Gt	**9** Gt	**8** Gt	**6** Gt
Energy	Industry	Agriculture	Transportation	Nature
41%	20%	15%	14%	10%

59 Gt
Total
100%

As I've noted, the world's greenhouse gas emissions amount to 59 gigatons of CO_2e per year. Business as usual will take us north of that figure, somewhere between 65 and 90 gigatons every year. (But if it's business as usual, we'll all be going out of business.) By standards of both logic and fairness, the nations responsible for the lion's share of the planet's emissions should be the first to cut them aggressively. As the developed world leads by example, it will drive down the costs of clean energy for the developing world as well.

Many lower "business-as-usual" projections assume that current policies are maintained. But as we've seen in the United States, there's no guarantee policies will stay intact.

Our targets conform with calculations by the Intergovernmental Panel on Climate Change, the United Nations Environment Programme, and the delegates who negotiated the Paris Agreement. All three bodies computed emissions levels that correlate with scenarios for warming of 1.5 degrees, 1.8 degrees, and 2 degrees Celsius over preindustrial levels. To simplify our objective, Speed & Scale has aligned its key results with the most ambitious target, a warming of no more than 1.5 degrees Celsius. That's our best chance to avert climate calamity—though scientists agree it is no sure thing. Which is all the more reason to move quickly.

So here is our plan: the Speed & Scale Plan to solve the climate crisis. Like FDR's pencil sketch, it contains but a handful of words. It barely hints at how hard our objectives will be to achieve. It truly could fit on a cocktail napkin:

Introduction

The first six items support our top-line objective: to solve the climate crisis by getting to net zero no later than 2050. All six are intricate worlds unto themselves, and each has its own chapter. They compose Part I of the book, "Zero Out Emissions." Beneath them you'll find a set of "accelerants" to speed the pace of climate action. That's Part II, "Accelerate the Transition." It contains four chapters, one for each accelerant.

To shape our key results, we've enlisted a team of policy experts, entrepreneurs, scientists, and other climate leaders who have given generously of their time and thoughtfulness. We've been inspired by the solutions and pathways recommended by the authorities at Project Drawdown, the Environmental Defense Fund, Energy Innovation, World Resources Institute, RMI (formerly Rocky Mountain Institute), and Breakthrough Energy.

In the spirit of FDR, we aim to be clear and concise:

By "electrify transportation," we mean switching from gasoline and diesel engines to fleets of plug-in electric bikes, cars, trucks, and buses (chapter 1)

By "decarbonize the grid," we mean replacing fossil fuels with solar, wind, and other zero-emissions sources (chapter 2)

By "fix food," we mean restoring our carbon-rich topsoil, adopting better fertilization practices, motivating consumers to eat more lower-emissions proteins and less beef, and reducing food waste (chapter 3)

By "protect nature," we're referring to interventions and protections for forests, soil, and oceans (chapter 4)

By "clean up industry," we mean that all manufacturing— particularly cement and steel—must sharply lower their carbon emissions (chapter 5)

By "remove carbon," we're saying we must remove carbon dioxide from the atmosphere and store it for the long term, using both natural and engineered solutions (chapter 6)

As for our four accelerants, we'll expedite these solutions by doing the following:

→ Implementing vital public policies (chapter 7)
→ Turning movements into meaningful climate action (chapter 8)
→ Inventing and scaling powerful technologies (chapter 9)
→ Deploying capital at scale (chapter 10)

Since we cannot afford to fail, each of these goals comes with its own set of measurable key results. We'll track our progress toward these milestones to show how we're doing and whether we need to pick up our pace or course correct.

Though I believe all of our goals are achievable, none are sure things. We may overachieve on some of our key results and fall short on others. And that's all right—as long as we land at net zero by 2050. That is our debt to future generations; it must be paid in full.

Our targets are informed by the work of a worldwide network of valiant climate researchers. For too long they have been voices crying in the wilderness; only in this eleventh hour have those with power, influence, and money begun to listen. Their work guides our estimates of the sources of carbon emissions, as of where and how the needed cuts might be made.

Speed & Scale: Countdown to net zero

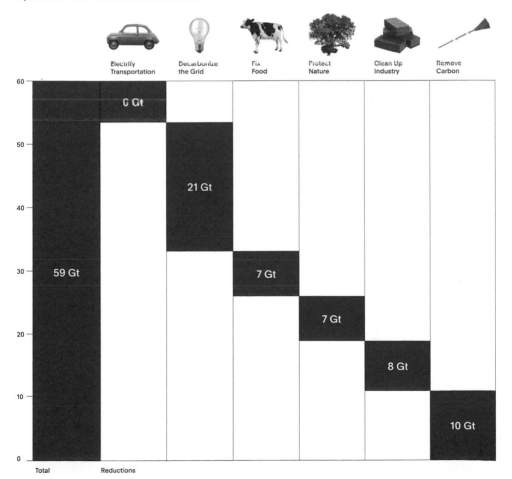

Introduction

In fairness, we must add a caveat. Though we know to a high degree
of accuracy how much greenhouse gas is in the atmosphere, our
calculations of present-day emissions—by country and by industry—
involve margins of uncertainty. Our targets for cutting those emissions represent one earnest view of how to tackle the crisis before
us, no more and no less.

In business, as I've learned, there are often several right answers.
The same holds true for public policy and climate solutions. The
Speed & Scale Plan isn't the only "right" plan for this emergency,
but we believe it strikes a practical balance. It is wildly ambitious
yet rooted in hard realities. In many ways, it's the ultimate application
of Objectives and Key Results. I've yet to see a bolder goal than
getting to net zero.

We're in peril, to say the least, and what's infuriating is that it didn't
have to be this way. Forty years ago, an Exxon scientist named James
Black connected the dots between fossil fuels, rising carbon levels, and
global warming.

VUGRAPH 18

SUMMARY

I. CO_2 RELEASE MOST LIKELY SOURCE OF INADVERTENT CLIMATE MODIFICATION.

II. PREVAILING OPINION ATTRIBUTES CO_2 INCREASE TO FOSSIL FUEL COMBUSTION.

III. DOUBLING CO_2 COULD INCREASE AVERAGE GLOBAL TEMPERATURE 1°C TO 3°C BY 2050 A.D. (10°C PREDICTED AT POLES).

IV. MORE RESEARCH IS NEEDED ON MOST ASPECTS OF GREENHOUSE EFFECT

V. 5-10 YR. TIME WINDOW TO GET NECESSARY INFORMATION

VI. MAJOR RESEARCH EFFORT BEING CONSIDERED BY DOE

Excerpt from internal Exxon presentation, 1978.

At the time, we might have gotten out of this jam with incremental
changes—say, emissions cuts of 10 percent or so per decade. But the
scientist's analysis was ignored and further research suppressed, as
Exxon (and later ExxonMobil, after a merger) went on to lead the

charge of climate change denial. Twenty-odd years ago, when Al Gore conceded and George W. Bush became president, we might have gotten away with aggressive actions to cut around 25 percent per decade.

But now we are out of time, and half measures will not suffice. To beat the odds and limit warming to 1.5 degrees, according to the IPCC, we cannot emit more than 400 gigatons. That's our carbon budget—and we are on pace to spend it within this decade. Nothing less than drastic, immediate action will do. We need to cut 50 percent of our emissions by 2030, and the rest by 2050. Because whether we're ready or not, irreversible climate damage is brewing.

President Obama framed the challenge before us with his customary eloquence: "We are the first generation to feel the effect of climate change and the last generation who can do something about it."

Let's consider the strategies that can unlock a net-zero future. In order of climate impact, they are:

1. CUT (slash emissions)

2. CONSERVE (get more efficient)

3. REMOVE (clean up what's left)

Avoiding greenhouse gas emissions in the first place—say, by electrifying transportation or decarbonizing the grid—remains our primary course of action. It's both the fastest and most reliable way to cut gigatons of greenhouse gases today. Next comes energy efficiency, which gets us more output per energy input.

The third strategy is the natural or technological removal and long-term storage of carbon. It addresses hard-to-avoid emissions, especially in transportation, industry, and agriculture. Even with the globe's best and most focused efforts, these emissions will be with us for the foreseeable future. We must add, however, that carbon dioxide removal is no substitute for avoidance or efficiency, but rather a critical complement. We'll need to pursue all three paths simultaneously.

The Speed & Scale Plan challenges leaders everywhere, in both government and business, to be guided by a deep sense of climate justice and equity. To ensure a just transition, we must acknowledge the differences between developing and developed countries. There are vast disparities in their economic ability to move away from fossil fuels and in how fast they can go. We must be mindful of the millions of everyday workers whose livelihoods are tied to fossil fuels. They deserve retraining and quality job opportunities in our green future.

Introduction

Finally, we must recognize climate-related inequities within countries as well. Fossil fuel pollution has a disproportionate impact on poor communities and communities of color. They're the least responsible for the crisis and least able to guard against its ravages. Those who are most damaged by carbon-intensive industries must receive their share of the benefits of the energy transition already under way.

Clean technologies can contribute to a fresh start. As coal-burning power plants close, we should seize the opportunity to revive down-wind communities and transition workers to clean energy jobs. We must stop dumping carbon, methane, and other greenhouse gases into our precious atmosphere as though it were a free and open sewer.

Keep in mind that our plan is designed to cut emissions to the bone. It is not intended to help us adapt to an ever-warmer world. Yes, climate change is already well under way. And yes, we do need to invest in securing our cities and farmlands against more ferocious hurricanes, cyclones, wildfires, floods, and droughts. But the more we do today to limit global warming, the less drastic our adaptations will need to be.

When asked why he robbed banks, Willie Sutton supposedly said, "Because that's where the money is." We need to go where the emissions are. **We must go for the gigatons.** That means tracking the twenty top emitters, the ones responsible for 80 percent of the world's greenhouse gases. It means targeting in particular the top five, which account for nearly two thirds: China, the United States, the European Union (plus the United Kingdom), India, and Russia.

As of June 2021, at least fourteen countries—including Germany, Canada, the United Kingdom, and France—had a law or had proposed legislation to ratchet down their carbon emissions to net zero by 2050. The problem is that all of these countries combined account for only about 17 percent of total global emissions.

Only recently have the very largest emitters begun to signal their ambitions. The Biden administration's plan for climate action calls for net zero by 2050, an impressive leap beyond previous U.S. policy. The European Union has committed to do the same. China has declared a national commitment to get there by 2060—ten years too late, in our view, but at least a basis for negotiation. India and Russia have yet to make any firm net-zero pledge. Still, there is finally some grounds for hope on the international front. What remains is the all-important question of follow-through.

Mitigating decades of reckless carbon emissions won't come cheap. But we know it will be vastly more expensive to defer aggressive action than to start today. In the eloquent words of the internationally recognized climate policy expert Hal Harvey: It is now cheaper to save the Earth than to ruin it. While betting on clean technology was once seen as risky or rash, it's beginning to be seen as the express route to economic growth.

As I write this, the coronavirus crisis is still with us, with horrific, unacceptable death tolls in many parts of the world. The pandemic reminds us how vital it is to act *before* disaster strikes. The same goes for our climate crisis, where every ounce of prevention will save us unimaginable pain.

In 2020, in the midst of a pandemic, life as we know it virtually ground to a halt. Yet all the restrictions forced by COVID took just 2.3 gigatons of carbon emissions off the top, around 6 percent of the world's annual greenhouse gas emissions. Soon enough, even that small reduction was gone; carbon pollution came roaring back. Short-term deprivation may help slow the spread of a plague, but it won't solve the climate crisis.

The task before us is clear. The need to act has never been more urgent. If we reach net zero in time, we can be justly proud of the planet we pass on to our children and future generations.

So let's do this, with speed and scale.

Electrify Transportation

Chapter 1

Electrify Transportation

There's an old venture capital axiom: *Never invest in anything with wheels.* In 2007, not long after committing to cleantech investing, Kleiner Perkins considered breaking that rule. Should we back an electric car company? Smart people warned me away. In a little more than a century, more than a thousand car companies have been launched, and nearly all are gone. Many failed spectacularly. Do you remember the DeLorean?

Kleiner was deep in discussions with a brilliant designer who'd made his mark at Aston Martin and BMW. Henrik Fisker hailed from Denmark but lived in Los Angeles. In our first meeting, he sketched a strategic plan to produce an electric car for luxury buyers, then to move down the price curve toward the middle of the market, where the real money is. Fisker Automotive would make only the car's frame, minimizing its risk. For the battery, the most expensive part, they'd contracted with the well-funded A123 Systems, with technology created by the highly regarded Yet-Ming Chiang at MIT.

Just around that same time, we were approached by a pair of engineers who'd named their startup after Nikola Tesla, the legendary inventor. They'd partnered with a wildly successful PayPal entrepreneur who'd put in so much of his own money that he was now chairman of the board. That is how Elon Musk came to us to pitch his idea.

Electrify Transportation

We liked Elon's three-step business plan. Tesla would begin with a high-end sports car, the Roadster, to show that electric vehicles (also called EVs) were feasible *and* cool. The company was ready to enter production as soon as it raised the cash. Next up would be a luxury sedan, the Model S, to compete with BMW and Mercedes. Finally, ten years or so down the road, Tesla would launch a lower-cost EV for the mass market.

The protracted time frame didn't bother me. In fact, nothing about Tesla's plan bothered me—it was strategically sound and beautifully structured. But even if Kleiner could have afforded to invest in both Fisker and Tesla, it wouldn't have been right. As competitors, they'd put us squarely in a conflict of interest. We'd have to choose one or the other.

We got the decision wrong—really wrong. By opting for Fisker, we lost out on one of the top-returning investments of all time. It still stings; Tesla would have been quite the ride. But even though we didn't get to take it, I'm thrilled with the outcome for the world. Elon steered the company through some of the tightest spots ever faced by a startup.

A seed-level investment of $1M in Tesla in 2007 would today be worth more than $1B.

Electric vehicles are growing in popularity

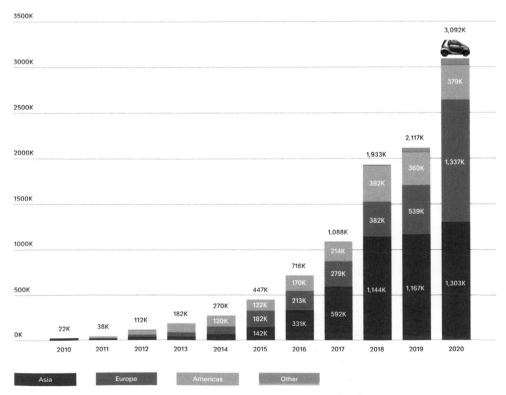

Adapted from data and visuals by BloombergNEF

Tesla thrived while pushing the car industry forward. To help boost the electric vehicle market, the company freely shared its patents with its competition.

By 2019, Tesla was selling one of every five electric vehicles worldwide. In 2020, it sold half a million. It boasts a stock market value of around $600 billion, more than its four closest rivals combined. Best of all, in a classic knock-on effect, Elon spurred global auto leaders to rev up their EV production. Every one of those sales is good news for our climate plan.

As for Kleiner's pick? The Fisker Karma made a magnificent debut for the 2012 model year. The car was sleek and gorgeous. But for reasons ranging from price ($100,000-plus) to performance, it didn't sell. Before Fisker's market could even materialize, A123 Systems, the sure-thing battery maker, folded. A pair of sedan fires triggered a recall. Any lingering hopes were washed away on a soggy day in October 2012 at the Port of Newark in New Jersey, when Hurricane Sandy flooded a $30 million shipment of Karma plug-in hybrids inbound from Europe. More than three hundred cars were totaled; sixteen of them blew up. Fisker was finished before it ever really started.

Counting Down in the Transportation Sector

Our global countdown from 59 gigatons to net zero covers five broad sources of emissions: transportation, energy, agriculture, nature, and industry. Our first objective, to electrify transportation, targets the 8 gigatons of emissions that come mostly from tailpipes. To achieve this goal, the world must replace all gasoline- and diesel-powered vehicles with a fleet of zero-emissions cars, trucks, and buses by 2050.

The electrification of transportation is already under way; as of January 2021, nearly 10 million EVs were on the road worldwide. But the technology we need to scale is behind schedule, and progress is frustratingly slow. We must accelerate. The world drives more miles each year. Over the next two decades, despite the growing popularity of EVs, the number of miles driven with combustion vehicles are projected to remain at their current level. We're not moving fast enough because EVs can't yet compete on convenience and cost with gasoline and diesel cars. With the average new car's life span of up to twelve years, the global fleet's turnover has slowed to a crawl. Combustion vehicles will keep spewing their carbon for a long time to come.

The impact of complete electrification cannot be overstated, and it goes beyond climate change. Each year, tiny particles from tailpipes and power plants cause 350,000 premature deaths in the United States alone and one of five worldwide. According to the Environmental Protection Agency, this pollution is linked to heart disease and lung cancer. Electrifying transportation is more than a cornerstone of our net-zero plan. It's essential to curbing deadly diseases that disproportionately plague poorer countries and communities of color. It is a matter of life and death.

Nine of ten cars on the road today are powered by fossil fuels.

In our effort to cleanse transportation of greenhouse gas emissions, we've framed a handful of key results. A good KR can be measured and verified against publicly available data. If we achieve all our key results, we'll be sure to meet our objective—in this case, cutting emissions from the sector to 2 gigatons per year.

Objective 1
Electrify Transportation

Reduce 8 gigatons of transportation
emissions to 2 gigatons by 2050.

KR 1.1   Price

EVs achieve price-performance parity with new
combustion-engine vehicles in the U.S. by 2024
($35K), and in India and China by 2030 ($11K).

KR 1.2   Cars

One of two new personal vehicles purchased
worldwide are EVs by 2030, 95% by 2040.

KR 1.3   Buses and Trucks

All new buses are electric by
2025 and 30% of medium and heavy trucks
purchased are zero-emission vehicles by 2030,
95% of trucks by 2045.

KR 1.4   Miles

50% of the miles driven (2-wheelers,
3-wheelers, cars, buses, and trucks) on the
world's roads are electric by 2040, 95% by 2050.

KR 1.5   Planes

20% of miles flown use low-carbon
fuel by 2025; 40% of miles flown are carbon-
neutral by 2040.

KR 1.6   Maritime

Shift all new construction to "zero-
ready" ships by 2030.

← For designated key results, the emissions cut is quantified in gigatons, e.g., KR 1.4 yields a 5 gigaton reduction.

Our Price KR (1.1) breaks a stubborn barrier for electric vehicles: parity on price and performance with combustion engines. If EVs are to capture the bulk of the passenger car market, they must be broadly affordable. When people spend more to buy a "green" product over one that emits more carbon, they're paying what's known as a "green premium," a term I first heard from Bill Gates. Markets have proven that when given a choice, most people won't pay or can't afford a premium for energy. "People are going to adopt the low-cost solution," says Eric Toone, technical lead at the Breakthrough Energy fund. "If it costs a penny a gallon more for cleaner fuel, versus petroleum from the world's dirtiest tar sands, many people won't pay for it." And even those willing to pay more will expect superior performance.

The green premium varies widely across sectors

	"Green" (no- or low-carbon) price	Traditional product price	Green premium
Electricity	$0.15 / kWh*	$0.13 / kWh**	$0.02 / kWh (15%)
Passenger EVs (U.S. prices)	$36,500 (Chevy Bolt)	$25,045 (Toyota Camry)	$11,455*** (46%)
Long-haul trucking/shipping transportation fuel	$3.18 / gallon (B99 biodiesel)	$2.64 / gallon (diesel)	$0.54 / gallon (20%)
Cement	$224 / ton	$128 / ton	$96 / ton (75%)
Aviation fuel	$9.21 / gallon	$1.84 / gallon	$7.37 / gallon (400%)
Round trip (economy) SFO to Hawaii	$1069 / ticket	$327 / ticket	$742 / ticket (227%)
Ground beef hamburger meat	$8.29 / pound	$4.46 / pound	$3.83 / pound (86%)

*Residential solar contract.

**Global average consumer price including distribution.

***Before incentives.

Source: Multiple sources. See endnotes.

Electrify Transportation

Early adopters and concerned citizens alone won't get us to net zero. To guarantee a market swing to electric vehicles, we'll need *better* performance at comparable prices. In this context, ==the green premium is a rough measure of the difficulty of each problem== —of how far we have to go to reach net zero, whether for electric vehicles or food products or cement.

Our **Cars KR (1.2)** calls for electric vehicles to account for the majority of new vehicle sales by 2030—a big stretch by any reckoning. Thanks to enlightened public policy, the future we need is happening today in parts of Europe. Norway is already at 75 percent EV market share for new car sales. China has passed 5 percent to become the largest EV market in unit sales. In large Chinese cities, one of five cars sold are EVs. The United States, despite being home to Tesla, the world's largest maker of electric vehicles, is at no better than 2 percent.

Miles driven by electric vehicles lags across categories

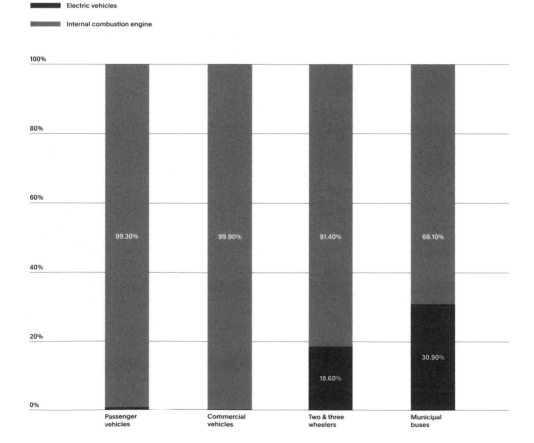

Electric vehicles

Internal combustion engine

	Passenger vehicles	Commercial vehicles	Two & three wheelers	Municipal buses
ICE	99.30%	99.90%	81.40%	69.10%
EV			18.60%	30.90%

Adapted from data and visuals by BloombergNEF

The big incumbent automakers can see the growth projections on the wall. Volkswagen is investing more than $85 billion in electrification by 2025. General Motors, Ford, and Hyundai are also placing big bets to electrify their fleets.

Our **Buses and Trucks KR (1.3)** focuses on two vehicle classes that get less attention than passenger cars, despite their outsize emissions. While buses and trucks represent 10 percent of the vehicles on the road, they generate 30 percent of the sector's global greenhouse gases.

Our **Miles KR (1.4)** ties most directly to emissions cuts. By focusing on total miles driven, it accounts for all vehicles on the road, from newly minted EVs to the oldest and dirtiest combustion vehicles. Globally, less than 1 percent of total passenger car miles were electric in 2020. Considering the sheer scale of the 13 trillion-plus miles driven worldwide each year, getting to 100 percent by 2050 will be an ambitious lift.

Our **Planes KR (1.5)** rallies the aviation industry to accelerate its adoption of sustainable aviation fuel. Our aim is for 20 percent of all airline miles to be flown with low-carbon fuels by 2025. Over a longer horizon, the industry will need to invent pathways to carbon-neutral flight with more efficient planes powered by synthetic fuels, electricity, or hydrogen.

Our **Maritime KR (1.6)** calls for more aggressive reductions in sea transport emissions by cargo and cruise ships. Heavy fuel oil generates large quantities of carbon dioxide and sulfur oxides. More than two thirds of these emissions are expelled within 250 miles of coastlines, exposing hundreds of millions of people to harmful pollutants.

Given the typical bulk carrier's fifteen-year life span, the maritime sector will be especially challenging to decarbonize. The path forward is to prod the industry to make or retrofit ships to be "zero-emissions ready" by using cleaner power sources. In the meantime, maritime emissions can be cut by slowing ships down, using more efficient engines, upgrading hulls and propulsion systems, and adding filters to catch those deadly small particles before they escape into the air.

As Goes General Motors, So Goes America

In 1953, Charles Wilson, the chief executive of General Motors, was nominated by President Dwight Eisenhower to be Secretary of Defense. When Wilson made it clear he would not sell off his substantial holdings of GM stock, a U.S. senator reasonably asked about the potential for a conflict of interest. Wilson famously replied, "I cannot conceive of one because for years I thought what was good for our country was good for General Motors, and vice versa." Over the years, Wilson's statement (with some license) has been invoked both to praise and pillory GM and business in general. But there can be no doubt that America's largest automobile maker has significantly shaped the nation's economy, even its identity.

After a false start or two, General Motors has embraced its leadership role in developing zero-emission solutions. Back in 1996, the company rolled out the first commercial electric car, the EV1, with a 50-mile range. As *Wired* noted, it was "impractical, dinky, and entirely doomed." GM leased about a thousand EV1s, mostly in California, before recalling and destroying them.

It took fifteen years for the auto giant to try again with the Chevy Volt, a plug-in hybrid priced for the mid-market. In 2011, the Volt won *Motor Trend*'s Car of the Year Award. Over the next four years, it vied with the Nissan Leaf for honors as the top-selling plug-in in the United States. The 2016 model year brought another mid-market Chevrolet entry, the Bolt, an all-electric EV designed to compete with Tesla's Model 3.

Still, GM's EV production plans lagged behind Tesla and global rivals—until March 2020, when the company surprised everyone with a string of turnabout announcements that capitalized on the company's economies of scale. The good news began with a sneak peek at the Ultium, a high-energy, large-format, EV battery platform. In November 2020, the company announced a new lineup of thirty EV models to be introduced by 2025. Even more stunning was the plan set by chief executive Mary Barra for 2035: to end GM's 112-year history of making internal combustion cars.

Mary Barra

It began with talking to customers across the country. We saw an inflection point in their view of EVs: *If it has the right range, and there's the right charging infrastructure, and the vehicle meets my needs, and I can afford it, I will consider it.*

We heard that everywhere. We came to believe we had a movement In the making. Given the importance of affordability, we also saw that GM had a critical role to play. If we want to electrify transportation, we've got to reach people who only buy one vehicle. It wouldn't be their family's second, third, or fourth vehicle. It would be their only one. So we decided to lead the transformation, and to do it at scale, worldwide.

It's a tremendous growth opportunity. We also want to supply EVs for zero-emissions autonomous ride-sharing and get the price down from three dollars a mile to one dollar a mile.

We developed an electric concept for potential use by the U.S. Defense Department. In commercial vehicles, we're selling electric delivery vans and last-mile solutions to FedEx Express and other fleets.

Ultimately, it's about execution. We have the know-how on our teams and in our plants. Electrification is now a core competency. We understand the customer. We have the resources to do this.

But first, we need to keep innovating—to take more cost out of batteries, for instance. We also need full-scale charging infrastructure. We're talking with the Edison Electric Institute about energy management, including ways to shift charging to between 2:00 and 5:00 a.m., when rates are lowest. There's a lot of innovation that still needs to happen.

I read that a small city in California banned the building of gas stations. Just two years ago, that would have been unthinkable. But especially with the Biden administration's EV adoption goals, it's clear that we need to accelerate. And we need to do it equitably, so there's no divide. EVs must be for everyone. We can't leave anyone behind.

Being a leader starts with a strong focus on customers. Then you need to consider corporate responsibilities as they relate to climate change and equity. You need to be willing to do the right thing—and frankly, your employees expect it.

It's not a choice between stakeholder capitalism versus shareholder capitalism; they're inextricably linked. Our constituencies are our employees, our dealers, our suppliers, our local communities, and our government, as well as our shareholders and customers. In making decisions, we need to understand the implications for all those stakeholders. As I've found in my time in this role, you make better decisions when you're focused on the right mission.

You need to be willing to do the right thing—and frankly, your employees expect it.

Accelerating Sales Through Policy

To meet our goal in the Cars KR (1.2), sales of EVs must ramp up in a hurry. To stay on track to reach a majority of sales by 2030, one of three vehicle purchases worldwide must be EVs by 2025, a huge advance in adoption over a very short time. New policy is paramount, as we'll discuss in chapter 7. But to spur on this transition, three existing policies must be strengthened, and soon.

First: We'll need more generous financial incentives, mainly tax credits or rebates, to bridge the gap between the up-front green premium on an EV purchase and the buyer's long-term savings on gasoline. That's exactly what a $7,500 U.S. federal tax credit, enacted in 2009, is designed to do. But we can do it smarter. Instead of credits being restricted to a model's earliest buyers, they would phase out only after EVs are well beyond sticker-price parity. As Mary Barra points out, "You shouldn't penalize the first movers for taking the risk."

Second: To speed the extinction of the internal combustion engine, owners need financial incentives to turn in combustion cars rather than resell them. A "cash for clunkers" incentive—better designed and more generously funded than the 2009 version—could clear millions of gasoline-powered cars off the road at a bargain cost.

Third: The ultimate transportation policy would ban all sales of internal combustion cars, politely known as "the electric vehicles sales requirement." This measure alone could achieve three quarters of the emissions cuts we need for the entire sector. At least eight European countries, along with Israel and Canada, say they will prohibit fuel-burning engines. China is working on a timetable. California's governor Gavin Newsom ordered a ban for 2035; eleven other governors appealed to President Biden to follow suit on a national basis.

As we wait for these policies to take hold, we need to raise the bar for fuel efficiency for all combustion and hybrid vehicles. If cars, trucks, and buses are going to burn carbon, they need to go farther per gallon.

How the E-Bus Market Moved the Fastest

Of all modes of transportation, buses are furthest along in adopting EV technology. Given the high level of air pollution from diesel-powered buses, it's an urgent matter, especially in the world's congested big cities. The rise of BYD, a manufacturer in the city of Shenzhen on China's central coast, shows how far a green company can go when savvy entrepreneurship gets rewarded with government support.

BYD's founder and chief executive, Wang Chuanfu, grew up in one of China's poorest provinces. Orphaned as a teen, he was raised by his elder siblings before making it to college to become an engineer. In 1995, he chose an acronym for Build Your Dreams as the name of his startup. A quarter century later, Wang's roaring success has landed him on lists of China's wealthiest individuals.

After starting out with cell-phone batteries, BYD expanded and began to make batteries for tablets, laptops, and solar energy storage. The company went public on the Hong Kong Stock Exchange. In 2003, Wang launched BYD's auto subsidiary—a significantly higher risk than building a battery business. But Wang had an ace up his sleeve: the backing of the Chinese government, which has enabled his company to compete with Tesla in the global EV market.

In many of China's largest cities, air pollution is a nightmare in plain sight. Under Wang's climate action leadership, BYD responded by developing electric buses at the same time as its budget-priced compact car. The company has succeeded in moving thousands of diesel-burning buses off China's teeming roadways. Shenzhen, with a population of 13 million, boasts a fleet of 100 percent e-buses and e-taxis and is closing in on 100 percent e-delivery vehicles.

In Shenzhen, a population of 13 million relies on an all-electric bus fleet.

The story of China's e-bus market shows how public policy can accelerate innovation and adoption. To overcome limited battery life and a shortage of charging stations, the government has channeled more than $1 billion in grants and subsidies to BYD, alongside financial incentives to EV consumers. The company has grown into a centerpiece of Beijing's $50 billion bid to become the world leader in electric vehicles, a core component of the Made in China 2025 strategic plan. Assured of public sector funding for R&D, tax exemptions, and financing for charging stations, at least four hundred companies have jumped into the EV business.

One investor who took notice was Warren Buffett, who snapped up an 8 percent stake in BYD. Buffett's stamp of approval opened other doors. In 2013, the mayor of Lancaster, seventy miles north of Los Angeles, invited BYD to build its first U.S. factory there. By 2016, Wang's company was delivering hundreds of electric buses to cities and towns throughout California. In 2017, when BYD expanded its California manufacturing footprint, the opening ceremony was ignored by the national media. But then Republican House Majority Leader Kevin McCarthy was on hand—the factory is in his district. McCarthy led the chorus praising BYD's promise to hire twelve hundred workers and build up to fifteen hundred buses per year.

Going the Distance: The Proterra E-Bus Story

The twentieth-century legacy of millions of noisy, dirty, diesel-powered buses—for municipalities, schools, and airports—must be eradicated as soon as possible. One U.S. entrepreneur leading the way is Dale Hill. He got his start in Denver by reviving a manufacturer of buses that run on compressed natural gas—cleaner than diesel, but still a CO_2 emitter. In 2004, Hill made the leap to exclusively building e-buses. He renamed his company Proterra: for the Earth.

The transition wasn't easy. In 2009, five years into the venture, Proterra's battery cost per kilowatt of power was stuck at twelve hundred dollars. To reach cost parity with diesel buses, Hill knew he'd have to drive that number down more than 40 percent, to around seven hundred dollars. As the technology kept getting better and cheaper, he began presenting his prototypes to municipal purchasing agents.

But making e-buses is capital intensive, with vehicle costs running into the hundreds of thousands of dollars. The market was growing slowly. Purchase decisions could take years.

In 2010, two of my Kleiner Perkins cleantech investment partners argued for backing Proterra. After researching potential applications for EV technology, Ryan Popple and Brook Porter found an opportunity. Since buses are highly utilized (lots of miles per year) and have notoriously poor fuel efficiency (less than 6 miles per gallon), they were an ideal fit for electrification.

Ryan was especially enthusiastic. A former army platoon commander in Iraq, his focus and discipline served him well as director of finance at Tesla, where he helped Elon Musk and his team survive the recession of 2008. Ryan understands firsthand what's required to make our global transformation to a net-zero economy. He was the perfect choice to step in as Proterra's interim chief executive.

Ryan Popple

There are tough breaks in the business world, but nothing like a bad day in Iraq. Risk in Iraq meant snipers, mortar fire, and roadside bombs. One of my best friends was killed in action while we were deployed together. I struggled to find the right words for a eulogy to honor his memory. My friend's helmet, boots, and rifle were posed with care in a desert far from home.

The experience left me with questions I couldn't answer. What did the war accomplish? Was it worth the sacrifice? Why were we so often drawn into conflict in that region?

One thing I knew for certain was that the Middle East wasn't going to settle down anytime soon, and yet the global price of oil depended on supply from there. In the ports of Kuwait, oil tankers came and went, even while ships arrived carrying tanks and heavy equipment for the invasion of Iraq.

I came home believing that the United States was crazy to think we could depend on imported oil from the Middle East, so I got very interested in helping to reduce the nation's exposure to oil. I was twenty-six years old.

I was admitted to Harvard Business School, and everything there seemed boring compared to cleantech. After graduating, I joined a biofuels startup. Electric vehicles weren't yet an industry. We were working with grain ethanol as a replacement for gasoline, but I didn't think it would fly. The conventional oil and gas companies still controlled distribution. At the end of the day, ethanol is still combustion. Most of its energy is wasted.

In May 2007, my wife, Jen, showed me the green issue of *Vanity Fair*. "Have you heard about Tesla?" she said. It sounded exciting, and electrification made sense to me. I sent my résumé over and became employee number 250 or so. It was the most impressive group of people, but we ran into obstacles nonetheless. Our first vehicle, the Roadster, hit production snags. People placing orders were losing patience, asking for their deposits back.

Then the Great Recession hit. Getting through it as director of finance was probably the hardest work I'll ever do. When you are selling a hundred-thousand-dollar sports car during a recession, every macro force is working against you. We had taken deposits up-front and spent most of that capital on development, not inventory. We knew that when we went into production, we'd be cash-flow negative.

We overcame the challenges by setting specific targets for vehicle cost structure, and we were fortunate that policy shifted in favor of EVs. By early 2010, we had stabilized production of the Roadster, announced the Model S, secured a Department of Energy Advanced Technology Manufacturing Loan, and filed to go public.

Then, out of the blue, a recruiter called me. She said, "There's a new position to create a cleantech portfolio at a venture capital firm." I said, "Unless it's with Kleiner Perkins, I'm not interested." She replied, "So I guess we should be having lunch, then."

That's how I joined the green team at Kleiner. I was part of running that first cleantech fund, and they asked me to focus on transportation. It was a great opportunity to have a lot of leverage in a space I cared about. The struggles with Fisker were actually a great impetus for shifting our focus away from passenger cars. None of us wanted to miss the rest of the EV sector.

So we asked, "What else is going to happen when batteries get cheaper?" I discovered that the value proposition for electrification was even better for municipal vehicles. Urban buses are the strongest example, based on miles and how inefficient the diesel models are.

China's National Energy Commission already understood this: "We electrify buses first." They threw a massive amount of money and incentives at urban buses. That's how BYD came into being. I started looking around. Who else was doing electric buses?

I have a ton of respect for founders. In 2010, I met Dale Hill, who'd started Proterra in Colorado on his AmEx card. Proterra had fewer than one hundred employees and just one customer when Kleiner invested in the Series A round. I served as a board adviser for two years.

The company got into the same spot as a lot of startups. The technology worked, but it was struggling to become a business. I came in as interim chief executive officer for a summer as they searched for a permanent replacement.

I understood the challenge and became determined to meet it. I didn't want to wake up ten years from now and see electric buses all over the place and be filled with regret. My kids would say, "Dad, didn't you used to work on e-buses?" So I stepped up.

What else is going to happen when batteries get cheaper?

After taking the reins as Proterra's chief executive in 2014, Ryan traveled far and wide to meet with purchasing managers in dozens of cities. The feedback was disheartening, but also honest and useful. Though everyone loved their Proterra e-buses, they viewed them as experimental. They wouldn't buy more until prices came down further and performance improved.

"Stay focused," Ryan kept telling his team. "Do one thing better than anyone else in the world can do it." Ryan believed they could build the best e-bus company on Earth.

As Kleiner learned with Fisker and again with Proterra, the most critical element in an electric vehicle is the battery. At Ryan's urging, we stepped up our investment and created a new battery manufacturing and R&D center in Burlingame, near the San Francisco airport.

We needed to recruit engineers who knew how to increase a battery's energy density. The problem was that we knew of only three such people in the entire country, and they all worked at Tesla. One of them, Dustin Grace, agreed to join Proterra as chief technology officer. It was all systems go on Proterra's next generation battery . . . until the project came to a screeching halt.

We need to bite the bullet.

Ryan Popple

Our new battery had sufficient energy for many heavy-duty applications, over 100 kilowatt-hours per charge. But to make our e-buses practical, we needed to reach 400. We spent most of the next year getting to the 250 level, when one of our battery packs started falling apart. Except I (the CEO) didn't know that yet.

One day in late 2015, I spotted two of our engineers pacing around in the back of the R&D area. They had that nervous look. I'm a glutton for punishment, so I asked them what was up. They looked at their shoes and said, "We don't know if we should tell you this, but we're concerned." They broke the news that it would cost us more money to fix the battery than to throw it away and start over.

I was staring at a huge dilemma. If we had to reboot our battery program, most of our revenue would be pushed out another two years. We'd be frustrating customers who wanted better buses yesterday. It would have been easier to keep the existing program on track and to try to work around the issues.

I couldn't make such a big call on my own. This was a board-level decision, the most important board meeting the company ever had. I let my engineers present directly. I didn't interrupt them. Dustin Grace said the team had 100 percent conviction that we could turn the bus market into 100 percent EVs. But the only way we could do it was if we threw everything in the trash, including eighteen months of capital, and started over.

I said, "Here's the deal: We need to bite the bullet, take down our revenue, and raise more capital." We had already raised close to $100 million, but we needed more. I thought the board might just get up and walk out of the building—and that might have been the end of the company, and the one chance we had to build a domestic e-bus manufacturing industry. But the board backed us. They voted yes.

In 2017, Proterra embarked on a pivotal road test: how far could an e-bus go on a single charge? With two drivers taking turns, a 40-foot Proterra electric bus began a range test on a closed track, with a third party measuring the results. By the time the bus fully depleted its battery, it had logged 1,101 miles without recharge, blowing away the previous e-bus world record.

That feat helped to seal more deals. Mayors of a dozen major cities—including Los Angeles, Seattle, London, Paris, and Mexico City—said they'd be buying zero-emissions buses exclusively by 2025.

Proterra's electric school bus wins on efficiency, operating costs, and zero emissions.

Rebooting Proterra's battery program was a tough call, but absolutely the right thing to do. The company's next-generation buses became 4,000 pounds lighter by swapping out steel for carbon fiber. With the team's new vehicle platform, called Catalyst, they exceeded 350 miles per charge. As bus batteries took a leap forward, the case for switching to an all-EV fleet was more compelling than ever. Municipal transportation officials focus on total costs of ownership. Over a typical twelve-year life cycle, by slashing maintenance and fuel costs, an electric bus saves between $73,000 and $173,000 over a diesel bus. It also removes the emissions equivalent of twenty-seven combustion cars.

For Proterra, the ride has just begun. Though its e-buses were operating in forty-three states as of 2021, the market is still wide open. Electric buses have but a tiny foothold in the United States, just 2 percent of the nation's public transit bus fleet—versus 25 percent in China. The overwhelming majority of municipal and school buses are still powered by diesel. But we believe that e-buses can consistently win competitive bids by 2025, and that the nation's bus fleet can be all electric by 2030. It's an audacious yet realistic key result that is essential to our Speed & Scale Plan.

Battery technology improvement will galvanize broader transport electrification. For passenger EVs, the first success came in high-performance sports cars. For commercial vehicles, buses were the first domino to tip. Now Proterra's technology is expanding to electrify delivery vans and heavy-duty trucks. The company recently entered into its first partnership—with Daimler, the largest commercial vehicle manufacturer in the world. As Proterra board member Brook Porter says, "Diesel's days are over."

Driving Down Costs While Improving Performance

Better, cheaper batteries top the list of breakthroughs we need. Despite significant progress, we're still early in the effort to drive down battery costs while improving performance. I saw something similar happen up close in the personal computer business. It was 1974, and I was fresh out of Rice University, looking to test what I'd learned about electrical engineering. I headed to Silicon Valley and landed a job at Intel, which was developing the first 8-bit microprocessor. The company hoped it would democratize computing by making microchips cheap and ubiquitous.

Moore's Law demonstrates exponential growth

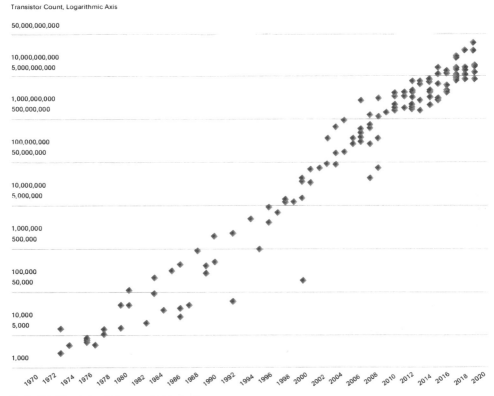

Transistor Count, Logarithmic Axis

Year in which different microchip types were first introduced

Adapted from data and visuals by Wikipedia and Our World in Data

Electrify Transportation

Intel's chairman, the great Caltech chemist Gordon Moore, believed we could sustain a compounding rate of improvement in computer chips indefinitely. He proposed that the number of transistors that could be packed onto a silicon wafer would double roughly every two years. The concept was astonishing even to us, the people trying to make it happen. "Moore's Law," as it came to be called, wasn't preordained; it became a reality through years of relentless, cumulative progress by thousands of engineers. It relied upon an ecosystem of innovation in physics, chemistry, lithography, circuits, design, robotics, packaging, and more.

As each generation of microprocessors affirmed Gordon's prediction, they paved the way for more powerful and affordable computers. Machines that once sold by the hundreds per year (UNIVACS), and then by the thousands (mainframes and minicomputers), soon were selling by the hundreds of thousands (Apple IIs), by the millions (IBM PCs and Macs), and ultimately by the billions (iPhones and Androids). Over the last half century, Moore's Law transformed the world's economy and changed nearly every aspect of business and daily life.

Wright's Law in Action: Solar
Solar module prices dropped as installations increased*

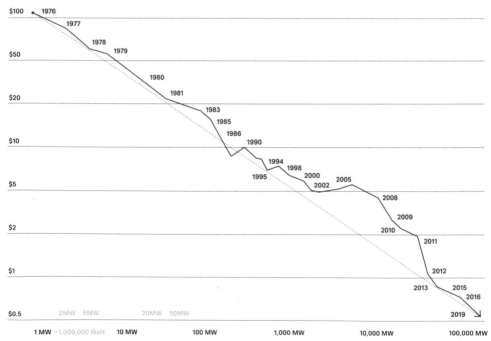

Cumulative installed solar PV capacity (logarithmic axis)

*Price per watt of solar photovoltaics (PV) modules (logarithmic axis)
The prices are adjusted for inflation and presented in 2019 USD
Adapted from data and visuals by Our World in Data
Licensed under CC-BY by the author Max Roser

Unfortunately, it does not apply to renewable energy. The materials and engineering challenges are very different. But could there be another way to anticipate the progress of batteries and other vital technologies?

In fact, there is. **It's called Wright's Law.** In 1925, the chief engineer at the Curtiss Aeroplane Company was an MIT grad named Theodore Wright (no known relation to Wilbur and Orville). He calculated that for every doubling of production, aircraft manufacturers could derive a reliable decline in costs. For example, if you have a thousand planes and the second thousand costs 15 percent less, the cost for the next doubling (to four thousand) should drop by 15 percent too. Wright's Law helps us forecast costs based on production. Wright became an aviation leader during World War II and later acting president of Cornell University. Though his rule of thumb wouldn't become as famous as Gordon Moore's, it was no less prophetic.

Wright's Law in Action: Batteries
Lithium battery prices over time

Consumer electric (cells)

Automotive (packs)

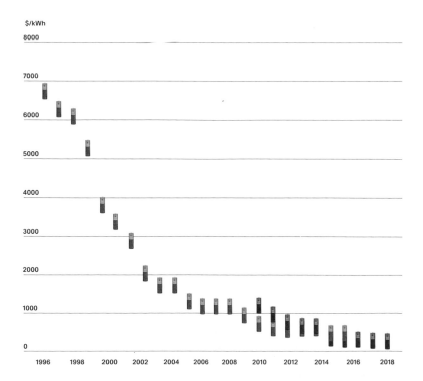

Adapted from data and visuals by IEA

Years later, a study by the Santa Fe Institute showed that Wright's Law could be applied to the cost curves across sixty-two different technologies, from televisions to kitchen appliances. Apply it to EV batteries, and you'll see something striking. In 2005, when early EV startups were first putting rubber on the road, a battery cost at least sixty thousand dollars. The only electric vehicles that could turn a profit were luxury or sports cars priced over one hundred thousand dollars, like the early models from Tesla and Fisker. In accordance with Wright's Law, each doubling of production lowered the battery packs' cost by 35 percent. By 2021, a similar-size battery pack cost only eight thousand dollars. All of a sudden, EVs were within reach of cost competitiveness with internal combustion vehicles.

Lightning Strikes at Ford

In May 2021, Ford unveiled the first electric version of its F-150 pickup, number one in sales among all U.S. vehicles for forty-four years running. After a $7,500 federal tax credit, the entry-level model will be priced under $32,500 when it goes on sale in spring 2022—and even less than that in states with additional EV rebates, like California and New York.

The new F-150 is named, with no hyperbole, the Lightning. It ranges up to 230 miles and rockets from zero to 60 miles per hour in 4.4 seconds. "This sucker's quick," said President Biden, after taking a spin (with a grin) in a prototype at a Ford test track. Since Ford sells around 900,000 F-150s per year, many see this as a Model T moment, a decisive turning point. "The future of the auto industry is electric," Biden said. "There's no turning back."

The Lightning isn't just a truck—it's an all-purpose, multitasking, highly portable generator. In a blackout, according to Ford, it could keep a house running for three days. With eleven power outlets, it can run an electric saw, a cement mixer, nighttime lighting for any worksite—or all of the above. After all, as The Atlantic noted, an EV is essentially "a giant battery on wheels."

Yet even as we approach price parity for passenger EVs, at least in the United States, we'll still need aggressive, innovative cost reductions to make affordable cars for the rest of the world. In India, the most popular car, the Maruti Swift, sells for between $8,600 and $12,600—around one third the average car price in Western Europe or the United States. What advances might emerge to erase the green premium in the developing world? Improved battery density is one. New materials and designs can pare a vehicle's weight, which translates to lower lifetime costs and longer range.

Spin with a grin: President Biden test-drives the electric Ford F-150 Lightning.

In the developed world, meanwhile, we still need to address the car buyer's fear of their batteries running out of charge—what some call range anxiety. Though the average American drives only twenty-seven miles per day, purchase decisions are guided by the *maximum* range someone might need for that long weekend holiday or summer road trip.

EV passenger cars and e-buses share a sweet spot for range: roughly 350 miles, or about six hours of uninterrupted highway driving. At the present pace of battery innovation, we are on track for ranges up to 500 miles. For longer trips and big rigs, we'll need even better batteries.

Electrifying transport is an ambitious undertaking in and of itself. But we'll gain a lot less from it if "clean" energy comes from a dirty source, from coal or a natural gas power plant. To put it simply, we can't decarbonize transportation without decarbonizing the grid — the topic of our next chapter.

Speed & Scale: Countdown to net zero

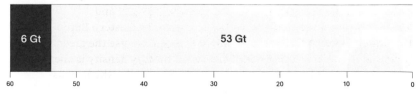

Objective	Reductions	Remaining
Electrify Transportation	6 Gt	53 Gt

60	50	40	30	20	10	0

Decarbonize the Grid

Chapter 2

Decarbonize the Grid

Long term, Thomas Edison once remarked, "I'd put my money on the sun and solar energy." But in Edison's day, there was no way to place that wager. Coal was the practical option for boiling water around the clock to make the steam that pushed the blades of giant turbines and generated electricity. Today it's a different story. Fossil fuels are but one of several ways to deliver energy into the power lines strung to people's homes and businesses.

Yet as late as the turn of this century, coal-burning power plants still furnished the largest share of the world's electricity. That's when Hermann Scheer, a long-serving member of the Bundestag, the German parliament, argued for Germany to become the first large country to scale up solar and wind power. Scheer powered his own home with a windmill. His vision for a society run on renewable energy was seen by some as quixotic. But the lawmaker had a plan: to use a special type of subsidy to drive down renewables' costs.

Through the 1990s, Scheer's call for a long-term phaseout of coal, as well as a gradual shutdown of Germany's nuclear power plants, was bitterly opposed by entrenched incumbents in the energy sector. The lawmaker refused to back down.

He founded the International Renewable Energy Agency and became president of Eurosolar, a coalition of energy entrepreneurs.

But none of that helped much in the scrum of domestic politics. A one-time member of the German modern pentathlon team, Scheer privately joked that getting his legislation passed would call on the political equivalent of all five of his sporting skills: swimming, fencing, equestrian show jumping, cross-country running, and pistol shooting. (He didn't say who he'd be aiming at.) In the year 2000, on the floor of the Bundestag, he declared: "Fossil fuels have created a climate disaster. The only real and realistic option is the total replacement of fossil and atomic energies by renewables."

When his colleagues seemed unmoved, Scheer brandished a metaphor: "Fossil and nuclear fuels amount to global pyromania. Renewable energy is the fire extinguisher." Thanks to his dogged persistence, the bill finally passed. What became known as Scheer's Law established the world's first large national marketplace for solar and wind power. The idea was simple, brilliant, and effective. Literally anyone—an ordinary homeowner, a farmer with spare acreage, a retailer with roof space—could install an array of solar panels or a bank of windmills and feed power into a utility's grid. In exchange, the utility paid for the homegrown power at a preset rate, locked in for twenty years. People could calculate their annual earnings in advance and secure bank loans for needed equipment.

Scheer's Law specified a pay rate as high as sixty cents per kilowatt hour, four times the going rate for electricity. The added costs were passed on to homeowners and some businesses as a surcharge on their electricity bills—for the homeowners, on average, less than ten dollars a month at the outset. A vocal minority opposed the premium on principle. But most German citizens backed the plan, which promised to generate thousands of jobs.

Money began flowing in new directions, and clean electric power with it. Soon hillsides were dotted with wind turbines. Photovoltaic panels blanketed neighborhood rooftops. Long stretches of the Autobahn were lined with sky-blue solar cells. A Bavarian livestock farmer named Heinrich Gartner borrowed 5 million euros to install ten thousand solar panels on his land. He calculated that they would generate more profit than his pigs.

Guiding Scheer's plan was a simple set of goals, or what I call key results: 10 percent of electricity from renewables by 2010, and 20 percent by 2020. While wind and solar energy were the plan's keystones, supporting roles would be played by other clean technologies: hydroelectric stations, geothermal power plants, biomass power. By 2006, the German renewables experiment was on track to meet Scheer's targets. But the politics grew fraught over the loss of coal mining jobs. As costs escalated, some consumers started fretting. Unbowed, Scheer kept citing climate change data and public opinion polls in the plan's favor.

As solar and wind energy scaled in Germany, cost reductions worked their magic. New business models emerged. The green premium—the extra cost for clean energy—began to shrink. For a time, skyrocketing demand for renewables set off a manufacturing boom. Germany's new "solar valley" industrial area, mostly in the former East Germany, created as many as three hundred thousand sorely needed jobs in solar panel design and manufacture. Several well-funded solar startups staged impressive debuts on the stock market.

In those more innocent days, it was easy to believe that starting a solar hardware technology company was a grand idea. I know this firsthand. Kleiner Perkins backed seven solar panel ventures around that time, to our subsequent chagrin.

As the market scaled in Germany, the United States got into the act. Though solar panels were invented there in the 1950s, the United States did little at first to push the technology to scale. As Scheer's Law kicked in, environmental advocates convinced a wave of American states to set modest requirements for renewable energy on their electric grids, and to adopt favorable pricing for solar power. Worldwide demand for wind and solar, a negligible niche market a few years earlier, started to boom.

Yet Germany's pacesetting policy failed to generate the new employment program the German politicians were banking on. The reason was simple: China saw a big new market for solar panels developing in Germany. With cash injections from their government, Chinese manufacturers marched into Germany and snatched the market away from the domestic competition. Cheap Chinese solar panels upended the American market as well, a big reason our investments at Kleiner Perkins failed to pan out.

As solar prices fell, demand soared

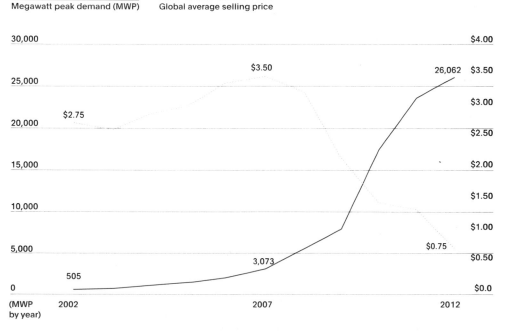

Megawatt peak demand (MWP)   Global average selling price

Adapted from data and visuals by Renewable Energy World

The Chinese solar expansion was spectacular. The government poured money not just into startup companies but into research and development, to capture a competitive edge. Every region of China, large and small, suddenly had its own solar panel startup. The Chinese government had decided that this was a strategic industry of the future, and they were going to own it. American and German manufacturers held some patents for technically advanced solar panels that might have given them a fighting chance, but neither government did much to help their solar companies stay afloat. The Chinese ultimately romped away with 70 percent of the global panel market.

Like other investors, I failed to foresee the German experiment's ripple effects or the impact of China's massive move into solar manufacturing. Once solar panels started selling in volume, it set off an all-out price war. Between 2010 and 2020, the panels plummeted from two dollars per watt of output to twenty cents. Six of the seven solar companies backed by Kleiner failed. For those of us who lost our shirts, it was another lesson learned: beware when investing in commodities, where price is king—especially when other governments are subsidizing them.

To get back to Germany: Even though most of their solar manufacturing jobs evaporated, solar installations were more popular than ever. The clean power subsidy for solar producers and installers stayed high even as the price of the panels plunged and the wholesale power price with it. By the time the Bundestag peeled back the subsidies, a number of German utilities woke up in the mid-2010s to see their profits vanish.

As more and more solar power was pumped into the grid, especially in the middle of the day, it undercut the most profitable hours for fossil fuel-powered plants.

Most of these utilities were forced to shed jobs and restructure, taking huge write-downs on their fossil power plants and turning their focus to clean energy. City governments with utility investments were forced to cut back on services. The collateral damage from the disruption was substantial.

But while Scheer's Law wasn't perfect, it showed that the right policy at the right time can be instrumental in helping clean energy technology scale—and get more affordable in the process. (It was also a cautionary tale for incumbents who ignore change.) Thanks to the German experiment, cheap solar panels are now available just about anywhere. Says Hal Harvey, chief executive of the think tank Energy Innovation: "It was Germany's gift to the world."

In 2010, Germany reached 16 percent renewables, well beyond its original target of 10 percent. Sadly, Scheer died that year at age sixty-six from heart failure. Three years later, his daughter, Nina Scheer, was elected to the Bundestag and named to its environmental committee. She backed a follow-on law to phase out the subsidies for renewables by 2021—they were no longer needed. She also helped lead the majority to pull the plug on coal power by 2038.

In 2019, 42 percent of Germany's electricity was generated by renewables. For the first time, a renewable "power majority" was at hand in a leading industrial country. In the summer of 2020, it happened. As overall demand dropped from the COVID-19 pandemic, solar ran near full capacity. Renewables became the leading energy source for 80 million Germans, powering on average 56 percent of the nation's grid.

Since Scheer's Law was passed, Germany has cut its grid emissions by nearly half. Nina Scheer says she only wishes her father were alive to see it.

Ramping Up Emissions-Free Energy

At 24 gigatons per year, more than one third of the global total, the power sector is the single largest source of carbon emissions. We count on it to heat our homes and offices, cook our food, and charge our electric vehicles. Remember, electricity is not a *source* but a *carrier* of energy. As long as energy is derived from fossil fuels, whatever we electrify will not be emissions-free. But of itself, electricity doesn't require combustion. It can be created by water, wind, or sunlight. Decarbonizing the grid and switching to clean energy is the single biggest step in making our plan a reality.

The problem with solar and wind power is that they don't work unless it's sunny or windy. To fully decarbonize the grid, we need layaway energy for when the sun has set and the winds have calmed. We need pinpoint forecasting to shift power from areas with an energy surplus to others with shortfalls. On-demand renewables, like geothermal and hydroelectricity, will have to fill in the gaps. In general, we need affordable short-term energy for hours to days at a time, plus storage to stock long-term reserves.

Our key results for the power sector call for even cheaper clean technologies in the future. To acknowledge the gap between rich and poor countries, they feature more flexible timelines for nations struggling with energy poverty.

Objective 2
Decarbonize the Grid

Reduce 24 gigatons of global electricity and heating emissions to 3 gigatons by 2050.

KR 2.1	**Zero Emissions** 50% of electricity worldwide comes from zero-emissions sources by 2025, 90% by 2035 (up from 38% in 2020).* ↓ 16.5 Gt
KR 2.2	**Solar and Wind** Solar and wind are cheaper to build and operate than emitting sources in 100% of countries by 2025 (up from 67% in 2020).
KR 2.3	**Storage** Electricity storage is below $50 per kWh for short duration (4–24 hours) by 2025, $10 per kWh for long duration (14–30 days) by 2030.
KR 2.4	**Coal and Gas** No new coal or gas plants after 2021; existing plants to retire or zero out emissions by 2025 for coal and by 2035 for gas.*
KR 2.5	**Methane Emissions** Eliminate leaks, venting, and most flaring from coal, oil, and gas sites by 2025. ↓ 3 Gt
KR 2.6	**Heating and Cooking** Cut gas and oil for heating and cooking in half by 2040.* ↓ 1.5 Gt
KR 2.7	**Clean Economy** Reduce reliance on fossil fuels and increase energy efficiency to quadruple clean energy productivity rate (GDP ÷ fossil fuel consumption) by 2035.

* This is the timeline for developed countries. For developing countries, this key result is expected to take more time (five to ten years).

Our top priority is the **Zero Emissions KR (2.1)**, which calls for zero-emissions electricity sources to exceed 50 percent of global power by 2025 and 90 percent by 2035. The solution includes nuclear power, which helps meet energy demands when wind and sun aren't abundant. While nuclear fuel yields virtually no emissions, it is not strictly renewable; it relies on limited quantities of radioactive elements. Nuclear plants last for decades and figure to remain part of the global mix. But as costs for this technology continue to rise and other options get cheaper, its role may be reduced in the future.

On the one hand, we need intensified research and development to make nuclear power safer. On the other, regulatory changes to speed plant construction can help us get to net zero by 2050. There's no time to waste on this front. Where renewables can be installed within weeks, nuclear plants can take a decade or longer to get up and running.

Each nation must choose its own path in constructing an emissions-free grid. Some, like Germany, will phase out nuclear power along with coal and gas. But France and China may choose a different tack. In the United States, twenty-eight states have nuclear plants; they generate one third of Virginia's electricity. In 2020, in enacting a carbon net-zero plan into law, Virginia counted nuclear as carbon-free. That's what our plan does too.

We're agnostic on which technologies are deployed, as long as they don't add greenhouse gases to the atmosphere. Hydroelectricity already accounts for 16 percent of global power, mostly from large dams. Wind and solar have global shares of about 6 percent and 4 percent, respectively. In the United States, the Southwest is dry and sunny, perfect for solar, while the gusty middle of the country is well suited for wind turbines. Iceland gets virtually all of its power from renewable hydro or geothermal sources. In making this momentous transition, each country and region will need to leverage its own geographic advantages.

We're already close to achieving the **Solar and Wind KR (2.2)**. New installations of these renewables are the cheapest energy sources in two thirds of the world, including the United States, China, India, South Africa, South America, and Western Europe. But our key result demands more by 2025: for solar and wind to be cheaper everywhere.

Our Storage KR (2.3) aims for energy dispatched from storage to be competitive with current electricity pricing. To get there, we'll need low-cost energy sources plus innovative storage technologies that meet specific price targets. The two must work in tandem to deliver energy to the grid.

Decarbonize the Grid

Renewables are winning as prices drop and installed capacity grows

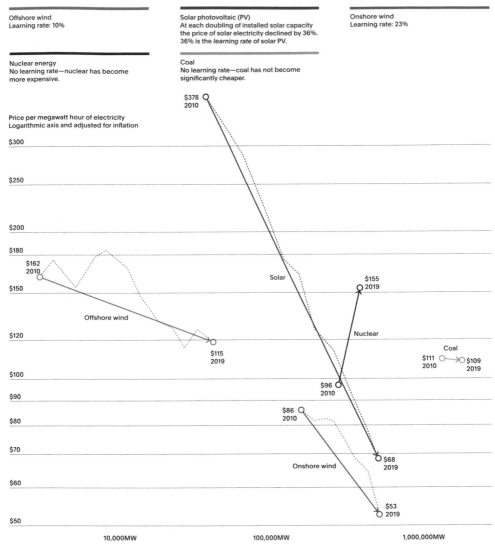

Offshore wind
Learning rate: 10%

Solar photovoltaic (PV)
At each doubling of installed solar capacity
the price of solar electricity declined by 36%.
36% is the *learning rate* of solar PV.

Onshore wind
Learning rate: 23%

Nuclear energy
No learning rate—nuclear has become
more expensive.

Coal
No learning rate—coal has not become
significantly cheaper.

Price per megawatt hour of electricity
Logarithmic axis and adjusted for inflation

$378
2010

$300

$250

$200

$180

$162
2010

$150

$120

Offshore wind

$100

$90

$80

$70

$60

$50

Solar

$155
2019

Nuclear

Coal

$111
2010

$109
2019

$96
2010

$86
2010

$68
2019

$115
2019

Onshore wind

$53
2019

10,000MW

100,000MW

1,000,000MW

Cumulative installed capacity (in megawatts)

Adapted from data by IRENA, Lazard, IAEA, and Global Energy Monitor
and visuals by Our World in Data

Transitioning out of fossil fuel power lies at the heart of our **Coal and Gas KR (2.4)**—a colossal challenge. Globally, new development of coal, oil, and gas must halt immediately. The world has enough supply for now, and we must begin to taper demand. In the developed world, we must stop building natural gas plants and continue to steer clear of new coal installations. Then the emphasis must shift to phasing out most existing fossil fuel plants. For those that stay in operation, removal technologies may be able to eliminate their emissions.

For the developing world, this key result will take more time, likely an additional five to ten years. In poorer countries that lack reliable access to electricity, clean energy portfolios may not meet their populations' immediate needs or stabilize their grids. In these cases, the construction of new gas plants may be justified—with the provision that they're retired or their emissions erased by 2040.

In general, the role for the developed world is to drive down the cost of renewables, eliminate the green premium, finance clean energy investments, and get its own house in order by decarbonizing first. Developing countries have the opportunity to leapfrog the outmoded fossil fuel model and move directly to clean, affordable energy sources. Investments from wealthier nations and the World Bank can accelerate the leap. It is easier and cheaper to build an energy infrastructure right the first time than to go back and correct past mistakes.

Though we don't yet know precisely how it will play out, a global farewell to coal-fired power is an idea whose time has come. In May 2021, in response to the International Energy Agency's clarion call for net zero and "a total transformation" of the world's energy systems, the seven largest advanced economies agreed to "stop international financing of coal projects that emit carbon" by the end of the year. Led by the United States and the European Union, this commitment could help to push renewable energy sources ahead of coal by 2026 and ahead of gas before 2030.

Our **Methane Emissions KR (2.5)** confronts "fugitive emissions" by eliminating 3 gigatons of methane from leaks and deliberate industrial releases by 2025. Existing regulations must be strictly enforced for better site management and the capping of old wells, mines, and fracking sites.

Our **Heating and Cooking KR (2.6)** aims to replace buildings' oil and gas installations with electric heating units and electric stoves. Modern electric heat pumps, which can boost heating efficiency by a factor of three or more, are reliable replacements for heating and cooling. Induction stoves have won the hearts of professional chefs while erasing the chief source of indoor air pollution: your stove. Rather than an appeal for sacrifice, this key result calls for modernization.

What is a zero-emissions economy? It's the way to sustain economic growth while turning away from fossil fuels. A country's clean energy productivity rate is its gross domestic product (GDP) divided by its consumption of fossil fuels. Our **Clean Economy KR (2.7)** looks to quadruple each country's rate by 2035.

Of the twenty top emitting countries, France sits at the top of the list, in large part because it derives 70 percent of its power from nuclear energy. Near the bottom are Saudi Arabia and Russia, two nations still addicted to oil and gas and that have yet to diversify their domestic energy sources. Countries can improve their clean energy productivity rate either by switching to cleaner sources of energy or by using fossil fuel resources more efficiently.

As with any well-constructed OKR, we need to attain all these key results to guarantee meeting our objective. Five out of six won't cut it. Thankfully, the sun delivers as much energy to our planet in an hour as we use in a full year. After decades of failed ventures, solar installations today are outpacing all other technologies, even wind power.

Europe generates more economic output with fewer emissions

Ratio of GDP to fossil fuel consumption

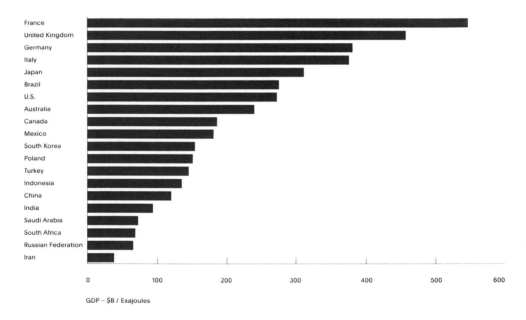

GDP – $B / Exajoules

Inventing a Solar Business Model: The Sunrun Story

A large part of solar's recent success rests on smart business models that scale aggressively. As chief executive of the San Francisco startup Sunrun, Lynn Jurich got in on the market's ground floor to create a new business model. By 2020, Sunrun counted three hundred thousand customer households across the United States.

Lynn Jurich

At a pretty young age, I read the entire biography section of the library. I was studying people who made a big impact over a long period of time.

Later I went to Stanford and stayed for business school, where I studied finance as a way to understand complex problems. You can't solve big societal challenges without knowing about the money.

I got a summer internship at a global bank, which took me to Hong Kong and Shanghai. It was 2006, and both cities were in the thick of huge construction booms, with cranes in all directions. Down at street level, I found myself walking through clouds of smog and breathing the pollution.

In the summer, the smog came from power plants burning coal to keep the air-conditioning going in all those offices. In the winter, it came from homes that were heating small spaces with coal. The system was built on fossil fuels, and it wasn't working.

While we had the technology to replace fossil fuels with renewables, a lot of work had to be done to make this transition economical. Envisioning the fifty-year career ahead of me, I knew I wanted to help. I wasn't necessarily focused on solar, but on distributed power.

At the time, a startup called SunEdison was gaining traction with a distributed utility approach for businesses. For no up-front cost, they'd install solar panels on the roof of a Whole Foods or a Best Buy or a Walmart to power part of the store's electricity and refrigeration needs. The store would pay SunEdison a fixed rate locked in over twenty years. Using that predictable revenue stream, the startup could raise or borrow enough money to fund its projects.

No one was doing this for the residential market, even though you could imagine there might be less friction there. I teamed up with two cofounders, Ed Fenster and Nat Kreamer, who were deeply experienced in consumer finance. Instead of going through long contract negotiations with gigantic, bureaucratic corporations, we thought about what it would take to go homeowner by homeowner.

Back in the 2000s, there was a lot of investment in solar PV hardware, for both panels and cells. We didn't want to be in that game; we wanted to invest in *how* solar is deployed. What's unique about solar is micro-ownership and its potential at scale. That's where grid parity could be achieved. We could empower homeowners to generate their own energy, instead of relying on a utility, and to create a distributed power system.

To launch Sunrun and scale our vision for residential solar-as-a-service, we raised money from friends and family. We put flyers on windshields in store parking lots. We talked to moms about solar as they shopped at a farmers market for their veggies and cheese.

From the early days, we benefited from solar's popularity. The public opinion polls told us that people believed in it—you just had to get in front of them. People didn't love their incumbent utilities, to say the least.

While our early adopters liked the climate benefits, most of all they liked feeling in control—they could be their own utility and save money. The up-front cost was zero, and they could lock in the price of their electricity for long periods of time. It was as if someone installed a gas pump in your backyard at no charge and said it was a dollar a gallon from then on. Would you want to keep it for one year or for many years? That's how we got homeowners to sign twenty-year contracts.

We raised the capital to cover the up-front solar system installation costs, about fifty thousand dollars per home—a number that has gone down significantly since. The customers signed power purchase agreements that covered maintenance and repairs.

We used our equity capital to buy the panels and install them, and to prove that people would sign up. It was a tough slog, but it was working.

The 2008 recession hit us hard. The mortgage bubble imploded. Who would keep funding a startup that lives on homeowner credit? For our customers, the biggest risk was our going bankrupt. Fortunately, we closed on a financing deal the day before Lehman Brothers failed. We got in right under the wire, with just enough capital to survive the Great Recession.

From a business model standpoint, going house by house seemed slow. But that's what we had to do, and we actually scaled faster than the commercial market. We covered as many neighborhoods as we could across ten states. Consumers got to create their own power, and more and more people wanted the opportunity.

By 2013, we finally were turning a steady profit. Our board felt the time was right for an initial public offering. It was tricky. My husband, Brad, and I had been married nine years and we weren't sure if we wanted children. He's also an entrepreneur, so we were always raising capital or dealing with crises at work. There was never going to be a good time. I was thirty-five and we said we have to do this if we're going to do it. I got pregnant, and, of course, it happened to perfectly coincide with our initial public offering.

We could empower homeowners to generate their own energy.

At the time, solar and wind were slowly expanding their market shares, the world over. The going was still tough, though. Most solar pioneers got rewarded with arrows in their backs from politically connected utility companies. But Sunrun's solar-as-a-service model quickly found a market. The company's 2015 initial public offering raised $250 million on the NASDAQ for funding operations across eleven states. Early investors were rewarded with a handsome return.

Sunrun's solar-as-a-service business model has made it the top rooftop installer in the United States.

Lynn Jurich

Because utilities are so entrenched, they don't feel much need to educate their customers about electricity. When we started out, nine out of ten people thought solar was the most expensive form of energy. Most people still believe solar is too expensive, even now, years after that's no longer the case.

The utility model is so broken, and the market is so fragmented. As a delivery vehicle, the grid was designed as a one-way street. When the sun is out, solar-powered homes produce more power than they can use. Under so-called "net metering" policies, utilities in thirty-eight states are required to buy that power and send it to other homes. But often grid operators say there is too much renewable power and they can't take it all. Utilities aren't incentivized to have a distributed grid, so it's up to the regulators to change the rules.

Utilities need to match supply with demand, and the intermittency of solar is their biggest challenge. They have to be smart about what to charge during peak and off-peak times. Consumers have to be smart about when to charge their car or dry their clothes.

One solution is to make utilities develop better demand-response management. Another is to include storage batteries with all residential systems, so homeowners can store their power for use at night or the next day. If we can do this at scale, we can build what we call "virtual power plants."

Hawaii is an interesting microcosm where about 30 percent of homes have solar and batteries. Having so much scale enables us to provide more affordable and reliable power than the utilities. We're now convinced that every home should have a battery, with grids designed to accommodate this model.

At this point, we're in twenty-two states. We've become the largest installer of rooftop solar in the United States—we've even surpassed Tesla, which owns another residential solar company. Anyone who's in our business is an ally to me. We all believe that electrifying everything is the way to go, and we are achieving that future together. That means distributed power. I have even more conviction about that today than when we started.

Decarbonize the Grid

In 2021, thanks to Lynn Jurich and other solar trailblazers, the United States reached 100 gigawatts of installed solar capacity. China, with more than four times the U.S. population, has reached 240 gigawatts. India set a target of 20 gigawatts by 2022, hit the milestone four years early, and is now targeting 450 gigawatts by 2030. Worldwide, solar is closing in on a historic milestone, its first full terawatt (1 trillion watts, or 1,000 gigawatts). Yet despite this rapid progress, we'll fall short of our net-zero target without fundamental changes in policy.

Lynn Jurich

In terms of global warming, we're already up against the 1.5-degree Celsius ceiling to avoid climate disaster. So we have to start making near perfect big decisions right now. In the 1950s, we built the national highway system incredibly fast. Eisenhower was a great leader. We need a general wartime approach to get solar on every roof where it can be useful.

That means incentives to help consumers, not just the incumbent energy companies. We need to power buildings with rooftop solar and storage batteries. The energy can be used to charge our EVs, and to switch us from natural gas or oil burners to cooling and heating with electric-powered pumps and compressors.

The technology is here. Battery costs keep coming down. Everyone talks about what China and India need to do. But if the United States leads, they'll do it even faster.

What constrains growth for solar? Inertia—it's always easier not to change. Incumbents defend their turf and make it complicated with red tape and paperwork.

Sooner than later, installing solar should be like installing a kitchen appliance. If a homeowner wants it next week, they should be able to get it. All new homes should be built with solar—it should be part of the price of a home, like granite countertops.

We have to make it super simple and inexpensive, which is happening. This new world of electrification is not about sacrifice. It doesn't cost more, and you can still have the house you want. But now you can have it while creating your own power from the sun.

The Way the Wind Is Blowing

Since solar and wind are the two fastest-growing sources of energy, you might think they'd be locked in competition for market share. But that isn't the case, because the two technologies are natural complements. Solar panels turn sunlight into electricity during the day, while turbines tend to be busier at night, when the wind picks up. In a sense, wind is solar energy too. The sun heats the Earth unevenly because of variations in terrain. Since warm air rises, it leaves pockets of low pressure behind it. The differential causes an imbalance. The resulting rush is the wind.

The two renewables have complementary business models as well. While solar accelerates the shift to a distributed grid, wind power is centrally sourced and managed. Wind allows utilities to keep doing what they do best: negotiate favorable purchase agreements on power sources and deliver electricity to customers. Wind is growing faster than any other utility-scale power source, including fossil fuels.

In the United States, wind has long enjoyed a higher market share than solar, due mainly to its embrace by big utilities. Sitting on the oil-rich Gulf of Mexico, Texas is home to the U.S. petroleum industry. Thanks to friendly state policies for entrepreneurs, it's also the longtime leader for wind. In 2006, the Horse Hollow Wind Energy Center went up in central Texas. Spinning 735 megawatts of power, it was the largest turbine farm in the world when it was built. (Since then it's been surpassed by larger installations, led by China's Gansu Wind Farm, which is twenty-seven times its size.)

Over time, wind technology improved by leaps and bounds. Blades got longer. Turbines got taller. As manufacturing capacity doubled, the cost of new turbines dropped by half. Once U.S. wind power became cheaper than train transport for coal, it simply made good sense to integrate it into the grid.

But future growth in onshore wind power faces several constraints: transmission bottlenecks, utility restrictions, a shortage of available land, local opposition to new sites. The new frontier for this sector is out into the sea, where a Danish visionary turned a financial crisis into a new green opportunity.

Ørsted's Offshore Revolution

The first offshore wind farm began as a small experiment. In 1991, off the coast of a little island in the Baltic Sea, Denmark's state-owned utility company, Danish Oil & Natural Gas (DONG), built eleven turbines. Named after the nearest shore town, the Vindeby Offshore Wind Farm could generate 5 megawatts when complete. It was a tiny enterprise in the scheme of things, meeting a fraction of 1 percent of Denmark's electricity standards.

In the 2000s, much of the growth in wind was left to more entrepreneurial ventures. State-owned DONG merged with rivals to fortify its position as a power company focused on oil and gas. But in 2012, the six thousand-employee utility verged on financial disaster. As the U.S. fracking boom boosted natural gas production to record highs, global prices collapsed by 85 percent, peak to trough, within four years. Though it seemed like good news for Danish electricity consumers, DONG's profit margin evaporated. Standard & Poor's downgraded the utility's credit rating to negative. The chief executive stepped down.

To replace him, the board tapped a leader from outside the energy sector. Henrik Poulsen, forty-five, was a former leader at famed Danish innovator LEGO, which pulled off a major turnaround during his tenure. At a time when DONG's future was cloudy at best, Poulsen was tasked with restoring the company's financial foundation and building a new strategy for growth.

In 2006, DONG merged with five other Danish energy companies to become an integrated oil, gas, and power company, still with fossil fuels at the core of the business.

In 2012, the offshore wind market barely existed. Costs were prohibitively high. To build platforms out in the ocean added risk, not to mention well-heeled opposition from the owners of beachside homes.

Henrik Poulsen faced some stark choices. Another CEO might have panicked, laying off workers until natural gas prices recovered. Not Poulsen. He seized the opportunity to make fundamental change.

The world's first offshore wind farm, built in 1991 off the coast of Denmark.

Henrik Poulsen

Soon after I joined DONG in August 2012, it found itself in a deep crisis. S&P downgraded the debt. Most other energy companies in Europe came under significant pressure too. The legacy business in conventional power production was quickly eroding. The liquified natural gas and gas storage business came under significant price pressure, driven by the shale boom in the United States.

It was clear to me that we needed a new action plan. We went through the business, asset by asset, to find out where we saw competitive strength and the potential for future market growth.

We settled on a plan to divest a long list of noncore businesses to reduce our debt. We needed to build an entirely new company. I believed strongly that we needed to shift from black to green energy to combat climate change. There was only one business that we decided to develop for our growth strategy, and that was offshore wind.

I believed we had a unique opportunity in offshore wind. We also had an early mover advantage. I believed we had no choice but to go all in.

Radical transformation is never easy. We looked at all existing wind farms out in the sea. Our leadership team reviewed all the costs and data. These operations were way too expensive; the energy cost more than twice as much as onshore wind.

We put a radical cost-down program in place to bring offshore wind to a level where we could beat fossil fuel energy production.

We broke offshore wind farms into their component parts, from the turbine to the transmission infrastructure, from installations to operations and maintenance.

We installed progressively larger turbines to increase the capacity of offshore wind farms. In collaboration with our suppliers, we kept bringing down the cost for each new installation.

In 2014, we put in a bid for a number of U.K. offshore wind projects, the largest auction ever in this sector, and won three of them. That secured enough volume to keep our cost reduction program going.

We had an installation cost target of one hundred euros per megawatt hour by 2020. We blew past it and reached sixty euros by 2016. Within four years, we'd reduced the cost of offshore wind by 60 percent, far beyond what we'd ever imagined. When we mobilized the whole industry and the whole supply chain behind our mission, it was really powerful.

We later scrapped the name DONG and rebranded the firm as Ørsted, after the legendary Danish scientist Hans Christian Ørsted, who first discovered that electric currents create magnetic fields.

There was only one business we decided to develop, and that was offshore wind.

When a new market opens, there's often little competition. First movers gain a big advantage—and for the wind industry, Denmark had two of them. Vestas, an industrial equipment manufacturer, became one of the first makers of wind turbines and is now the largest in the world.

But Ørsted was the first to see that offshore locations could be scaled far larger than land-based sites. Each of its early offshore wind farms generated about 400 megawatts. The new business was growing and profitable. In 2016, the company went public with a market value of $15 billion. Four years later, it reached $50 billion. As European governments showed rising interest in offshore wind, other companies jumped in and helped drive down costs for the entire industry.

Henrik Poulsen

Our global pipeline of new projects enabled a more industrialized approach to the design and construction of offshore wind farms. Rather than thinking of them as one-off projects, it became a standardized conveyor belt. That gave us the momentum and conviction to push forward aggressively in waters around Europe, into Asia, and across to North America.

We had no idea whether there would be a significant offshore wind market in the United States. But we opened our North American headquarters in Boston. Then we acquired the company that won the bid for the first plant, the Block Island Wind Farm off the shore of Rhode Island. Now it lowers carbon dioxide emissions by 40,000 tons per year. That's the equivalent of taking 150,000 cars off the road.

Since then, offshore markets have opened up around the world, and we've won a fair share of those bids. What's especially amazing is how we've repurposed our people to learn new skills. The business has been turbulent, but the mission has infused a new purpose into the company.

By 2020, 90 percent of the energy generated by Ørsted was renewable. After cutting its own CO_2 emissions by 70 percent, it was named the world's most sustainable company at the World Economic Forum. Today it's the world's largest offshore wind developer, with one third of a growing global market. For any fossil fuel company looking to escape its past, Ørsted is a role model par excellence.

Size Matters: Ørsted's Wind Turbines Produce More Power as They Get Larger

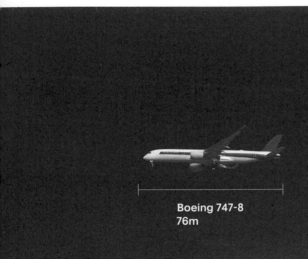

Boeing 747-8
76m

↓
Vindeby

Year: 1991
Diameter: 35m
Height: 35m
Capacity: 0.45MW

↓
Middelgrunden

Year: 2001
Diameter: 76m
Height: 64m
Capacity: 2.00MW

↓
Nysted

Year: 2003
Diameter: 82m
Height: 69m
Capacity: 2.30MW

↓
Horns Rev 2

Year: 2010
Diameter: 93m
Height: 68m
Capacity: 2.30MW

Anholt

Year: 2013
Diameter: 120m
Height: 82m
Capacity: 3.60MW

Westermost Rough

Year: 2015
Diameter: 154m
Height: 102m
Capacity: 6.00MW

Burbo Bank
Extension

Year: 2017
Diameter: 164m
Height: 113m
Capacity: 8.00MW

164m

The Dirty Secret About Natural Gas

In April 2020, the Environmental Defense Fund reported an emissions emergency in the Permian Basin in West Texas, the largest mining and drilling zone in the United States. The EDF's chief scientist, Steve Hamburg, was startled by the images and data rolling in from their methane emissions monitoring network: satellite feeds, surveillance drones, helicopters with infrared cameras.

Then he grew alarmed. "These are the most emissions ever measured from a major U.S. basin," Hamburg reported. The Permian site was leaking 4 percent of the basin's total natural gas production—a gush of climate-punishing methane.

The monster emissions flare-up showed how accurate, real-time measurement strengthens our efforts to cut and contain emissions. Within days, the Environmental Defense Fund was firing off legal memos to the responsible oil and gas companies and their regulators, insisting that they move to shut down the leaks.

A one-time professor of environmental science at the University of Kansas and Brown University, Steve Hamburg's life took a turn when he served as lead author for a series of frightening reports for the Intergovernmental Panel on Climate Change, work that earned his band of scientists the 2007 Nobel Peace Prize. When the EDF reached out to him the following year, Hamburg saw it as his chance to take direct climate action. He left his endowed chair at Brown to go out in the field and up in the sky.

The EDF's marquee project is MethaneSAT, a satellite dedicated to tracking and measuring methane emissions. By 2023, in collaboration with SpaceX, a joint U.S.-New Zealand space mission will launch the satellite into low-Earth orbit with a Falcon 9 rocket. MethaneSAT will be on the prowl for worldwide fugitive emissions from oil, gas, and coal sites, while tracking additional gigatons of methane from livestock farms and landfill food waste. It seems sure to raise the profile of real-time global measurement in our effort to solve the climate crisis.

Natural gas consists of up to 90 percent methane, which traps more than 30 times as much heat, pound for pound, as carbon dioxide.

Steve Hamburg

The MethaneSAT effort started with collecting the data and doing studies to quantify methane emissions. The snapshots of data we had were better than anything before. But we needed a motion picture, a continuous data stream, and not just in some places but everywhere.

We still can't collect data in lots of parts of the world right now. We need to be able to get a plane in or a crew on the ground, and we just can't get those permissions.

So satellites are the way to go. For many years, I said okay, if we want to do this at a scale and with precision, we need a dedicated methane satellite.

It started with the data we needed to collect and how much precision we could get by tracking from space. We went to faculty at both the Harvard and Smithsonian Astrophysical observatories and asked: Could you build this? Is this technologically possible? And they took a little bit of a deep breath and said, You know, there is new technology coming along.

So we said, Let's figure this out. We put everything on whiteboards and said: I think we can. This is buildable. So then we had a project.

The urgency of hunting down fugitive methane stems from its potency in warming the planet—and its relatively brief shelf life. In preindustrial times, methane was present in the atmosphere at 722 parts per billion (ppb). Today its concentration has more than doubled. If we succeed in cutting human-caused methane emissions by 25 percent by 2025 and 45 percent by 2030, it could help limit global warming in our lifetimes.

From methane sensors flown on aircraft to research conducted in homes, we know that methane leaks occur not only in oil and gas production, but at every step of the supply chain to the gas appliances we use. The sooner we replace our gas appliances with electric ones, the sooner we'll eliminate these added sources of methane leaks.

Getting out this message—as both a code-red emergency and a unique opportunity—has long been a focus for Fred Krupp, EDF's president since 1984. His organization has a long and storied history in protecting the environment, with more wins than losses. It played a prominent role in the removal of lead from gasoline and the banning of the dangerous pesticide DDT. Today the global nonprofit has seven hundred full-time employees and an annual budget of $225 million.

The urgency of hunting down methane stems from its potency in warming the planet.

Fred Krupp

Methane has this extraordinary, immediate impact that is warming the planet right now. Scientists and policy makers are waking up to the reality that mitigating methane matters in its own right, completely apart from decarbonization. Methane diffuses from the atmosphere much faster—after about ten years versus more than one hundred years for CO_2. So meeting our methane milestones by 2025 and 2030 would reduce warming and even have a *cooling effect* soon after that.

The urgency is especially apparent with the Arctic summer sea ice. We know of no way to keep it from disappearing without driving down methane emissions. That includes the methane from beef and dairy cows. But at EDF, we're focused on the oil and gas industry. We have an immediate opportunity to reduce methane and prevent those feedback loops. If we don't, we could lose the Arctic summer sea ice almost completely.

The good news is that the industry now recognizes the problem. The big players have set forth their commitments, including ExxonMobil and Chevron, Shell and BP, Saudi Aramco, Petrobras in Brazil, and Equinor in Norway. These companies formed an investment consortium called the Oil & Gas Climate Initiative, which has committed to reducing methane intensity, and supports the elimination by 2030 of routine flaring, the burning of wasted methane.

The bad news is that the consortium includes only publicly held companies, not the state-owned enterprises in Russia and Iran or in Mexico, Indonesia, and China, which will have to be addressed with the help of diplomacy. Together, all those companies need to shut down leaks and all other fugitive emissions from their operations, now.

Beyond the industry's medium-term goals, we need a *plan* for fossil fuel companies to pick up their pace and eliminate methane emissions by 2025. Oil and gas companies need a strategy for field measurement and monitoring and equipment upgrades. Many companies still operate with valves that are *designed* to release methane, based on the pressure of gas running through the valve. This legacy equipment can easily be replaced with modern valves that don't bleed gas. The technology exists today and costs only three hundred dollars per valve.

As industry begins to meet its responsibility, MethaneSAT will be out to detect leaks that go unplugged and unreported. We'll need strong legal prohibitions and enforcement against these fugitive emissions.

In 2016, in the waning months of the Obama administration, the Environmental Protection Agency finalized a methane pollution rule for leak detection and repair at all new oil and gas sites. But with the Trump administration suspending enforcement, leaks from U.S. fracking sites surged between 2018 and 2020. In April 2021, the Biden administration targeted methane leaks as a core element of its infrastructure package, with a focus on abandoned wellheads.

To get mining and drilling operations under control, nations and their regulatory bodies must be more vigilant. Laws must be enforced and leaks properly capped at both new and existing operations. "Some of these leaks are intentional," says Fred Krupp. The practice of flaring, where waste gases are burned on-site for both safety and economic reasons, must be strictly regulated as an unnecessary emission and a waste of natural gas. Open venting, which releases methane directly into the atmosphere, must be prohibited entirely.

Electrifying Everything

As a proud resident of California, I'd like to give a shout-out to Jerry Brown, our four-term governor. Long before climate became a global cause célèbre, Brown was leading the charge to a cleaner environment. He signed into law an astounding series of world firsts. In 1977, California enacted an unprecedented tax incentive for rooftop solar. The following year came the first-ever energy efficiency standards for buildings and appliances. In 1979, Brown signed the world's strictest antismog laws, a mandate to ban leaded gasoline, a nuclear power moratorium, and a prohibition on offshore oil drilling.

The world's fifth-largest economy has never let up. In 2002, its climate leadership turned bipartisan with the election of Arnold Schwarzenegger. California became the official world leader on climate when the

Republican governor signed the Global Warming Solutions Act, which targeted an 80 percent reduction in greenhouse gases by 2050.

Returning to office in 2011, Jerry Brown picked up where he left off in signing landmark legislation. In 2018, the year before he left office, he signed a mandate for 60 percent clean electricity by 2030 and 100 percent by 2045. California became the world's largest jurisdiction to commit to completely clean electricity.

From a national standpoint, however, we have a long way to go. Roughly half of American homes and restaurants still rely on gas stoves and ranges, and many cooks are reluctant to switch to electricity. Gas utilities are crafting marketing campaigns to exploit people's bias toward fossil fuels.

Yet according to extensive testing by Consumer Reports, all-electric induction ranges outperform their gas counterparts for most cooking tasks, including boiling, simmering, and broiling. Instead of a flame, these stoves use magnetic fields to produce energy to heat your cast iron or steel pots and pans. With no burners, they are safer than gas ranges and throw fewer toxic fumes into the air. "I love our big induction range," says James Ramsden, co-owner of the Michelin-starred London restaurant Pidgin. "I'd never go back to gas." Superstar chefs Thomas Keller, Rick Bayless, and Ming Tsai are also on board. Building codes everywhere should stipulate induction stoves for new construction. Incentives for existing buildings can ease switch-outs over time.

All-electric induction ranges outperform gas stoves for most cooking tasks.

Our Majestic Energy Future

As we change the way we cook, the way we heat our homes, and the way we drive, electrical demand—long flat in many states—will rise again. **Our electrical grid needs upgrading to support the energy loads of the future** and the growing influx of variable sources like solar and wind. American grids might be charitably described as antiquated. To meet demand in real time and move energy over long distances via high-voltage transmission lines, the grid needs to get a lot smarter.

Some utilities do a better job of this than others. A few have installed "demand response" systems with software connected to chips built into thousands of thermostats. In a regional power crunch, "smart" grids credit consumers who agree to have their air-conditioning reduced to cut peak-power usage. Another approach is "net metering," where rooftop solar feeds energy back into the grid whenever a home produces more than it consumes. Beyond shaving solar owners' utility bills, net metering is a win for the planet.

As more zero-emissions power surges through energy grids around the world, an even greater challenge looms. The 27,000 terawatt-hours of electricity generated by the world today will soon fall well short of our needs. By 2050, according to the International Energy Agency, we'll need at least 50,000 terawatt-hours of capacity for tens of millions of additional electric cars, among other things. Once solar panels are powering homes, as Elon Musk and Lynn Jurich understood from the start, people's garages will become their gas stations.

With solar homes, people's garages become their gas stations.

The Power of Energy Efficiency

For much of history, national economies grew in lockstep with their energy usage. Many believed that the energy needed to generate a given dollar of gross domestic product was more or less fixed. But an Oxford-trained physicist named Amory Lovins took a contrary view: that we can use far less energy and still have our economies grow. In 1982, Lovins cofounded the Rocky Mountain Institute to promote energy efficiency. His solar-powered house in Old Snowmass, Colorado, with 99 percent passive heat, became a showcase for efficient design. It features a greenhouse where Lovins grows bananas year-round without a furnace.

Lovins says that leaps in efficiency are so predictable that they can reliably accelerate our transition to a zero-emissions future. LED lighting, for one example, uses 75 percent less electricity than traditional bulbs. More efficiently designed pipes and ducts can cut up to 90 percent of the friction of pump and fan systems.

In 2010, consulting on a retrofit of New York City's Empire State Building, the Rocky Mountain Institute showed the way to a 38 percent energy savings. Other famous buildings followed. "What's improved aren't necessarily the technologies," Lovins says. "It's the design, the choices you make in combining technologies already invented." At the Empire State Building, gains came from insulating windows, adding radiant barriers, and optimizing the heating and cooling system. Any office building or home could do the same.

In the United States, buildings use nearly 75 percent of our electricity. They must be heated and cooled—most often, for the moment, with two separate appliances—a furnace powered by natural gas or oil and an air conditioner powered by electricity. The next leap is to get rid of the old appliances entirely and install an electric heat pump that provides both services in one device. These clever systems both heat and cool and turn one unit of electricity into three units or more of heat, with industrial versions available for large buildings. While this technology still needs to drop in price, it's ready and waiting at your nearest authorized dealer.

Most people will consider a heat pump only when their old equipment breaks. But utilities could offer incentives to replace gas appliances, just as they've incentivized efficiency measures like weather stripping, insulation, and Energy Star appliances. In 2019 alone, the Energy Star program helped Americans cut energy costs by $39 billion and reduce greenhouse gas emissions by 390 million tons, or 5 percent of the nation's total.

Decarbonize the Grid

As of 2018, the United States ranked an uninspiring tenth in energy efficiency, behind Germany, Italy, France, and the United Kingdom. Had the rest of the United States kept pace in energy efficiency with California, we would have cut our current CO_2 emissions by 24 percent.

Most of the potential in this arena remains untapped. For Lovins, the next generation of efficiency gains could dwarf the savings achieved since the 1970s.

The big incumbent utilities are acutely vulnerable to disruption. Gathering forces, from regulatory shifts to smarter, more efficient grids, could gobble up half the industry's revenue by the end of this decade. Every rooftop solar installation means less revenue for conventional utilities.

The grid is slowly moving away from the old, inefficient, fossil fuel model: centralized, one-way, supply-focused, and brittle in periods of peak demand. The smart renewables grid of the future will be distributed, two-way, and customer-focused. It will be both more efficient and more resilient. One hurdle is to amass enough storage for on-demand electricity, to compensate for the variability of solar and wind. But with continued cost reductions for renewables, grids can continue to add cleaner sources and storage that suit the demands of their customers and the constraints of where they operate. Every procurement, efficiency improvement, and metering policy change is one step closer to a clean grid.

Humanity's transition to a new energy model over the next three decades will be a majestic achievement. Ultimately, every fossil fuel power plant will need to be shuttered. Natural gas must be phased down, and coal consigned to the past—there is no way around it. The objective is to make the electrical grid 100 percent emissions-free in as many countries as possible, as quickly as possible.

Speed & Scale: Countdown to net zero

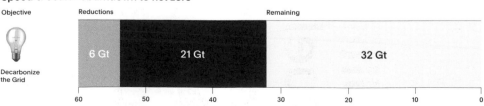

Objective	Reductions		Remaining

Decarbonize the Grid

6 Gt — 21 Gt — 32 Gt

60   50   40   30   20   10   0

But to safeguard our climate, we need to broaden our sights beyond energy—in particular, to how we feed people. What we eat and how it's grown accounts for an alarming share of the world's greenhouse gas emissions. In the next chapter, we'll examine how we can transform our food and agriculture system to help us arrive at our net-zero future.

Ultimately, every fossil fuel power plant will need to be shuttered. Natural gas must be phased down, and coal consigned to the past — there is no way around it.

Fix Food

Chapter 3

Fix Food

After *An Inconvenient Truth* drew unprecedented attention to the climate crisis, the search was on for grand-scale solutions to decarbonize electricity and transportation. For far longer than I've invested in cleantech solutions, Al Gore has publicly called out emissions offenders and rallied support for promising technologies. Decades after convening the first U.S. congressional hearing on climate change as a twenty-eight-year-old freshman in the House of Representatives, Al returned to his roots. He's refocused on what may be the most promising climate solution of all: growing food in better ways.

Al spent his early summers working on Caney Fork Farms, the family's acreage near Carthage, Tennessee. Albert Gore Sr. walked with his son around the farm, to teach him where the best soil lay. When they came to a river bottom, the soil was black and moist. The son held it in his hands. "The dark soil is the best soil," Al says. "My dad taught me that."

Al never forgot that lesson. But he admits to embarrassment that it took him another fifty years to understand *why* that rich soil was black. The reason was carbon. "Higher concentrations of carbon help feed all the life in the soil," Al says. "And the darker soils retain moisture better, because the carbon builds a latticework that holds it in place."

What happens at the microscopic level determines what happens at the planetary level. The Earth's soil, Al notes, contains 2,500 gigatons of carbon, more than three times the amount in the atmosphere. To reach net zero, we need soil to absorb even more of it. The potential is huge—but we're moving in the wrong direction. Our topsoil is in danger. Over the past century, fully one third has been depleted.

Al Gore at Caney Fork Farms in Carthage, Tennessee.

Back at Caney Fork Farms, Al is implementing what should be standard practice for the farms of the future. "The entire farm, the barn, the food production, the house—all of it runs on one hundred percent renewable energy," he says. "But the most critical part of the farm is the topsoil." Whether growing lettuce or squash or melons, the challenge is the same: keep the soil as carbon rich as possible to fuel more interaction between plants and microbes.

In the 1930s, depleted farm soils turned the American plains into the Dust Bowl.

In the 1930s, poor farming practices so depleted the plains of Texas, Oklahoma, and Kansas that much of the region's topsoil blew away with the wind. Hot brown gusts rose taller than buildings and blotted out the sky. Since the days of the Dust Bowl, we've learned a lot about crop rotation and the importance of cover crops for holding our fragile soil in place. This hard-won wisdom spawned the regenerative agriculture movement.

In traditional farming, plows tear apart the soil's connective tissue, disrupting the natural ecosystem and pouring carbon dioxide into the air. Nitrogen-rich fertilizers try to coax more productivity from the damaged dirt. Then come the pesticides and herbicides, shedding chemicals into our streams and water tables and killing valuable microorganisms. The nitrous oxide that rises from industrial fertilization traps heat at three hundred times the rate of CO_2 and stays in the atmosphere for a century and more. Fertilizers alone account for 2 gigatons of CO_2-equivalent emissions.

All told, more than 15 percent of the entire emissions emergency, around 9 gigatons a year, can be attributed directly to our food system—to industrial farming, livestock (especially beef cattle), rice production, and emissions from fertilizers and food waste. To reach net zero, we must change the way farming and our food system works, from the ground up.

By 2050, the global population will expand to nearly 10 billion people, up from 7 billion today. A growing middle class will increase demand for meat and dairy products. For everyone to have enough to eat, we'll need to produce up to 60 percent more calories than we did in 2010. We've embedded the speed and scale required in the following OKRs:

Objective 3
Fix Food

Reduce agricultural emissions from
9 gigatons to 2 gigatons by 2050.

KR 3.1   Farm Soils: Improve soil health through
           practices that increase carbon content in
           topsoils to a minimum of 3%.

KR 3.2   Fertilizers: Stop the overuse of nitrogen-
           based fertilizers and develop greener
           alternatives to cut emissions in half by 2050.

KR 3.3   Consumption: Promote lower-emissions
           proteins, cutting annual consumption of
           beef and dairy 25% by 2030, 50% by 2050.

KR 3.4   Rice: Reduce methane and nitrous oxide
           from rice farming by 50% by 2050.

KR 3.5   Food Waste: Lower the food waste ratio
           from 33% of all food produced to 10%.

To supply enough food for all while cutting our emissions from agriculture, we must tackle all five of these drivers. Our **Farm Soils KR (3.1)** targets improved soil health as measured by the carbon content in our topsoils. By accelerating the adoption of regenerative agriculture, we can boost that carbon content. Applied broadly, this practice could absorb 2 gigatons of CO_2 each year.

Our **Fertilizers KR (3.2)** calls for limiting the use of nitrogen-based fertilizers, the source of 2 gigatons of CO_2e. With new delivery methods and precision techniques for fertilizer timing and placement, farmers can cut emissions with no harm to yields. In addition, we must invent ways to produce fertilizers without using fossil fuels. Together, these actions could cut carbon dioxide and nitrous oxide emissions in half.

Our **Consumption KR (3.3)** aims to cut emissions from the production of livestock—and cattle in particular—by reducing consumption of beef and dairy products. To do so, we need to improve and scale plant-based alternatives to compete with beef and dairy products, and to shift demand from high-emissions foods. Carbon labels and dietary guidelines can guide consumers to better choices.

Our **Rice Methane KR (3.4)** reduces methane emissions from paddies while still growing enough rice, a staple food in much of the world.

Our **Food Waste KR (3.5)** reins in emissions from foods discarded in production and transport or by retailers and consumers. Worldwide, one third of all food produced today is wasted. Most of it wends its way to landfills, where it generates nearly 2 gigatons of CO_2-equivalent emissions, mainly methane gas. Reducing food waste also eases the burden on production. Every pound of wasted food is a waste of energy and water.

The Unrivaled Potential of Topsoil

To see why soil is crucial, we need to understand how it works. Soil is created over time, as carbon-rich plant and animal residues are broken down by insects and millipedes, then by bacteria. The organic matter that's left is a storehouse of carbon and also nutrients for plants. Healthy, undisrupted soil contains a network of subterranean pores, the work of plant roots, fungi, and earthworms. These micro-tunnels allow roots to extend more deeply into the soil and help the soil retain water, making it more drought resistant.

Regenerative agriculture is a coordinated set of farming and grazing practices that enhance the soil's ability to retain carbon.

It rebuilds organic matter and restores soil biodiversity, the variety of life. The regenerative movement limits traditional plowing and tilling, which expose buried organic matter to oxygen, hasten decomposition, and hurl carbon dioxide into the air. By contrast, no-till farmers punch thousands of shallow holes—no wider than a kernel of corn—into the dirt. Seeds are implanted with minimal disruption to the topsoil. Roots grow deeper, tapping more nutrients and moisture along their way. Limited to less than 7 percent of croplands worldwide in 2004, no-till farming has expanded to 21 percent in the United States and to the majority of croplands across South America.

These farmers are following proven, centuries-old practices. But as populations grew in the industrial era, as Vaclav Smil notes, it became less labor intensive to extend cultivation than to intensify planting on existing cropland. In the nineteenth century, the trend accelerated. Instead of enriching depleted soils, farmers slashed and burned woodlands and grasslands to add room to plant. In the twentieth century, industrial agriculture fostered higher yields on existing croplands and more profits per acre, but at a price: more emissions.

Regenerative agriculture challenges the modern reliance on chemical fertilizers and pesticides. These farmers use cover crops like clover to

Less tilling creates healthier roots and soil

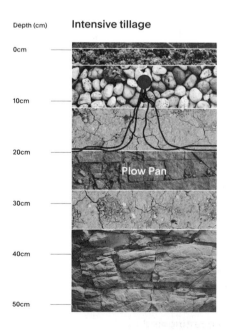

Depth (cm) — Intensive tillage — 0cm, 10cm, 20cm, Plow Pan, 30cm, 40cm, 50cm

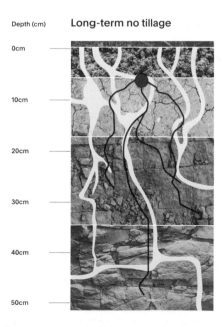

Depth (cm) — Long-term no tillage — 0cm, 10cm, 20cm, 30cm, 40cm, 50cm

Adapted from information and visuals by Ontario Ministry of Agriculture and Food

Regenerative agriculture explained

Increasing biodiversity
to boost nutrients, natural
decomposition and attract
insect predators of pests

Cover crops
that are grown in the soil
after the commercial
harvest and can be grazed
or harvested themselves

Rotating crops
to naturally balance what
is being taken out and put
into the soil

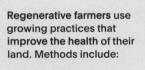

Regenerative farmers use
growing practices that
improve the health of their
land. Methods include:

Integrating livestock
to combine animals
and plants in a circular
ecosystem

Minimizing chemical inputs
that destroy biodiversity
and pollute waterways
due to runoff

Minimum till systems
that improve soil health and
prevent erosion thanks to
minimal soil disturbance

Adapted from visual by Eit Food

nourish the soil and guard against weeds. At the end of the cover crops' life cycle, they are left to compost, leaving behind a natural layer of mulch and nutrients. If 25 percent of the world's farmlands used cover cropping, they could remove nearly half a gigaton of carbon dioxide per year from the atmosphere while helping to prevent droughts. In 2019, in the United States alone, 20 million acres lay fallow after floods knocked them out of commission. By holding more topsoil in place, regenerative farms can thrive again after the waters recede.

Crop rotation, an age-old regenerative practice, restores essential nutrients to the soil. Well-managed grazing can replace chemical fertilizers with manure. Combine the two practices and you have rotational grazing, where a portion of pastureland is emptied of livestock in off years. Another regenerative mainstay is silvopasture, the integration of trees into livestock pasturelands. "You can see it when you take a walk through the pasture," says Al Gore. "The trees are just better for the land."

When spaced properly, trees mitigate summer heat stress on grazing animals while permitting enough sun to penetrate to allow for "understory" grass growth. Silvopasture also diversifies farmers' income; the trees can be harvested for timber, while the grass can be turned into hay or biofuel.

All things considered, regenerative agriculture can be more profitable than industrial farming. Even so, many large industrial farms remain invested in the status quo. Near-term costs of the transition can exceed what many financially strapped farmers can afford. To speed the shift to regenerative agriculture, governments must provide incentives to farmers and entrepreneurs to deploy these new solutions.

Stop the Overuse of Fertilizers

Due to its outsize impact in trapping heat in our atmosphere, nitrous oxide is an especially noxious greenhouse gas. Though present in relatively small quantities, it accounts for 5 percent of total global emissions. Most nitrous oxide derives from fertilizers that are especially popular among corn farmers in the United States. Since many natural fertilizers have their own nitrous oxide profile, they aren't much better as an alternative.

According to the World Resources Institute, nitrous oxide emissions can be reduced with cover crops like legumes, which cultivate microbes that capture nitrogen from the air in a form plants can use. They can be further cut with nitrification inhibitors, the farming equivalent of time-release capsules.

Modeled after car fuel-economy standards, government nitrogen efficiency standards could push fertilizer companies to clean up their products' performance in the field—especially if the standards come attached with financial incentives.

Creating synthetic fertilizer is a carbon-intensive process that produces ammonia by fusing hydrogen from fossil fuels with nitrogen from the air. It requires high heat and extreme pressure. As a cleaner alternative, companies around the world are exploring the use of solar or wind to make "green ammonia." In the short term, using less fertilizer will lead to fewer emissions. In the longer run, we need cleaner ways to produce synthetic fertilizer to reduce emissions at scale.

The Menace of Methane

When I was in high school, I worked one summer at a fast-food restaurant called Burger Chef, flipping hamburgers. I did such a fine job that the manager encouraged me to consider a future in the business: "Doerr, you'd make good hamburger material." (My siblings got a kick out of that one.) That job taught me two important things: the rigor of doing a task right each time, and Americans' deep love for burgers.

What I wouldn't learn until years later was that feedlot beef has the highest emissions footprint of all popular foods, by far.

After Argentina, the United States eats more beef per capita than any other country. Typical Americans annually consume more than their own weight in red meat and poultry, about 220 pounds—a bonanza for the fast-food industry, which brings in $648 billion in revenue worldwide, one third of it from the United States. Convert that booming business into emissions and you begin to get a sense of the scale of the problem.

Any serious discussion of the climate crisis must focus on atmospheric methane, much of it from livestock and food waste. Together they generate 12 percent of all greenhouse gases, or 7 gigatons of CO_2e per year. As you might have guessed, cattle is king, at 4.6 gigatons. If the world's 1 billion cows were a country, they'd rank third in greenhouse gases after China and the United States. Accounting for nearly two thirds of total livestock emissions, beef and dairy cattle dwarf the climate threat from all other farm animals combined, including pigs, chickens, lambs, goats, and ducks.

Emissions by kilogram of food

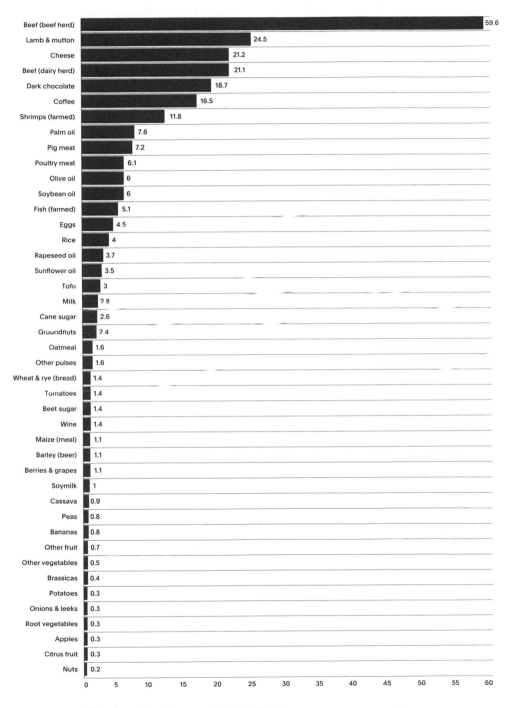

Food	Emissions
Beef (beef herd)	59.6
Lamb & mutton	24.5
Cheese	21.2
Beef (dairy herd)	21.1
Dark chocolate	18.7
Coffee	16.5
Shrimps (farmed)	11.8
Palm oil	7.6
Pig meat	7.2
Poultry meat	6.1
Olive oil	6
Soybean oil	6
Fish (farmed)	5.1
Eggs	4.5
Rice	4
Rapeseed oil	3.7
Sunflower oil	3.5
Tofu	3
Milk	2.8
Cane sugar	2.6
Groundnuts	2.4
Oatmeal	1.6
Other pulses	1.6
Wheat & rye (bread)	1.4
Tomatoes	1.4
Beet sugar	1.4
Wine	1.4
Maize (meal)	1.1
Barley (beer)	1.1
Berries & grapes	1.1
Soymilk	1
Cassava	0.9
Peas	0.8
Bananas	0.8
Other fruit	0.7
Other vegetables	0.5
Brassicas	0.4
Potatoes	0.3
Onions & leeks	0.3
Root vegetables	0.3
Apples	0.3
Citrus fruit	0.3
Nuts	0.2

GHG numbers reflect emissions generated across supply chain.

Adapted from data by Joseph Poore and Thomas Nemecek and visuals by Our World in Data

Most of us don't think much about emissions as we bite into a burger or a slice of pepperoni pizza. But these everyday meals create huge amounts of emissions in every phase of their production cycle, from the fertilizers used to grow cattle feed to the cow's digestive process (belching, mostly). In addition, the 80 pounds of daily manure from a 1,000-pound dairy cow has an emissions profile of its own.

More than 75 percent of farmland is dedicated to raising and feeding animals for our food supply. Yet these animals supply only 37 percent of our global protein and just 18 percent of our global calories. Aside from their considerable impact on greenhouse gas emissions, they're an underperforming, inefficient food source.

As the global demand for calories grows, available land can only get scarcer. As major segments of the population see their incomes rise, so too will their demand for meat and dairy products, adding pressure to clear more land for farming and cattle ranching—a chief driver of deforestation. Our net-zero goal is undercut twice—through methane emissions from livestock and from the release of carbon as trees burn or rot. (Our Speed & Scale Plan calls for an end to all deforestation, a topic addressed in the next chapter.)

Most writings on climate change are pessimistic about the prospects for wholesale emissions cuts from food production. I don't minimize the difficulties. By 2050, nearly 10 billion people will need to eat—and they'll want to eat what they like to eat. And as we've observed, Americans in particular like to eat beef and cheese. A lot.

So how can we possibly satisfy people's appetites while paring gigatons of livestock and farming emissions? Though we may never fully eliminate these greenhouse gases, we can certainly accelerate reductions and reach a manageable annual budget of 2 gigatons by 2050. But even that partial mitigation will demand action through the collective power of markets, innovation, education, policy, and measurement.

The shift is already in progress. Plant-based proteins have entered the market in a big way as a viable alternative to meat. These low-emissions food products are improving in taste all the time; the best of them approach the experience of eating beef or pork. They're widely available in supermarkets and restaurants, and are scaling fast.

On the supply side, there's promising R&D for cutting enteric emissions (the belching problem) by mixing natural additives into cattle feedstock. According to research at the University of California, Davis, small amounts of seaweed can cut emissions by an impressive 82 percent.

Education is one more powerful tool. Rolled out in 1994, the Food and Drug Administration's Nutrition Facts labels are correlated with healthier eating: a 7 percent drop in average U.S. calorie intake and a 14 percent rise in vegetable intake. In similar fashion, climate guidance food labels could lead consumers to planet-friendly choices and expand markets for low-emissions foods. Consumers "underestimate the emissions associated with food but are aided by labels," says a Duke University study.

In 2019, Denmark became the first country to propose "environmental price tags" in food stores. "Food is the strongest lever we have as individuals for fighting climate change," says Sandra Noonan, chief sustainability officer at Just Salad, a restaurant chain that has adopted climate labels.

Clear climate labels can guide consumers

Grass-fed Beef

100% natural, raised without antibiotics
Use by 08/23

Climate Footprint
27kg CO_2e
Per kg

Climate Footprint
27kg CO_2e
Per kg

In 2020, Panera Bread became the largest chain to designate "climate-friendly" foods. In collaboration with the World Resources Institute, the restaurant awarded "Cool Food Meal" badges to low-emissions menu choices. Adding quantified carbon footprint data to easy-to-understand labels would do even more to help customers' decision making.

Back in 1992, the U.S. Department of Agriculture introduced the country's first official dietary guidelines, the "Eating Right Pyramid." By 2011, the pyramid had morphed into a plate; in 2020, it added more emphasis on vegetables and grains. These guidelines have exerted significant influence over what people eat, from school lunches and corporate cafeterias to individual choices. By steering consumers away from beef and dairy and toward plant-based proteins, policy makers can boost demand for lower-emitting foods.

What would the climate-optimal diet look like? Botanist and author Michael Pollan has a simple prescription: "Eat food. Not too much. Mostly plants." Research from Johns Hopkins University concluded that a "two-thirds vegan diet," restricting meat and dairy to a maximum of one serving apiece per day, would cut livestock emissions by as much as 60 percent.

With tasty plant-based meat and dairy, we need not limit servings at all.

Reinventing the Burger: The Beyond Meat Story

In 2010, a young Kleiner partner named Amol Deshpande, alert to reports of global food shortages, began looking into technologies that could use plant proteins to replicate the texture and flavor of meat. Later that year, around the time I began to grasp the magnitude of cattle emissions, Amol escorted a giant of a man named Ethan Brown into our office for a presentation. Standing six foot five in jeans and a T-shirt, Ethan made a powerful impression. More than anything else, I was blown away by his vision for "a plant-based McDonald's," and his passion for making an all-natural, plant-based burger that could compete on taste with the real thing.

A climate-friendly diet: lots of fruits and
vegetables, limited animal-based proteins

Ethan Brown

I was raised in Washington, D.C., and College Park, Maryland, where my father was a professor at the University of Maryland. Not being one for city life, my dad spent as much time as he could on a farm we owned in the mountains of Maryland. Though he bought it for recreation and conservation purposes, being entrepreneurial himself, we soon had a hundred head of Holstein cattle and a milk operation.

As a child, I was fascinated by the animals around us, whether in our home, the barn, or streams or woods. My earliest career ambitions were to be a veterinarian.

I grew up eating meat, and at my size, probably more than most. One of my favorite fast-food items was the Double R Bar Burger with ham and cheese at Roy Rogers. As I got older, reflecting on time spent on our farm, I found it more difficult to disassociate the products (ham, cheese, beef) from the animals from which they came.

Fast-forward to my early twenties. I was sitting with my father in his office at the University of Maryland and discussing my career path. He asked me an important question: What's the most important problem in the world? I thought it must be climate change. If the climate collapsed, nothing else mattered.

And so, after finishing school and a stint working overseas, I narrowed my focus to clean energy in service to climate. I progressed quickly in the field, got married, had children, a mortgage. And then two things happened. One, by my early thirties, my discomfort intensified as I realized that the food system that my kids were growing up with was largely unchanged, and they would face the same dilemmas and narrow choices I had. Two, my personal interest in animals and agriculture and a career focused on energy and climate started to merge. Specifically, I can recall attending cleantech conferences where thousands of professionals would come together to discuss how to increase the efficiency and density of a fuel cell or lithium-ion battery and then go out for a steak dinner. Having learned of the vast emissions associated with livestock, I couldn't help but think that a massive solution was awaiting us right there on the plate.

By my late twenties, my diet was fully plant based. In my mind, the earliest iteration of what is now Beyond Meat was a plant-based McDonald's. As I soon realized, however, what we needed more than a venue were better products. And to get better products, we couldn't think about "meat substitutes" as a culinary exercise. We needed to apply science and technology and big budgets—what I had seen in the energy sector—and get away from "alternatives" and "substitutes." We needed to go toward building meat itself directly from plants: plant-based meat.

For me, a real breakthrough occurred when I stopped thinking about and defining meat in terms of its animal origin (e.g., chicken, cow, pig), but instead in terms of its composition. At a very high level, meat is really five things: amino acids, lipids, small amounts of carbohydrates, trace minerals, and, of course, water. The animal eats plants and turns them into muscle tissue, or what we call meat. But with today's technology, instead of using a biological bioreactor (animal), we can harvest those core inputs directly from plants themselves. We can use other systems to assemble them in the familiar architecture of meat.

I began to look around the globe for technologies that could be part of the solution. I ultimately found two researchers at the University of Missouri who were working on a method to break the bonds in plant protein and restitch the proteins into the striated structure of muscle. In 2009, the year I founded the company, I called them and introduced myself, and thankfully they agreed over time to partner with me. I then reached out to the University of Maryland for additional research support. Between the two universities, over several years, we were able to get a viable prototype.

Fix Food

By the time Ethan shared his vision with us at Kleiner, he'd raised money from family and friends for an experimental kitchen in an old hospital building. As I came to know him, I found Ethan to be one of the most authentic people I'd ever met. He was committed to giving people what they love—the experience of grilling and tasting meat— but with peas and lentils and seed oils subbing for animals. He chose the most sustainable crops and pulled out their proteins to create the biochemical essence of beef—no cow required. While Ethan might have looked like a latter-day hippie, he had a sound business plan, backed by science and consumer taste testing. Plus we loved the name: Beyond Meat. Kleiner Perkins became the first big investor in Ethan's fledgling company.

Ethan Brown

The years went by, with ups and downs. I'd put in something like $250,000 of my own money, but we needed millions to turn Beyond Meat into something real. The team at Kleiner stuck their necks out for us. Once they got involved, others came aboard and we really started to move.

Though we had brought a beef product to market in late 2009, it wasn't until 2012 that we offered consumers what I then considered to be a breakthrough in muscle structure and sensory experience, our plant-based chicken strips. Whole Foods sold them in their prepared-foods section to great fanfare, including a feature article by Mark Bittman on the cover of the Sunday Review of *The New York Times,* complete with an artist's sketch of a chicken with broccoli for a head. It was a great moment for us.

In 2016, we launched the Beyond Burger in the meat case, right next to beef from cows, first at Whole Foods and then throughout the nation, and now globally. Made from all-natural ingredients and presented in raw form for the consumer to cook, this product was our breakthrough. Even today, as we go to market with our 3.0 version, we have miles to travel before we close the gap between the Beyond Burger (and our other products) and their animal protein equivalent. We're getting there with our Beyond Meat Rapid and Relentless Innovation Program. The good news is that we don't see a material obstacle to someday achieving that perfect indistinguishable build.

In 2019, we reached another massive milestone. McDonald's began testing a burger we developed for them in a small number of stores in western Ontario, Canada. Late one evening, after meetings in Toronto, I had the opportunity to drive a couple of hours to the stores, walk in, and eat our product—it was delicious, and I savored the entire experience. Out in the parking lot, I felt both immense gratitude and a sense of relief. What began as a dream was now a reality.

Growing meant we needed more capital; it was time to take the company public. Our public offering in May 2019 took everyone by surprise, including me. We opened at more than double the offer price, and the stock quadrupled in value over the next few months. Suddenly everyone knew about Beyond Meat.

Our ancestors began consuming meat from animals over two million years ago. This dietary choice, and later the discovery of fire for cooking, delivered higher nutrient density. It was like finding a Clif Bar on the savanna versus consuming high volumes of grasses and other plants. No longer needing to process so much material, our stomachs shrank. Energy was freed up to power our ancestors' rapidly growing brain, which doubled in size. Today, we can use that brain power and technology to separate meat from animals and realize the attendant benefits to our health, the climate, natural resources, and animal welfare—for ourselves and future generations. That seems like change at an evolutionary level, and it's endlessly energizing to my colleagues and me.

Beyond Meat is aiming for price parity with beef burgers by 2024.

Beyond Meat has surpassed 118,000 points of distribution in over eighty countries, including the vast China market. It has signed strategic global agreements with McDonald's and Yum!, two of the world's largest restaurant brands. But that's only the start. A recent consumer study shows that more than 90 percent of plant-based burger customers are neither vegan nor vegetarian. The broader market confirms the staying power of plant-based meat; the category grew 45 percent year-over-year in 2020. The growth curve is showing no signs of flattening. Beyond Meat's new target: price parity with beef by 2024.

Ethan Brown is a climate crusader who has persevered in a highly competitive industry. Beyond Meat is going head-to-head with Impossible Foods, which makes its own burger with heme, a bloodlike molecule derived from soy. In 2019, Burger King started selling the Impossible Whopper worldwide. They were joined that year by Tyson Foods, which came out with chicken-like nuggets made with pea protein. Rather than buck the trend, the largest U.S. meat producer was choosing to fight for market share. Plant-based proteins have captured nearly 3 percent of the packaged meat market—about where plant-based milk was ten years ago.

Cultured meats—also known as synthetic, lab-grown, or cell-based meats—are another future facet of the alternative proteins market. After biopsies of an animal's muscle, fat, and connective tissue, cells are cultivated in a nutrient-rich serum. While synthetic meats are not vegan or vegetarian, and are still priced higher than the natural variety, their production has the potential to cut emissions. Uma Valeti, a Mayo Clinic cardiologist and the CEO and cofounder of Upside Foods, says their self-renewing cell technology could "entirely remove the animal from the meat production process."

The Dairy Dilemma

If shifting consumers from meat to plant-based alternatives still sounds like a long shot, consider recent developments in your super-market's milk section. Fifteen percent of total milk sales in the country now come from oats, soy, almonds, or other plant-based sources. With the green premium on plant-based milks nearing zero, their market share is growing by the year. No matter which type consumers choose, they're all better for the environment than dairy milk, based on three critical metrics: emissions, land use, and water use.

We should point out, however, that milk represents a small fraction of dairy-related emissions. The larger problem is cheese, the third-highest-emitting food item after beef and lamb. In global sales, the big cheese is mozzarella. It takes ten pounds of dairy milk (about five quarts) to produce one pound of mozzarella, enough for two traditional pizzas. That's equal to a productive dairy cow's daily output—and that single cow emits about 250 pounds of methane per year.

Where milk alternatives have taken off, we are still in search of a great cheese alternative. For my taste, dairy-free cheese substitutes made of nuts and soy have yet to meet the mark. But I have no doubt that food innovators will cook up something better in short order.

Rethinking Rice Cultivation

While most debate around food and climate centers on meat consumption, a seemingly innocuous staple generates hefty emissions of its own. Rice—a dietary cornerstone for over 3 billion people—provides 20 percent of calories consumed worldwide. It also accounts for 12 percent of the globe's methane emissions, with some estimates running even higher.

Rice is commonly cultivated by flooding paddies, a practice that prevented weeds from growing and was presumed to increase yields. Unfortunately, flooded rice paddies make an ideal environment for methane-producing microbes, which feed on decomposing organic matter in airless conditions.

It's a knotty problem, but solutions are afoot. Improved rice production methods are employed today by millions of small-scale farmers. They abate methane with intermittent flooding, a planet-friendly alternative to the continuous variety. Besides eliminating up to two thirds of methane emissions, these practices can double a rice farmer's yield and sharply boost profits. But they come with a catch: a drastic increase in nitrous oxide emissions, which pack a planetary-heating punch three hundred times more powerful than carbon dioxide.

To contain the problem, it's essential for water levels to be closely monitored and managed. Shallow flooding, together with nitrogen and organic matter management, can limit this seesaw effect and reduce greenhouse gas emissions up to 90 percent.

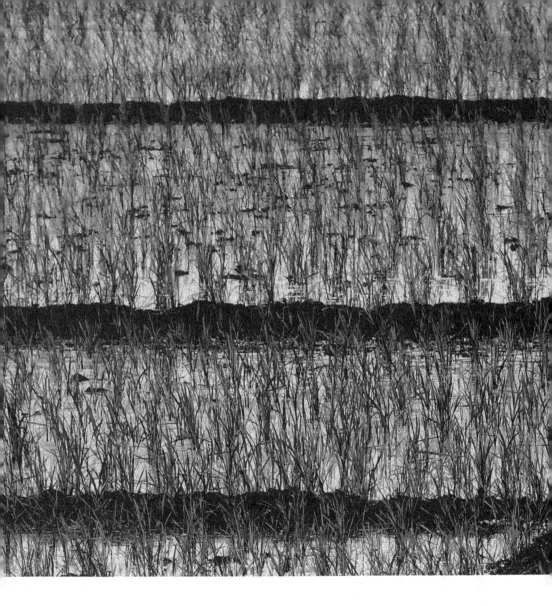

More sustainable rice farming comes down to avoiding big swings of water content. Leading suppliers are buying more grain from farms that no longer engage in continuous flooding. Uncle Ben's parent company, Mars, Inc., reached the 99 percent mark in 2020. The United Nations-backed Sustainable Rice Platform has issued a verified logo to guide consumers toward the right grain for both farmers and the climate.

*Rice is a staple
for more than
three billion
people
worldwide.*

Shifting to lower emissions cultivation is no simple check-the-box exercise. Promoting shallow flooding will require close work with hundreds of millions of growers. If asked to change long-held practices, they'll need to be convinced by the promise of greater yield and profitability. More research, education, and measurement are needed. But of the many solutions needed to ease the climate crisis, this one offers tremendous rewards at a relatively low cost.

Redirect Supply-Side Subsidies

Despite substantial progress, we're not yet on track to meet our objective in cutting agricultural emissions. We see both hopeful signs and serious obstacles. Since 2010, the global cattle population has remained steady even as the human population has grown. Milk prices have declined in the United States, eroding dairy farm profits. Some farmers are scaling down; others are selling or converting their land to other uses.

But our beef and dairy emissions problems won't solve themselves. Most countries subsidize their farmers. In 2019, U.S. government support for agricultural producers totaled $49 billion, including heavy subsidies for the dairy industry. China ($186 billion) and the European Union ($101 billion) spent even more.

Let's use this moment to disrupt the status quo—to push the food industry to lower emissions. Farmers need help to switch to new crops. As a first principle, we must shift government subsidies toward more sustainable agriculture. "A wider movement of farmers wants to get on board with regenerative agriculture," says Amol Deshpande of the Farmers Business Network. "It's both more profitable and it helps preserve their land—and ninety percent of their wealth is in their land." In shifting more land from pasturing to high-demand crops, he adds, farmers can further increase their farms' value.

In the end, fixing our food system will be more profitable and better for the planet. "Food is the mother of all sustainability challenges," says Janet Ranganathan, the World Resources Institute's vice president of science and research. "We can't stay below even a 2-degree centigrade increase without major changes to this system."

Focusing on Food Waste

To achieve a drastic reduction in agricultural emissions, we must also take on one of the biggest problems in the food system: food waste. A staggering 33 percent of the world's food is wasted every year, with even more thrown out in higher-income countries. Collectively, the food we waste is responsible for more than 2 gigatons of global emissions. Meanwhile, more than 800 million people worldwide are undernourished. In short, too much food is left unconsumed, and too many people don't have enough food.

In lower-income nations, food waste is mostly unintended—a result of improper storage, substandard equipment and packaging, or bad weather. Most waste occurs early in the supply chain, with food rotting before it's harvested or spoiling en route to buyers.

In the United States, by contrast, consumers throw out 35 percent of their food. Annual waste amounts to $240 billion, or nearly $2,000 per household. It's driven by misleading expiration dates on food labels, which prompts premature discards of safe and edible items. The waste is compounded by food items rejected at the retail level, often for superficial reasons.

This global unevenness calls for a range of solutions. In wealthier countries, our strategy includes standardized labeling, municipal composting programs, and public awareness campaigns. And we need more effective waste-reduction programs between retailers and food banks and their supply chains.

In 2015, France barred large grocery stores from disposing of unsold food that could instead be donated to charities. Each day, more than 2,700 French supermarkets send items nearing their expiration date to 80 warehouses around the country, rescuing 46,000 tons of food per year. Donations to food banks are up more than 20 percent.

But with due credit to these efforts, we need earlier interventions in the supply chain. In developed countries, the usual suspects for waste include slaughterhouses, farms, and distribution sites. In poorer nations, the problem can best be solved with upgraded infrastructure for food storage, processing, and transport. Inexpensive measures like better storage bags, silos, or crates can go a long way. Clearer communication and consistent coordination between producers and buyers is another must.

Working with the private sector, governments will need robust measurement and reporting to track progress. As avoiding food waste becomes a higher priority, we can meet our objective to cut one full gigaton of emissions.

Land in Limbo

While decarbonizing the food sector will be challenging, the impact can be huge. By fixing food, we will substantially cut agricultural emissions, a giant step toward climate stability, healthier lives, and less hunger.

To help people make better choices, we need more education on the links between food and greenhouse gas emissions. We need more innovation in regenerative agriculture. And we need more money— a lot more—to promote sustainable choices by farmers, suppliers, food companies, retailers, restaurants, and, not least, consumers.

Part of the solution lies in enhanced agricultural productivity and efficiency. To grow everything we'll need without devastating valuable

By fixing food, we will substantially cut agricultural emissions.

resources, we'll need to harvest more food and more calories per acre. U.S. farmers have trended in this direction over the last half century. Efficiency gains have reduced the amount of land and water used to produce a pound of beef or a gallon of milk. Worldwide, the public sector has a big role to play. More research and development can unlock greater productivity; better policy can dissuade farmers from converting forest or grasslands into croplands.

By the year 2050, humankind will have some large bills coming due. As we've noted, our growing population will require more than 50 percent more calories. Meeting people's needs while cutting emissions and protecting vital ecosystems will be a tall task. But by embracing more productive and regenerative agriculture, shifting to precision fertilization, encouraging lower-emission diets, and reducing waste, **we can both help out the climate and enjoy our food, as we should.**

Speed & Scale: Countdown to net zero

Fix Food

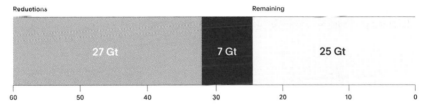

| Objective | Reductions | | Remaining |
| 27 Gt | 7 Gt | 25 Gt |

Protect Nature

Chapter 4

Protect Nature

I like to approach problems as an engineer—looking at the whole, then dissecting the parts. As a student at Rice University, I would take apart old audio equipment and repurpose it for KTRU, our campus radio station. Sometimes we'd use it for concerts as well, which drove home to me a certain principle. When setting up the stage, a common mistake was to place a live microphone too close to a loudspeaker. The result would be that familiar, high-pitched noise, so sharp and piercing that it's painful.

That problem is known as a "feedback loop." The audio from a speaker is picked up by the microphone, amplified out of the speaker, picked up by the mic again, and sent out of the speaker even louder. When feedback strikes, you need to either switch off the mic or cut out the speakers. Otherwise, the unchecked amplification can blow out the equipment—and your ears.

In our climate today, we're seeing several dangerous feedback loops. It's terrifying and uncharted territory; even the best climate models don't fully account for them. To better understand the damage being done, think of Earth as a gigantic, supremely intricate machine. High concentrations of atmospheric carbon warm the globe. High ambient temperatures suck the water from forests. The dry heat ignites and spreads wildfires, which throw the trees' stored carbon into the atmosphere, which drives temperatures still higher. That's the quandary we're in now.

Feedback loops can get stuck on repeat. If the fundamental cause of carbon emissions doesn't get switched off, the global warming loop will become a runaway disaster. Whole ecosystems will be destabilized—forests, to be sure, but also farmlands and savannas, river deltas and oceans. And when a disturbance goes on long enough, it can reach a point of no return. One of the most vulnerable regions in this respect—with terrifying planetary implications—is the permafrost, the frozen ground beneath the surface of the Arctic lands. As it thaws because of warming temperatures, microbes break down plant matter that had been frozen in the soil for eons. CO_2 and methane are

Protect Nature

Carbon moves through the land, atmosphere, and oceans

Adapted from data and visuals by U.S. DOE Biological and Environmental Research Information System

Human emissions   Natural fluxes   Stored carbon

Photosynthesis

Human emissions

Plant respiration

Plant biomass

Soil carbon

Microbial respiration
& decomposition

Fossil carbon

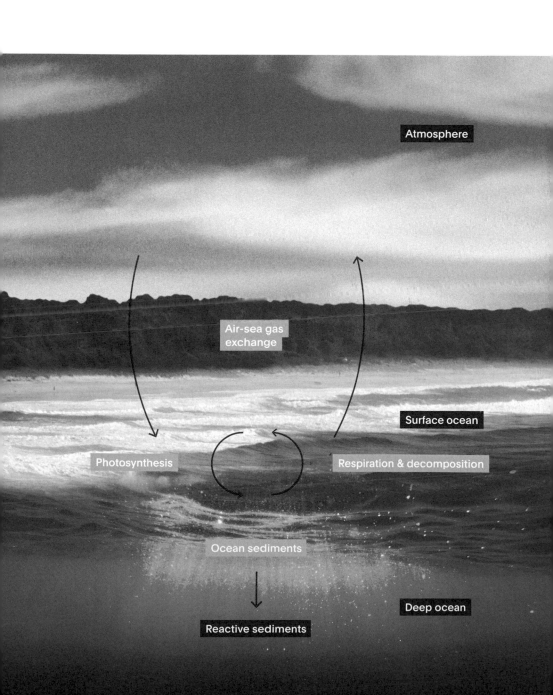

Atmosphere

Air-sea gas exchange

Surface ocean

Photosynthesis

Respiration & decomposition

Ocean sediments

Deep ocean

Reactive sediments

discharged into the atmosphere. There's no way to hit the pause button or get that carbon back under wraps. Left unchecked, a thawed permafrost could flip the Arctic from a greenhouse gas sink to a massive source of emissions.

If we're going to stop feedback loops from making Earth uninhabitable, we must stabilize the carbon cycle. Our planet has natural ebbs and flows. Trees breathe in carbon dioxide and emit oxygen. Oceans and soils absorb huge quantities of carbon; rocks do the same. Before the industrial era, when the natural carbon cycle was in balance, the atmosphere contained about 280 parts per million of carbon dioxide. Then humanity began to burn coal for heat, steam power, and electricity. Then we burned petroleum for transportation. Soon enough, our CO_2 emissions exceeded what Earth could absorb and store, and the concentration of the gas in the atmosphere started rising. It is up 50 percent since the mid-1700s, and the rate of increase accelerates each year.

Today's rising carbon level portends a storm of planetary emergencies. Earth's "carbon sinks"—our lands and forests and oceans—take in carbon from the atmosphere. But our sinks are at risk of being overloaded. Industrialization, fossil fuel pollution, and damaging practices threaten their ability to absorb our emissions. Unless we change our ways, we could be killing any chance we have of reaching net zero. If we're serious about avoiding climate catastrophe, we need to restore all three of Earth's sinks to function as nature designed them.

———————————

How can we solve this monumental problem? We might look to the vision of biologist E. O. Wilson, the towering figure known as "Darwin's natural heir." In 2016, toward the end of his illustrious seventy-year career, Wilson gave the world a gift, a book titled *Half-Earth*. As a desperate measure to preserve Earth's rich variety of life, it proposes to commit half the planet's surface to nature. "The Half-Earth proposal offers a first, emergency solution commensurate with the magnitude of the problem," Wilson writes. "I am convinced that only by setting aside half of the planet in reserve, or more, can we save the living part of the environment and achieve the stabilization required for our own survival."

Wilson's bold proposal would protect 50 percent of all oceans, forests, and lands as a means to an all-important end—normalizing Earth's carbon cycle and pulling the plug on our climate's escalating feedback loop.

Objective 4
Protect Nature

Go from 6 gigatons of emissions
to -1 gigaton by 2050.

KR 4.1   **Forests**

Achieve net-zero deforestation by
2030; end destructive practices and
logging in primary forests.

-2.6 Gt

KR 4.2   **Oceans**

Eliminate deep-sea bottom trawling
and protect at least 30% of oceans
by 2030, 50% by 2050.

-1 Gt

KR 4.3   **Lands**

Expand protected land from 15%
today to 30% by 2030, 50% by 2050.

Protect Nature

We have only one planet to share. Any fundamental rethinking of our relationship to the Earth will disrupt ingrained patterns of land use and development. Local populations will need to be considered; issues of climate equity must be addressed. There will be inevitable tradeoffs between the global demand for resources and the need to protect our environment. None of this will be easy. But if we're going to prevent a full-blown climate disaster, we need to start working with nature, not against it.

To rebalance the carbon cycle, we must hit all three of our targets. Our **Forests KR (4.1)** calls for drastically reducing human-caused deforestation by 2030, and for planting more trees than we log or burn. Protecting our forests starts by providing political and economic support to halt the clear-cutting of trees, keeping them intact for centuries to come. Tighter regulation, monitoring, and certification are needed to ensure that only sustainable wood enters the market.

Our **Oceans KR (4.2)** demands an end to the devastation of our seas. Our coastal oceans, home to immense underwater meadows of carbon-absorbing marine vegetation, must be protected from pollution and destructive fishing practices. The deep ocean is layered with marine sediments, Earth's single largest stockpile of stored carbon. In these watery realms, the commercial fishing industry's aggressive bottom trawling—dragging heavily weighted nets across the sea floor—releases carbon dioxide into the seawater, and ultimately, a portion into the atmosphere. We need to end deep-sea trawling and protect 50 percent of all oceans by 2050.

Our **Lands KR (4.3)** protects 30 percent of all lands—from tundras and ice caps to grasslands, peatlands, and savannas—by 2030, and 50 percent by 2050. That's a prodigious stretch goal from the 15 percent protected in 2020.

Taken together, our three KRs in this chapter would reduce emissions by 7 gigatons, or 13 percent of our current crisis. To achieve them, we need bold government action, heroic private-sector innovation and investment, and focused philanthropy.

The pressing subtext of this chapter is climate justice. Humans often disrupt nature for legitimate reasons—for fuel, food, or shelter. For many people today, practices like tree clearing are the best ways to sustain themselves. Any just nature protection plan must offer viable alternatives for people to earn income and feed their families. We cannot simply ban our way out of the climate crisis, nor can we solve it from on high. Affluent nations, corporations, and philanthropists must bear the global cost of these restorative measures, in poorer countries as well as their own.

It's tempting to believe that we can plant our way out of deforestation. Some factors to consider: where trees can best thrive; their impact on adjacent land; their longevity; and what happens if they're not planted.

A Future for Forests

The calamity of the world's forests has claimed headlines for decades. As we're barraged with nonstop bulletins on the burning Amazon or the latest California wildfire, it's easy to grow numb. But too many of us know too little about the importance of trees and what happens when we lose them.

Let's begin with the basics. Trees take in carbon from the atmosphere, storing small amounts when they're young and more as they mature. They hold this carbon for the rest of their lives, often a hundred years or longer. When trees are burned, intentionally or not, they release their carbon stores. The CO_2 they once held is expelled into the atmosphere. Deforestation is both a cause and a result of global warming.

The numbers are damning. The world loses a football field of forest every six seconds. All told, deforestation—the sum of all chopping and burning—accounts for 6 gigatons of annual CO_2, or about 10 percent of the globe's total emissions.

Tropical deforestation is especially problematic because of the immense amount of stored carbon in the tropics, not to mention the rich diversity of wildlife. The Amazon rainforest alone houses 76 billion tons of carbon. If ranked among nations, according to the World Resources Institute, tropical deforestation would place third in carbon emissions, behind only China and the United States.

It's a time-sensitive emergency, says Nat Keohane, former top climate officer at the Environmental Defense Fund: "If we don't protect the tropical forests now, we won't have the opportunity ten years from now."

Trees are cut to make room for livestock and crops, or to produce wood and paper, or to build roads and dams. Deforestation is in part a product of the demand for agricultural land to feed the world's growing population. To put it mildly, we have a land crunch. To feed 10 billion people in 2050 without runaway tree clearing, we need to increase the productivity of existing farmlands and shift away from consumption of emissions-intensive foods like beef. We must implement policies to cut emissions from food production and invest in innovations like feed additives and precision-based fertilizers, as described in chapter 3. But even if we succeeded in doing all these things, they wouldn't be enough.

To end deforestation, we need financial incentives to shift behavior toward forest protection. Crucially, these funds must make their way to the people whose livelihoods depend on cutting down trees. Without replacing—and exceeding—the income earned through deforestation, we cannot expect our forests to meet a kinder fate. Not last, we need to enforce national policies that protect forests.

Deforestation inflicts damage on both local ecosystems and the planet's carbon cycle.

Protect Nature

Tropical deforestation drives global forest loss

Temperate forests (the sum of "boreal" and "temperate" areas)

Tropical forests (the sum of "tropical" and "subtropical" areas)

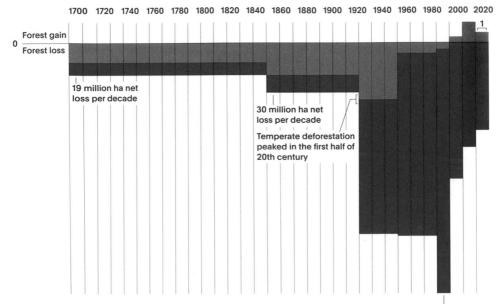

1700 1720 1740 1760 1780 1800 1820 1840 1860 1880 1900 1920 1940 1960 1980 2000 2020

Forest gain

0

Forest loss

19 million ha net loss per decade

30 million ha net loss per decade

Temperate deforestation peaked in the first half of 20th century

[1] Temperate regions have had a net gain in forest since 1990—passing the forest transition point.

Global forest loss peaked in the 1980s. 151 million ha net loss. This is equivalent to area half the size of India.

Location of temperate and tropical forests

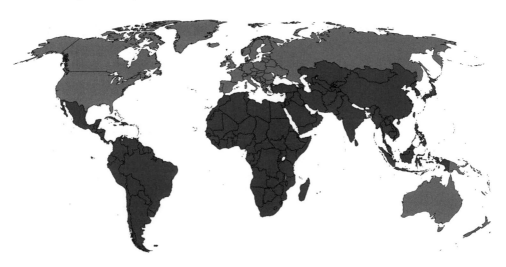

Adapted from data in Williams's *Deforesting the Earth* and UN's 2020 Global Forest Resources Assessment and visuals by Our World in Data

In 2007, the United Nations announced REDD+, an initiative to reduce emissions from deforestation and forest degradation in developing countries. The idea seemed straightforward: Wealthy nations would pay developing countries to leave their forests standing. But because of funding gaps and the lack of a global price on carbon, the program has been largely ineffectual; emissions from tropical forests have risen since it was launched. Until wealthier nations deliver on promised contributions, programs like REDD+ cannot move the needle.

In the absence of effective large-scale public policy, the private sector has attempted to fill the gap, building programs in which corporations pay public or private entities to keep trees intact. These avoided emissions, known as offsets, are purchased to help neutralize a company's carbon footprint. While we'll address this topic in detail in chapter 6, we'll note that these programs have been challenged on grounds of additionality (would the emissions have been avoided anyway?), durability (how long will the trees live?), and verifiability (is the program fulfilling its promises?).

To curb deforestation in time, we need more funding for high-quality protection efforts. Greater transparency is essential to validate programs. Improved satellite data and certification programs are critical. Finally, governments must make stronger commitments to end deforestation with two-pronged policies: financial incentives and tightly enforced legal protections.

For an example of pioneering leadership, we need look no further than an international organization that created market-based solutions to the deforestation crisis.

The Rainforest Alliance: Creating a Market for Sustainable Harvesting

In the year 2000, conservationist Tensie Whelan became president of the Rainforest Alliance, an organization dedicated to protecting biodiversity and the livelihoods of those who rely on it. When Tensie took charge, climate change was mostly a nonstory, ignored by the media and governments alike. She set an audacious goal: to safeguard the rainforests of the world by building a multibillion-dollar market.

Tensie Whelan

I grew up in New York City. But my family also had a farm in Vermont, where I went camping and fishing. I learned a lot about nature from my dad, who worked at the Museum of Natural History. My mom spoke Spanish and worked in criminal justice reform. And my grandparents on her side lived in Mexico City, so I saw poverty up close.

As a young adult, I found myself editing an international environmental journal in Sweden and then went to work as a journalist in Latin America, covering sustainable development issues, including deforestation. I saw firsthand that people cut down trees due to economic pressures, not because they are bad people.

One big flashpoint at the time was McDonald's practice of sourcing beef from Costa Rica. It kept U.S. hamburger prices down, but the added grazing also led to deforestation. Environmental boycotts led McDonald's and others to stop sourcing beef from there, but that did not stop the deforestation. People turned to slash-and-burn agriculture to put food on the table. That got me interested in how we could help people pursue sustainable livelihoods.

The Rainforest Alliance was founded by Daniel Katz. He was moved to act after reading that fifty acres of rainforests are destroyed every minute, and two dozen species become extinct each day.

Daniel used his winnings from a casino to sponsor a conference of experts. They came up with two ideas that worked together: First, they replaced boycotts with buycotts—positive campaigns to promote purchases of sustainable products.

The Rainforest Alliance seal: a badge of sustainable sourcing practices.

Second, they created a symbol to certify sustainability in the marketplace. They chose a frog for the consumer-facing seal because healthy frog populations are emblematic of healthy ecosystems.

The Alliance's first victory was to certify banana farms in Costa Rica as sustainable. Then they did the same with coffee farms in Guatemala, and they started a cocoa program in Ecuador.

The 1990s saw considerable progress on deforestation, but it wasn't fast enough for Tensie. It was a long, hard slog to go through the developing world farm by farm to collaborate with local growers and Indigenous peoples. Tensie promised money for people to protect the rainforests instead of destroying them. She won farmers' trust, and sign-ups rose each year. By mandating safe working conditions and fair pay, the program also caught on among farmworkers.

Then came a breakthrough. Chiquita, one of the world's largest banana companies, decided to do the right thing—to work only with suppliers who renounced forced labor practices. All Chiquita-owned farms became Rainforest Alliance Certified. They committed to harvesting ingredients from the forests without harming the trees or exploiting the workforce.

Getting one of the world's most famous brands on board made all the difference. Its partnership with Chiquita made the Rainforest

I saw people cut down trees due to economic pressures.

Alliance one of the highest-profile nonprofit organizations in the environmental movement. Scaling the sustainable practices program became Whelan's top priority.

Tensie Whelan

The Alliance had thirty-five employees and a $4.5 million annual budget spread across too many programs. We needed to figure out how to scale it up, especially once large companies and big brands started paying the way.

It took about a decade to get all of Chiquita's production certified. Other banana companies began to change their practices and work with the Alliance, under my watch. We were scaling coffee and cocoa certification at the same time. To raise money and awareness, we began charging to license use of the Rainforest Alliance seal.

We also worked to help Indigenous forest-dwelling communities to get cash for carbon credits and other ecosystem service payments [cash transfers from brands and businesses]. We had to get more money flowing from big corporations to small producers.

We invited the CEO of Kraft Foods, a company with more than $30 billion in revenues, to see what we were doing. When Roger Deromedi came to El Salvador, we had the farmers talk to him about changes they'd made—what was working and what wasn't. He saw how sustainable production made a positive impact—economically, environmentally, socially.

As a result, Kraft committed to sourcing from Rainforest Alliance–certified growers for many of its brands. Within five years, the company was certifying 25 percent of its products as sustainably sourced. It reduced its carbon footprint by 15 percent.

By 2006, the Alliance hit a key milestone: a billion dollars per year in sales of certified brands.

Then Unilever invited me onto their sustainable agriculture advisory board. Their own standard had a lot of alignment with ours. They were particularly interested in tea—they had about twenty brands, including Lipton. They were actually persuading me to expand into the tea market, rather than me persuading them.

That had a big impact on thousands of small tea producers in Kenya. We also saw Argentinian farmers improve their environmental practices for tea sold in the American market. Then we went on to source sustainable cocoa with Mars, Inc.

So we went way beyond bananas. We were certifying 20 percent of the world's tea, 14 percent of cocoa, and 6 percent of coffee—

sustainable sourcing for five thousand companies through entire supply chains. All told, we were helping local growers ensure sustainability for about 7 percent of the world's agricultural forests.

By channeling cash payments to growers to sequester carbon and protect their ecosystems, the Rainforest Alliance spread the pay-for-performance gospel. In essence, the organization put a price on carbon. In 2015, Tensie Whelan left to direct the Center for Sustainable Business at New York University's Stern School of Business. By that point, thanks in part to the Alliance's staunch efforts, deforestation was on the wane, at least temporarily. Between 2000 and 2015, carbon emissions dipped 25 percent, from 4 gigatons to 3.

Since moving to NYU, Whelan has led development of a powerful metric to justify corporate spending on sustainability. It tracks progress in employee retention, customer loyalty, and company valuations.

Third parties are following the data. The International Institute for Sustainable Development publishes an annual report on certified products. New York University has partnered with a firm that compiles product barcode data at the retail level, gleaning powerful insights on consumer behavior.

Tensie Whelan

When I came to NYU, I was keenly interested in the "green gap," the difference between what consumers say they value in terms of sustainability and what they actually buy. Instead of conducting surveys, I wanted to look at real-world numbers.

We obtained data for five years of consumer-packaged retail goods across thirty-six product categories and more than seventy thousand personal care and food items. We looked at everything that was labeled sustainable, such as plant-based, organic, or non-GMO, versus conventional products.

We found that sustainability-marketed products accounted for 55 percent of growth in consumer-packaged goods over those five years—and earned on average a 39 percent premium. There wasn't any green gap after all. It was actually a green premium.

While most consumers may be unwilling to pay a green premium for clean energy or electric cars, many are willing to pay a little extra for sustainable foods.

In 2015, the Paris Agreement addressed the forests of the world with unprecedented emphasis: "Parties should take action to conserve and enhance, as appropriate, sinks and reservoirs of greenhouse gases . . . including forests." The treaty went so far as to specify "results-based payments" to national governments and nongovernmental organizations that protect rainforests and nature's other carbon sinks.

Stopping deforestation will require far more investment. On a global basis, deforestation funding outpaces forest protection funding by a ratio of forty to one. There are early signs, however, that the trend may be shifting greenward. In April 2021, at President Biden's Leaders Summit on Climate, private and public sectors came together to pledge to raise at least $1 billion within the year for large-scale forest protection and sustainable development. Benefits will be channeled to Indigenous peoples and forest communities. This just might be the winning formula to save our forests: more funding, with high transparency and standards, overseen by the world's leading stewards of the land.

Indigenous Leadership

Perhaps the most underestimated force for averting climate disaster is the protection of Indigenous rights and lands and way of life. While Indigenous peoples are only 5 percent of the global population, their lands contain 80 percent of the world's biodiversity. They include at least 1.2 billion acres of forest, which store 38 billion tons of carbon. But the role of Indigenous peoples transcends these numbers. Their traditions are rooted in the care and relationships of natural ecosystems. Indigenous wisdom and practices, honed over centuries, will be indispensable to mankind's efforts to mitigate a warming planet—and to adapt to it as well.

In quantitative terms, the power of Indigenous practices is indisputable. When forests are managed by Indigenous communities, deforestation rates are often two to three times lower than in surrounding forests, even in those that are nationally protected. According to the World Resources Institute, "tenure-secure indigenous lands in the Amazon store carbon, reduce pollution by filtering water, control erosion and flooding by anchoring soil, and provide a suite of other local, regional and global ecosystem services."

To keep Indigenous peoples' stewardship intact over centuries to come, their lands must be legally protected and recognized as their own. Within climate circles, the principle is known as secure land tenure.

In stretches of the Amazon across Brazil, Bolivia, and Colombia, tenure-secure lands generated a net benefit of up to four thousand dollars per acre, or a total benefit of more than a trillion dollars over a twenty-year period. The cost of securing the lands was no more than 1 percent of the gains. Making Indigenous lands legally secure could keep 55 percent of the Amazon's sequestered carbon under wraps. It's one of our most cost-effective mechanisms to prevent added emissions and protect the planet.

Replenishing the Oceans

The world's oceans provide half the oxygen we breathe and abundant fish to eat. As climate regulators, their capacity is unmatched. We need them to thrive—and to survive.

But the truth is that our oceans aren't what they used to be. Decades of pollution have taken their toll. As far back as the 1960s, scientists could measure small but distinct declines in oceanic carbon absorption. Those data points, along with evidence of increased ocean acidity, convinced Roger Revelle to study the problem. As one of the world's original climate scientists at the Scripps Institution of Oceanography in La Jolla, California, Revelle concluded that the planet was suffering from anthropogenic—man-made—global warming. As a guest professor at Harvard, Revelle taught the young Al Gore, who came to see him as a mentor. And the ocean's growing resistance to absorbing CO_2 would become known as the Revelle factor.

Oceans naturally exchange carbon with the atmosphere; they've been giving and taking for as far back as we can tell. But in the centuries since atmospheric carbon began to grow unabated, oceans have mostly functioned as a receptacle, taking in more than they release. In addition to the deluge of fossil fuel emissions from the air, overfishing, overdrilling, and overdevelopment have released tremendous amounts of carbon that wind up stored in aquatic ecosystems and ocean sediment.

As a result, the seas nearest our coastlines, home to most marine life, are in trouble. Sea grasses, coral reefs, and mangroves have all suffered from humanity's abuse. If we halt this destruction, we'll prevent 1 gigaton of carbon emissions from entering the atmosphere each year.

Then there's the second ocean zone, the deep seas that lie beyond the continental shelves and cover 50 percent of the surface of our planet. Their sediment floors contain thousands of times more carbon than in all of our lands. Deep-sea mining and fishing disturb the sediment, releasing its carbon and contributing to increased ocean acidity, to the point where it dissolves certain shellfish. Bottom trawling is especially destructive. Dragged along the seafloor, enormous nets release 1.5 gigatons of aqueous CO_2 emissions, though researchers aren't yet sure how much of that enters the atmosphere.

On average, the ocean's acidity is about 25 percent higher than in the pre-industrial era.

Our oceans are under assault on two fronts. They're being pressed to take in carbon from both the air above and their floors below. Meanwhile, marine life is choking on plastic that has sullied the furthest reaches and depths of our waters. Overfishing has spread from 10 percent of the world's stocks in 1980 to 33 percent today, with China, India, and Indonesia the worst offenders. Tragically, 90 percent of the globe's coral reefs could be gone by 2050, killed by warmer waters and acidity. China has already forfeited 80 percent of its barrier reefs. Australia's Great Barrier Reef, the largest system in the world, has lost more than half its corals to mass "bleaching" events, telltale signs of warming waters.

By far the biggest contributor to ocean acidification is CO_2 from the air, which gets absorbed by the ocean. By achieving net zero and reducing atmospheric carbon dioxide, we can stem the tide on ocean warming and acidification. In the meantime, we can cut ocean-related emissions by expanding the seas' protected zones.

A Miracle in Mexico

Leading this charge is marine ecologist Enric Sala, one of the world's top experts on ocean protections. Sala came by his life's passion naturally, growing up on the Costa Brava of northern Spain. After studying biology at the University of Barcelona and earning his doctorate in ecology, he became a professor at the Scripps Institution of Oceanography.

In 1999, on Mexico's Baja Peninsula, Sala visited Cabo Pulmo, a once-rich ecosystem that had become an underwater desert. The fishermen could no longer catch enough to make ends meet. The marine vegetation that once fed their fish—and captured many tons of carbon—had disappeared. Desperate, the fishermen did something no one expected. As Sala explained in a TED talk, "Instead of spending more time at sea, trying to catch the few fish left, they stopped fishing completely. They created a national park in the sea, a no-take marine reserve."

Ten years later, this underwater barren zone had become a kaleidoscope of life and color. Even the large predators—groupers, sharks, and jacks—returned. As Sala noted, "We saw it come back to its pristine level. And those visionary fishermen and towns are making much more money, from economic growth and tourism."

Sala quit his academic job to work full time as a conservationist for the National Geographic Society. In collaboration with the naturalist Mike Fay, he persuaded the president of Gabon in Central Africa to create a network of national sea parks. In 2008, Sala and Fay launched the Pristine Seas initiative to document the wild places left in the ocean and to work with governments to protect them. These spectacular refuges, sprawling over an expanse half the size of Canada, are now fully protected by government laws or regulations.

"This showed what the future ocean could be like," Sala says. "Because the ocean has extraordinary regenerative power. We just need to protect many more places at risk so they can become wild and full of life again." It's a story with a simple moral: When business aligns with conservation, miracles can happen.

While coastal marine reserves have recently grown in number, only 7 percent of ocean waters are fully protected from overfishing and other destruction. For our plan to work, we need at least 30 percent of the ocean under protection by 2030, and 50 percent by 2050. "The jury is in on marine reserves—they work," National Public Radio reported. "Research has repeatedly shown that fish numbers quickly climb following well-enforced fishing bans, creating tangible benefits for fishers who work the surrounding waters. In fact, many experts believe fishing will only be sustainable if marine reserves are expanded significantly."

While the vast majority of fishing is in our coastal oceans, the high seas host their fair share as well. Fishing in the deep oceans is largely unregulated. Regional authorities can police practices in coastal areas. But the rules get muddier—and enforcement spottier—the further out into the seas you go.

Sala has set his sights on the spectacularly destructive practice of bottom trawling. "Super trawlers, the largest fishing vessels in the ocean, have nets so large that they can hold a dozen 747 jets," he says. "These huge nets destroy everything in their paths, including deep corals on sea mounds, which can be thousands of years old." Satellite data shows that Russia, China, Taiwan, Japan, Korea, and Spain account for nearly 80 percent of high-seas fishing. These governments subsidize trawling with cash incentives for buying bigger ships.

Based on Sala's analysis, more than half of high-seas fishing grounds depend on these subsidies, which total $4 billion per year. The Pristine Seas project advocates an international ban on bottom trawling. Backed by leading marine scientists, with discussions convened by the United Nations, the ban would not harm the world's fish supply.

In our warming world, the people striving to protect our oceans are fighting uphill. Coral reefs and plant and marine life remain in jeopardy. The feedback loop of climate change has yet to be switched off. To put it plainly, we need a more serious global commitment. In 2016, twenty-four countries plus the European Union agreed to protect the Ross Sea in Antarctica with a thirty-five-year ban on commercial fishing. The signatories included the fishing-dependent nations of China, Japan, Russia, and Spain. If we could expand these protections, we'd have a chance to help oceans resume their rightful role as vibrant homes to a myriad of species.

Net loss: bottom trawling releases stored carbon into the oceans.

Cultivating Kelp

Though the oceans can still be saved, the mission is pressing. My lifelong hero Gordon Moore, the aforementioned author of Moore's Law, has thrown his talent and capital into the Bay Foundation's efforts to restore the underwater kelp forests of California's Monterey Coast. Sea kelp absorbs twenty times more CO_2 than a comparably sized land forest. The Monterey Bay's sea otter population was decimated by the fur trade of the seventeenth and eighteenth centuries. In the sea otters' absence, one of their favorite snacks—sea urchins—multiplied and devoured much of the kelp. Beginning in the 1980s, protection efforts have helped to revive the sea otter population. With the sea urchins once again in check, the Monterey kelp forests have gradually returned to their former glory. The ecosystem has been restored.

Kelp forests could be one way to expand the seas' annual carbon absorption capacity. Their near-surface seaweeds capture CO_2 through photosynthesis. One of the fastest growing forms of plant life, kelp expands by up to two feet in length per day. As Charles Darwin commented upon his kelp encounters during his mid-nineteenth-century voyage on the HMS *Beagle*: "I can only compare these great aquatic forests with terrestrial ones. The number of living creatures of all orders, whose existence depends on kelp, is wonderful."

In 2017, the prospect of growing hundreds of thousands of acres of new kelp was cited by Project Drawdown as a "coming attraction," a promising climate solution for the future. As the nutrient-rich leaves of kelp die and fall to three thousand feet or more below sea level, their carbon "can be considered permanently sequestered," says Dorte Krause-Jensen, a professor of marine ecology at Aarhus University in Denmark.

Kelp forests are "an efficient conversion pathway that doesn't use up valuable cropland which should be reserved for growing food," says Timothy Searchinger, technical director of the WRI's food program. Historically, CO_2-absorbing biomass is cultivated on land. Kelp is one potential path for carbon removal that doesn't compete with agriculture.

Enter "marine permaculture arrays," the invention of Brian Von Herzen, a planetary scientist based in Woods Hole, Massachusetts, and Queensland, Australia. Von Herzen's nonprofit Climate Foundation has developed these kelp-growing grids—irrigated by renewably powered pumps—to draw nutrient-rich waters from the lower ocean depths. By 2020, the organization was running small pilot arrays in the Philippines and designing thousand-square-foot grids for the Australian coast. It is raising funds to build some at least ten times that large in the Philippines.

Though the size of Von Herzen's ambition dwarfs the small amount of territory in play, his business model makes sense. By harvesting and selling some of the seaweed and restored schools of fish, operators can recoup their cost of installation within four years. If innovators like Von Herzen could scale this practice to 1 percent of the world's oceans, or about half the size of Australia, kelp forests would absorb 1 gigaton of CO_2 annually.

Kelp forests are seaweed farms that can absorb carbon and safely store it at the bottom of the ocean.

It's an alluring vision. But we're going to need more proof on a larger scale before we can begin to count on kelp forests to help get us to net zero.

The Power of Peatlands

The Earth's massive carbon storage chests span from wetlands to tundras and polar ice caps and the frozen soil—the permafrost—that lies beneath them. These natural sinks capture and sequester carbon for tens of thousands of years.

Perhaps the least known but most important are peatlands. A thick, muddy, waterlogged substance, peat historically has been harvested for fuel. After oceans, peatlands hold the second-largest store of carbon in the world. When they are damaged, the repercussions reach the world's atmosphere. Drained peatlands, for example, represent only 0.3 percent of the world's land area but generate 5 percent of anthropogenic CO_2 emissions.

In parallel with forest protection efforts, moratoriums are already in place to prohibit peatland draining. In Indonesia, the early returns are promising. Incentives, consistent monitoring, and rigorous standards will be required to scale protection and restoration.

Biodiversity as a Measure of Carbon Resilience

In restoring lands, forests, and oceans, we are protecting the planet's biodiversity, the millions of species that make ecosystems thrive. Life forms are interdependent; the loss of one threatens another. Whenever a species goes extinct, the ecosystem may change in unforeseen and invisible ways. When worms or microorganisms are killed off by industrial farming, they can't consume organic matter. They no longer deposit carbon into the soil. In Yellowstone National Park, small packs of wolves regulate the elk population, which feeds upon aspen seedlings before they become trees. If the wolves die out, the elk grow too abundant. The seedlings are decimated, and over time the tree cover disappears. The wolves help the trees just by being wolves.

The welling threat of "ecocide"—the wholesale destruction of an ecosystem and its biodiversity—is a central strand of our climate emergency. Earth contains an estimated 8 to 9 million plant or animal species. Before the onset of humanity, just one species in a million went extinct per year. In the early twentieth century, annual extinctions rose to around a dozen, and the rate of increase kept accelerating. Today, more than 1 million species are threatened with extinction. The worst, as E. O. Wilson writes, may be yet to come: "If extinction mounts, biodiversity reaches a tipping point where the ecosystem collapses."

In the United States, one of the main instruments for protecting lands and species is the National Park Service. Novelist and environmentalist Wallace Stegner called our park system "the best idea we ever had." A multitude of countries has tried to follow America's lead.

Yet despite impressive gains by conservationists, extinction rates remain distressingly high. Wild habitats are under assault. How can we help our ecosystems to bounce back?

Our Grandest Challenge: 30x30 and 50x50

I often find myself admiring the methodical magic of the Earth. The carbon cycle is a measurable process that desperately needs rebalancing. To avert climate disaster, as E. O. Wilson would tell us, we must be just as aggressive in restoring nature as we are in making the transition to clean energy.

In 2018, in response to Wilson's Half-Earth challenge, the National Geographic Society formed the Campaign for Nature, a coalition of scientists, entrepreneurs, Indigenous peoples, and environmental leaders. Where Wilson's vision aspires to safeguard half the globe by 2050, the Campaign for Nature has a nearer-term goal: the protection of 30 percent of the planet by 2030. They believe that's the least we need to do to address climate change and "prevent a mass extinction crisis." In 2021, the Biden administration endorsed this target for the United States.

The Speed & Scale Plan is geared to both timelines, the urgency of 30×30 and the make-or-break objective of 50×50. When you're piloting a plane, you cannot lift off before reaching a minimum speed. Only then can you confidently chart the course for your final destination.

All three realms of nature—lands, forests, and oceans—are essential to our net-zero mandate. Our carbon sinks lie at the fault line between climate stability and climate catastrophe, between biodiversity and mass extinctions. We are among the fragile species at risk. On the tree of evolution, humanity is now out on a limb, an imminent danger to itself. But before you may be tempted to despair, remember this: time and again, we've seen that restoration efforts can actually work, no matter how far gone an ecosystem might have seemed.

All things considered, we've had it pretty good since the end of the last Ice Age, eleven thousand years ago. We've flourished on a Goldilocks planet, neither too cool nor too warm. "The beautiful world our species inherited took 3.8 billion years to build," Wilson writes. "We are the stewards of the living world. We've learned enough to accept this simple and easy-to-use moral precept: do no further harm."

Speed & Scale: Countdown to net zero

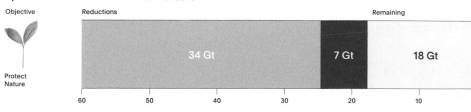

Objective

Protect Nature

Reductions

34 Gt | 7 Gt

Remaining

18 Gt

60 | 50 | 40 | 30 | 20 | 10 | 0

Clean Up Industry

Chapter 5

Clean Up Industry

At age twenty-five, James Wakibia set off on a path that would make him "the vision-bearer of banning plastics." A photojournalist from Nakuru, Kenya's fourth-largest city, Wakibia got his start by patrolling the streets with a zoom-lens camera. He snapped pictures of daily life: stores opening in the morning, kids heading to school. And, of course, nature—Kenya is justly renowned for its natural beauty. Wakibia posted the photos on his blog, which soon brought him paid assignments. "Photography has helped me to see," he says. "It has given me the courage to stop and shoot, and later question."

One day in 2011, Wakibia passed the local garbage dump. It overflowed with plastic supermarket bags that blew onto the roads nearby. He photographed plastic bottles clogging the shores of the city's lakes and ponds—and hated what he saw. In 2013, he launched a petition to relocate the dump site. City officials were unmoved and rejected the proposal.

For Wakibia, that was just the first skirmish. He gained thousands of followers through his Twitter hashtag #BanPlasticsKE. He took to quoting a celebrated American folk singer and environmentalist: "If it can't be reduced, reused, repaired, rebuilt, refurbished, refinished, resold, recycled, or composted," Pete Seeger sang, "then it should be

restricted, redesigned, or removed from production." Wakibia's own rallying cry is shorter: "Less plastic is fantastic." Four years after passing that garbage dump, he threw his energy behind a more audacious goal. He petitioned Nakuru to become the first city any-where to ban all single-use plastics, including bags, water bottles, and throwaway utensils. Once again, the city turned him down.

James Wakibia took on plastic pollution in Kenya—and won.

This time around, though, the campaign caught the eye of Kenya's cabinet secretary for the environment and regional development. Judi Wakhungu loved Wakibia's idea so much that she pushed to enact it nationwide. She cast it as an economic imperative: half of Kenya's cattle had plastic in their stomachs, depressing the nation's milk supply. In 2017, the country passed the world's strictest ban on single-use plastics. "I didn't expect this," said a jubilant Wakibia. "It is great news."

Kenya's penalties are severe. Anyone found making, importing, or selling plastic carrier bags can be fined up to forty thousand dollars or be imprisoned for up to four years. Anyone using a banned bag could be fined five hundred dollars or jailed up to a year. The government meant business. Hundreds of violators were fined and more than a dozen put behind bars.

Less than eighteen months after the law was passed, Kenya announced that 80 percent of the population had stopped using plastic bags. The BBC reported on the triumph to an international audience. Wakibia was hailed by the U.N. Environment Programme.

A single small country passing one green law will go up against powerful forces, especially when a banned material in question is made by huge multinational companies and used in hundreds of thousands of ways. In a famous scene in *The Graduate*, a businessman tells the young character played by Dustin Hoffman, "I just want to say one word: plastics. There's a great future in plastics." In 1967, when the movie was made, plastics were just beginning to become the ubiquitous material we know today: overwhelmingly popular, undeniably useful, and ultimately a nightmare.

Plastic pollutes twice: when it's made and when it's discarded. Not only are plastics fabricated from petroleum and natural gas, but they emit carbon dioxide in the process. Half of all plastic in human history has been made in the past fifteen years, a fact that carries grim implications for the future. The problem is getting worse and worse, faster and faster, at industrial scale.

At the point of creation, almost everything in our built world throws off emissions —from the plastic banned in Kenya to concrete bridges on our roadways to the steel in our tallest buildings. Even if we succeed in decarbonizing our grids and electrifying our modes of transport, the manufacture of the materials we need to make them releases greenhouse gases. Some emissions are derived from the fossil fuels used as a direct heat source; others are created by the chemical reactions in the manufacturing process.

There's a playbook that can be reliably applied across every industrial sector to draw down emissions. The first strategy is to use less. Structures designed to use less concrete add less carbon dioxide to the atmosphere. The second is to recycle and reuse. By salvaging textiles and creating recycled fabrics, we bypass the need to produce them from scratch. The third is alternative heat sources. By using electricity for the heat needed to melt steel, we can tap into emissions-free energy sources. And last but not least: invention. By creating new types of containers that can be composted, you prevent them from ending up in the landfill.

Objective 5
Clean Up Industry

Reduce 12 gigatons of industrial emissions to 4 gigatons by 2050.

KR 5.1   **Steel**

Reduce total carbon intensity of steel production 50% by 2030, 90% by 2040.

↓ 3 Gt

KR 5.2   **Cement**

Reduce total carbon intensity of cement production 25% by 2030, 90% by 2040.

↓ 2 Gt

KR 5.3   **Other Industries**

Reduce emissions from other industrial sources (i.e., plastics, chemicals, paper, aluminum, glass, apparel) 80% by 2050.

↓ 2 Gt

At 12 gigatons of emissions, around 20 percent of the global total, the things we manufacture place a heavy burden on our atmosphere. Our **Steel KR (5.1)** targets the single largest emissions source in this category, at 4 gigatons. It calls for the world's steel companies to identify practices and technologies that reduce the use of fossil fuels in the production of steel. The **Cement KR (5.2)** applies to the fabrication of concrete, which emits almost 3 gigatons. The production of steel and cement today is dominated by China and its breakneck pace of urbanization and construction. For these two mega-industrial sectors to decarbonize, the cost of new approaches and technologies must make sense for the developing world.

The manufacture of plastics, chemicals, paper, aluminum, glass, and apparel all use fossil fuels for direct heat. Many of these goods end up incinerated, producing even more emissions. Our **Other Industries KR (5.3)** applies our industrial playbook—use less, recycle, swap heat sources, and invent—to slash these emissions.

This will be easier said than done, of course. In this chapter, we'll begin with the sectors that touch consumers most directly, plastics and apparel. Then we'll look at new ways of generating heat, an approach that can help to broadly decarbonize this sector. Finally, we'll take on the two giant emitters, concrete and steel, where the challenges are most daunting.

There is no way around it. If we don't cut industrial emissions, they'll follow in lockstep with the world's growing population and demand for energy—through the roof.

The Scourge of Plastics

Plastics are a part of the chemicals industry, which is responsible for 1.4 gigatons of annual carbon emissions. Just twenty global polymer producers account for more than half of all single-use plastic waste. ExxonMobil and Dow in the United States and China-based Sinopec top the list. At this decisive moment in human history, these companies have a choice: to continue using fossil fuels at our climate's expense, or to lead the transition to a sustainable future.

A ban on single-use plastics must provide functional replacements. Plastic bags can be replaced with unbleached paper or reusable bags made from recycled bottles. Compostable fibers can be used to craft sturdy utensils, straws, and containers. For drinks, it's a simple move to glass and aluminum for shelf goods and reusable bottles and compostable cups for orders to-go. The biggest recycling success story ever? The aluminum cans that pack the single-serving shelves in your

Greener alternative: PLA polymers—made from plant starch—emit 75 percent less greenhouse gases than petro-plastics.

neighborhood supermarket or convenience store. Nearly 75 percent of all aluminum ever produced in the United States remains in use today, proof of the potential of the "circular economy."

With plastics, the current state of emergency reflects the failure—at least to date—of two once-promising solutions. The first is plastics recycling. Despite widespread adoption and high compliance levels, the market has broken down. Longtime recipients of America's plastic trash, notably China and Malaysia, are less willing to serve as dumping grounds. Not so long ago, China was importing 7 million tons of scrap per year. It closed the door after seeing its waterways despoiled by plastics dumped there by recycling companies. Rather than take a hit to their bottom line, the companies blighted China's rivers with mounds of plastic that didn't meet their quality standards. To be blunt, global recycling is a mess.

In the United States, much of the blame can be laid at the feet of the petroleum and plastics industries. What's more, their missteps were deliberate. Back in 1989, they lobbied for recycling symbols to be added to *all* plastics, knowing it would subvert the program. (The ubiquitous 1-7 code is confusing, at best.) As a result, confused U.S. consumers took to chucking just about anything into their blue bins, including nonrecyclable items and others contaminated by food.

To make the programs more effective, people need to be able to identify what's actually recyclable. New labels designed by the Sustainable Packaging Coalition can help. For example: Cookies are typically packaged in a plastic tray surrounded by a thin plastic wrap within a cardboard box. A proper label would tell us that the cardboard is recyclable, the plastic wrap is not, and the tray needs to be cleaned before going into the blue bin. More precise labeling would also expose companies that might otherwise feign ignorance of their nonrecyclable materials.

Instructive labels can help consumers make the right recycling decisions

Cookies

Freshly baked gingersnap cookies.

Rinse Before Recycling

PAPER	PLASTIC	PLASTIC
BOX	WRAP	TRAY

*Not recycled in all communities

Adapted from guidelines by How2Recycle

The second unmet promise is bioplastics. To understand why they have failed so far, we need to ask two questions: Is the material organic? And is it biodegradable? The ideal bioplastic, an organic resin that is decomposed by bacteria, checks both boxes. But as it turns out, neither one is easy to check.

In a very gradual shift away from fossil fuels and petrochemicals to renewable sources, Coca-Cola experimented with bottles using 70 percent petroleum and 30 percent ethanol, derived from sugarcane. While this hybrid marginally reduces the carbon footprint, it still results in a plastic bottle that takes hundreds of years to decompose. Other bioplastics may be even worse for the environment—and our health—than traditional petroplastics. Some approaches generate more pollutants, deplete more ozone, and swallow more precious land.

What we need is an approach that's more degradable yet still scalable. Chemists are coming closer with a polymer called polylactic acid, or PLA. Fabricated from corn or tapioca plant starch, PLA is sturdy—it feels like a normal plastic cup. But it cuts greenhouse gas emissions by

Clean Up Industry

75 percent as compared to traditional plastics. What's the catch? The polymer is biodegradable only in specialized industrial composting facilities, where it takes ten to twelve weeks to break down, limiting its utility. In a conventional landfill, or if dumped into the ocean, it struggles to decompose.

Still, wherever possible, using PLA for single-use food servings would be a win. It reduces the tonnage of plastic sent to landfills, prevents contamination of recycling bins, and ensures that food waste is composted and returned to the soil instead of rotting in a landfill and releasing methane. For bioplastics to meet the moment, more innovation is needed for them to break down in our home compost bins.

The plastic pollution lifecycle shows where we are today, and it's not a pretty picture. **Only 9 percent of the world's plastic waste is recycled.** What happens to the rest? Twelve percent is incinerated, emitting carbon dioxide, and the rest ends up in landfills and ultimately our oceans. Plastic pollution has exploded by a factor of ten since 1980. Due to waste stream mismanagement, 8 million tons of it enters the oceans each year. It kills up to a million seabirds and harms hundreds of species of fish, sea turtles, and marine mammals. When animals get entangled in plastic debris, they can suffocate or drown. When they ingest it, they can starve to death—the plastic fills their stomachs.

The push to prohibit these deadly materials has reached far beyond Kenya. In 2018, the British Parliament enacted a twenty-five-year plan

Plastic pollutes at every stage of its lifecycle
Global plastic production and use, 1950–2015, billions of tons (BT)

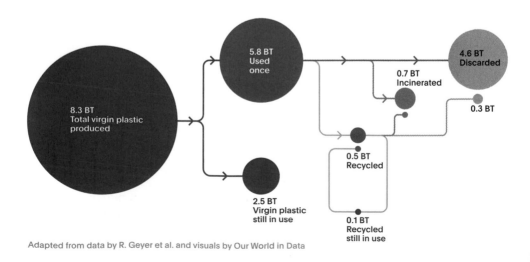

Adapted from data by R. Geyer et al. and visuals by Our World in Data

to phase out certain plastics and to conform to U.S. retail prohibitions on microplastics, the tiny beads used in detergents, cosmetics, and skin-care products. As of 2021, the European Union has banned single-use straws, plates, and cutlery, and set "a 90 percent collection target for plastic bottles by 2029." All told, 127 countries have some kind of regulation to limit plastic use.

We know we cannot ban all uses of plastic, at least not yet. Plastics are irreplaceable for certain medical supplies, home appliances, and multiuse containers. But there's nothing stopping us—save for a lack of political will—from clamping down on packaging, grocery bags, and other single-use items that account for nearly half of plastic-related emissions.

Sartorial Solutions

While clothing and footwear are a smaller source of greenhouse gas emissions, their cultural importance transcends the numbers—we wear these products every day. Over the last two decades, the advent of "fast fashion" has accelerated both production and consumption of apparel—62 million tons of it per year. Chasing rapid trend cycles, manufacturers produce low-quality garments that are worn briefly and discarded, piling landfills with textile, rubber, leather, and plastic waste.

According to a study published by Harvard Business School, materials used for clothing sold by Zara are designed to last for no more than ten wearings.

The industry has been slow to take climate action, despite a recent *Vogue* report of a budding industry consensus to cut 50 percent of emissions by 2030. To get there, fashion brands, manufacturers, and retailers need to accelerate their deployment of both upstream and downstream solutions. The upstream challenge—textile operations—is more straightforward. By using renewable energy and improved energy efficiency in garment production, we can directly cut emissions. Downstream, solutions are more diffuse. We need brands to choose lower-emissions transport and cleaner retail operations while also reducing overproduction. To drive change at scale, we need these enterprises to boost consumer adoption of garment rental, resale, refurbishment, collection, and reuse. The broad goal is to find ways to lengthen the life span of each sweater, coat, shoe, or handbag.

One high-impact strategy is to manufacture apparel from recycled materials. Patagonia became one of the first to make this leap back in 1993, fabricating fleece from recycled plastic bottles. Now the movement has been joined by scores of leading brands and market insurgents.

Clean Up Industry

Other industry leaders are turning to materials that are greener from the start—including a San Francisco-based shoe manufacturer that uses OKRs to track progress on its net-zero goals. Within a crowded, hypercompetitive market, Allbirds was founded in 2014 with a unique aspiration: to build comfort and high-quality design into the "most sustainable footwear on the planet." Two years after launching, they sold their millionth pair of shoes: their signature merino wool sneakers, with midsoles made from Brazilian sugarcane.

As Joey Zwillinger, cofounder and chief operating officer, says, "We treat the environment as a critical stakeholder in the success of our business." To guarantee that their financial objectives stay in sync with their net-zero goal, the company sets OKRs for all 250 employees. These efforts have established Allbirds as more than just a shoe company—today it's an environmental company too. In 2021, to accelerate the sustainable footwear market, Allbirds publicly shared its carbon footprint calculator with others in the industry.

One top-level company objective stipulates an ironclad promise to customers: All Allbirds products are strictly carbon neutral through their lifecycle. The corresponding key results track emissions targets across all facets of the operation, from the supply chain to manufacturing to shipping to retail. To implement this OKR, the company trains every employee to quantify—and reduce—carbon emissions at each step.

Allbirds is one of a growing cadre of green fashion leaders. Companies like Reformation source and rank fibers by sustainability. Brands like Stella McCartney are showing that high fashion can make the transition too.

On the consumer side, we're witnessing a boom in used and vintage clothing. Young people in the United States, Europe, and Asia are quickly turning the secondhand apparel category into a bona fide fashion trend. Services like Trove and Tradesy are creating new online resale markets to encourage shoppers to buy quality garments and resell them, countering the negative impact of fast fashion. Now the rest of the industry needs to catch up with the new vogue: net-zero emissions.

Electrifying Industrial Heat and Our Hopes for Hydrogen

Heat for industrial processes accounts for nearly one fifth of global energy use. It's also the largest contributing factor to CO_2 emissions from across the industrial sector. In manufacturing, varying degrees of heat are needed to produce everything from paper and textiles to steel to cement. The heat is generated on-site from natural gas, coal, or oil, all high-emissions sources.

It is now technologically feasible to electrify the manufacturing processes that account for at least half of industrial fuel consumption. Electric heat pumps, electric boilers, and recycled waste heat are established alternatives for low- and medium-heat needs. While some electric furnaces can reach temperatures higher than 1,000 degrees Celsius, a necessity for steel manufacturing, they're limited by high costs and excessive energy demands. As industry searches for more practical ways to achieve high heat without carbon emissions, it has turned to something literally elemental.

Replacing fossil fuels in many industrial processes is possible

Share of fuel consumption for process heat	%	Example of processes	Technology status
Very-high-temperature heat (>1000°C)	32%	Melting in glass furnace, reheating of slab in hot strip mill, and calcination of limestone for cement production	Research or pilot phase
High-temperature heat (400–1000°C)	16%	Steam reforming and cracking in the petrochemical industry	Available today
Medium–temperature heat (100–400°C)	18%	Drying, evaporation, distillation, and activation	Available today
Low-temperature heat (≤100°C)	15%	Washing, rinsing, and food preparation	Available today
Other (potential not assessed)	19%		

Adapted from data and visuals by McKinsey and Company

Clean Up Industry

In recent years, the most intense climate debate in industry has centered around the prospective role of hydrogen. Like electricity, hydrogen is a carrier of energy, not a source. It holds huge promise because it's nearly omnipresent; if you have water and an electric current, hydrogen is there for the making. A process called electrolysis uses electricity to split water—two parts hydrogen, one part oxygen—into its atomic components. The hydrogen component then can be stored, used directly on-site, or condensed into a liquid and transported to generate heat or electricity.

The industry uses a color code for various classes of hydrogen:

Brown or Black Hydrogen:
produced from coal

Gray Hydrogen:
from natural gas

Blue Hydrogen:
from natural gas, with emitted CO_2
captured and sequestered

Green Hydrogen:
from a zero-emissions energy source

Most hydrogen produced today is for the manufacture of chemicals, and 95 percent of it is made from natural gas. Our ambition over the next decades is for clean hydrogen to become the standard for generating high-intensity heat. Then we'd be able to decarbonize tough-to-tackle sectors like cement and steel.

While the cost of producing clean hydrogen will fall as more factories are built, it could take twenty years or more to become cheaper than dirty hydrogen in most parts of the world. And regardless of sources, hydrogen faces a cost barrier to adoption in industries that haven't used it in the past. Pressurization and refrigeration are expensive. Liquid hydrogen is difficult to transport, and pipelines need to be rebuilt to prevent hydrogen gas from leaking—or exploding. The sum of these costs is a hefty green premium.

Green hydrogen must compete with batteries and other clean energy sources that are trending down in cost. Where can green hydrogen win? One opportunity is to replace dirty hydrogen in generating ammonia for fertilizers. Down the road, green hydrogen could prove economical for steel fabrication and other high-heat manufacturing. Costs might be contained by producing the hydrogen on-site, with solar panels or wind turbines powering the electrolyzers.

In other sectors, green hydrogen might work for grid storage and maritime transport. But it's unlikely to power cars, buses, trucks, or trains, where denser and lighter batteries are becoming a more practical option. As an emissions-free choice to make the materials to build our cities, however, green hydrogen could usher in a new industrial era.

Cementing Carbon

Concrete has been used for over two thousand years. But it wasn't until the dawn of the industrial era, in the mid-1800s, that builders of cities learned how to use it at scale. Joseph-Auguste Pavin de Lafarge ran a limestone quarry in southeastern France. Seeking new ways to exploit the quarry's white rock and mineral-rich clay, Lafarge scaled a recently patented process for Portland cement, so named because it resembled the famed limestone on the Isle of Portland in the English Channel.

Cement releases carbon dioxide at different stages of production.

By the 1830s, Lafarge could see the emissions billowing from his smokestacks. He had no way to know he was contributing to a future climate crisis—or that the twenty-first-century cement industry would

Emissions from cement production

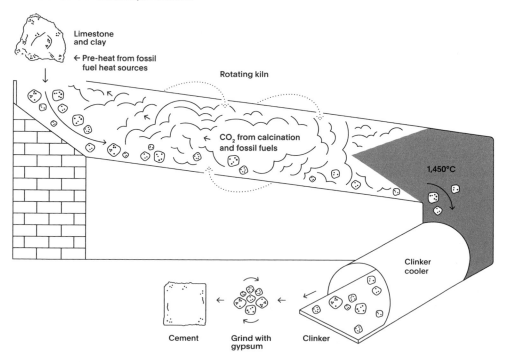

Limestone and clay

← Pre-heat from fossil fuel heat sources

Rotating kiln

← CO_2 from calcination and fossil fuels

1,450°C

Clinker cooler

Cement

Grind with gypsum

Clinker

one day spew around 3 gigatons of annual CO_2 emissions. Making cement has always been a fiery endeavor. Limestone and clay are heated to 1450 degrees Celsius in a fossil fuel–fired kiln, spawning half of cement's carbon dioxide emissions. As the materials are rotated and heated, limestone splits into calcium oxide and carbon dioxide—which generates the other half. From the kiln emerges pebbles of "clinker" that are ground with gypsum and other materials to create cement. Cement is essentially a glue that is mixed with water, sand, and gravel to form concrete. For the nineteenth-century construction industry, it was a winning formula. Lafarge grew into one of the world's largest cement makers and remains so today, with $25 billion in revenue.

For every ton of concrete that's produced, nearly the same amount of carbon dioxide goes into the atmosphere. Barring radical change, the industry's carbon emissions—5 percent of the globe's total greenhouse gas output—will continue to rise with economic growth. Pressure for radical change is mounting. In 2019, the Institutional Investors Group on Climate Change moved to leverage their $33 trillion in money under management to compel the industry to reach net zero by 2050. In a strongly worded letter to four gigantic European cement makers (including LafargeHolcim, now a publicly traded Swiss multinational with 67,000 employees), they urged adoption of both short- and long-term goals on emissions.

"We don't take this challenge lightly," said Lafarge CEO Jan Jenisch. He cited the company's use of 20 percent renewable energy to power its plants and promised to redouble their efforts to develop carbon-neutral cements. Climate-concerned investors weren't impressed; LafargeHolcim wasn't moving fast enough to make a big enough difference. Finally, in September 2020, the company committed to net-zero emissions by 2050.

Eric Trusiewicz has spent a decade in the cement and concrete business. Jan Van Dokkum was an operating partner at Kleiner Perkins when we developed our green investing strategies. Together, they serve on the board of Solidia, a young company that has developed a new chemical recipe: cement that absorbs CO_2 during the hardening process.

Eric Trusiewicz

Concrete is the invisible underlying material of all civilization. If you think of any urbanization, any industrial activity, any energy production, any transportation infrastructure, any building—all are based on concrete.

Concrete is 30 billion tons per year of magic. You take a little bit of this white powder and you mix it up with whatever you find around, and you make a rock that will last more than fifty years. Forty percent of the world does this with little more than a shovel. The other 60 percent uses industrialized equipment to make larger buildings and structures. It is unimaginable to have civilization in any form without this thing.

Concrete is thirty billion tons per year of magic.

Jan Van Dokkum

Here's the challenge in decarbonizing the cement industry. Today, there are around two dozen companies in the world that drive most global cement production. As of now, there's really no incentive to adopt new technologies as they relate to CO_2 reduction.

With companies that produce cement, there is a basic misalignment between economics and the environment. Their carbon footprint is significant. The margins in the cement industry are razor-thin because it is a highly competitive commodity business. So it's challenging for companies to spend money on innovation when other producers don't adopt it. Innovation has been seen as a hindrance, not a necessity.

The industry needs a clear mandate to clean up. Only then will innovation and change become a necessity for them.

Eric Trusiewicz

Yes, this is a hard problem. But already you can get to a 50 percent emissions reduction or more with existing techniques and technologies. First, you can design your building using half as much concrete. Second, you can resort to CO_2-reducing solutions like supplemental cement-like materials and fillers. These alternatives aren't as flashy as a branded form of concrete, but they work. The challenge is to scale their availability.

Yes, this is a hard problem.

Making change is not easy. There is a green premium with new concrete technology. While the materials are sometimes even cheaper, the approach often costs more in terms of the need for technical expertise and supervision. People have spent decades building structures in a particular way; hundreds of millions of workers, many of them low skilled, work in the construction industry worldwide. Plus there are regulatory and public safety issues. We need to ramp up education on how to use these new materials effectively and safely, and how to design buildings using less of them.

Governments can create incentives or mandates to accelerate the pace of change. The cement and concrete titans are not to blame for the fact that cement is high in CO_2. If governments and people want a sector to change, they have to change the rules for the sector. These companies have resisted change because until now it's been cheaper to resist.

But we're at the point where it's no longer going to be cheaper for these big companies to ignore their emissions. They see what is happening in the coal, oil, and gas industries. Rather than be left behind, they're asking how they can innovate.

Jan Van Dokkum

The burden is not just on governments. The investment and financial community can also put pressure on the cement industry to clean up its act. The cement industry needs to feel pressure to adopt change. Financial pressure affects their stock price, their access to the capital markets, and their return on investment. Shareholder pressure is the greatest driver for the adoption of new technologies and practices.

Eric Trusiewicz

How do we get this industry to net zero? We need to scale existing innovations and continue to bring viable technology to market. To be accepted by industry, a new approach needs to use raw materials that are widely available and cheap. The final product needs to be simple to use and cost competitive and reliable. Ideally, it would use existing infrastructure. That's a high bar, but there is a lot of entrepreneurial energy going into solving this problem.

Cement innovations need to target the two primary points of emissions: the fuel for the heat, and the chemical reaction within the kiln. For heat, there are promising approaches to replace fossil fuels with electricity or clean hydrogen. As to actual production within the kiln, several companies are redesigning the process to capture the carbon dioxide coming off the limestone. Groups like Jan and Eric's company, Solidia, are deviating from cement's original recipe with new chemistries that change how concrete is hardened. In altering the chemistry, Solidia's concrete absorbs CO_2 as it sets. Other groups are testing new materials and additives that can help form concrete with less cement— and less emissions.

As Eric says, concrete is a hard problem. If these new and cleaner approaches are to be deployed, large multinationals and innovative startups will need to work together. With the cost of a new cement facility ranging up to $400 million, entrepreneurs and university research teams can't do this on their own.

There is no guarantee that we'll find a way to eliminate all emissions from this sector. But our civilization isn't about to stop building and so we must keep trying. The payoff will be a giant step toward net zero.

Steeling Ourselves for the Future

Skilled craftsmen have forged superstrong metals for centuries. But in the 1880s, Andrew Carnegie, a Scottish-born industrialist in Pennsylvania, made a discovery that changed the world. He scaled a process that used blazing temperatures to burn off impurities in crude iron and create something harder and more durable. Steel soon became indispensable. With steel girders, buildings could be flung skyward, far beyond the old limit of four or five stories. Cities became vertical. And when the automobile industry came along, sheet steel became the material of choice.

From a climate perspective, the problem with steel manufacturing is that it accounts for close to 4 gigatons of annual carbon emissions, or about 7 percent of the global total. Like cement, steel production requires extreme heat—temperatures approaching 4,000 degrees Fahrenheit. Those blast furnaces are fed by coal, adding to the carbon pollution from the metallurgy itself.

To subtract emission from steelmaking is no easy task. It's a complex, multistep process, from reheating semifinished products to rolling the steel into sheets, with fossil fuels required at every turn. One partial solution is to use an electric current to melt recycled scrap steel from recycling, a process deployed in nearly two thirds of steel production in the United States. It's less popular in China, however, which accounted for more than half of the 1.8 billion tons of global production in 2020.

Emissions-free steel must meet three criteria. First, the furnace must be powered by a zero-emissions source. Second, the iron going into the furnace must either be produced without fossil fuels, or replaced with scrap steel. Finally, the heating step before rolling needs to be conducted with green hydrogen or some other clean source.

Before a green steel solution is feasible at scale, performance and costs need a real-world test. In 2020, in a joint effort with hydrogen producer Linde Gas, the Swedish steelmaker Ovako installed a green hydrogen system at its scrap-based mill in Hofors. According to project leader Göran Nyström, it marked the first time that hydrogen was used to heat steel in its rolling mills. The trial's success enabled Ovako to raise needed capital and switch over all of its factories, slashing its carbon footprint from cradle to grave.

Ovako proved that no-emissions hydrogen could be used in the fabrication process without any negative impact on steel quality. Predictably, the fossil fuel sector pushed back. Liquefied natural gas (LNG) companies, in particular, feared a competitive threat to the use of "blue" hydrogen from natural gas. Nils Rokke, chairman of the European Energy Research Alliance, called it "ridiculous" to skip LNG hydrogen and move to 100 percent green. "You need to do both," he argued.

While Rokke had a point, at least for the transition, the green genie was out of the bottle. The HYBRIT consortium of Swedish steelmakers came out in favor of green hydrogen at an even larger scale. In August 2020, the first large-scale green hydrogen steel plant was inaugurated by consortium member SSAB, Sweden's leading sheet steel company. Swedish prime minister Stefan Löfven was enthusiastic: "We are embarking on the biggest technological shift in steel manufacturing in a thousand years."

The Swedish government effectively created an OKR. They set a clear objective: zero carbon emissions in the Swedish steel industry by 2040. One key result was large-scale, production-level testing through 2024; another was a broader rollout. As the chief executive of SSAB declared: "We must seize this chance."

As we've seen, industry may be the most complex area to decarbonize. Even so, new technologies and business models are making remarkable headway in plastics, apparel, cement, and steel. All told, they contain the potential for eight gigatons of emissions reductions. If you're keeping count, you know that ten more gigatons hang in the balance, in our atmosphere . . . and in our next chapter.

Steel manufacturing accounts for about 3 gigatons of annual CO_2 emissions.

Speed & Scale: Countdown to net zero

Objective	Reductions		Remaining

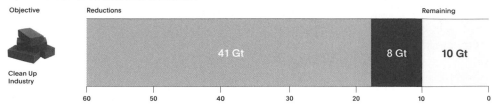

Clean Up Industry

41 Gt   8 Gt   10 Gt

60   50   40   30   20   10   0

Remove Carbon

Chapter 6

Remove Carbon

Let's imagine we reach the objectives of our first five chapters. We clean up transportation and electricity; we transform agriculture and reinvent our methods for making cement and steel. More than likely we'll underachieve on some of these big goals and outperform on others, but let's say our numbers on balance hold up. By our own math, we'll still be left with 10 billion tons of heat-trapping gases per year.

That's what keeps me up at night. Our **Carbon Dioxide Removal KRs (6.1 & 6.2)** is a truly devilish key result; we must somehow find a way to shed those 10 gigatons of CO_2e each year. Anything less would constitute failure—not just for our plan, but for humanity.

And so the question becomes: Should we focus on reducing our emissions, or should our priority be to remove them? Given the fierce competition over limited climate action funding, it's more than an academic debate. Here is where we stand: *The world needs to do both.* The two efforts are interwoven. Without carbon removal at scale, emissions cuts would need to double each year through 2040 for us to reach net zero in time. In sectors where clean alternatives don't yet exist, the burden would be crushing.

So what exactly is carbon dioxide removal—or, for short, carbon removal? It's a range of activities that would capture CO_2 molecules from the atmosphere, then store them. The CO_2 can be embedded in industrial products or in reservoirs underground, soils, forests, rocks, or oceans. In practice, carbon removal is composed of engineered and natural solutions. A prime example of the first is direct air capture, where CO_2 is separated from ambient air and permanently stored. Nature-based solutions include reforestation (replanting trees where they once flourished), afforestation (encouraging new growth), and agroforestry (integrating trees and shrubs into farmland).

Winnowing out CO_2 from our atmosphere will be no small technological feat. But what makes this challenge really difficult—almost implausibly so—is the colossal scale of the job. According to the World Resources Institute, an environmental think tank whose specialty is real-world mitigation, we're not nearly on track to pull down billions of tons of carbon each year.

As things stand, our net-zero objective—10 gigatons of annual carbon removal, about 17 percent of the world's total emissions—is a truly audacious goal. As JPMorgan's Michael Cembalest observed, only partly tongue-in-cheek, the highest ratio in the history of science is the number of academic papers written on carbon removal versus the amount of carbon removed in real life. Of all the tough tasks before us, according to Cembalest, engineered carbon removal may be "the steepest climb of all."

I look at it this way: We'll need to be spectacularly innovative and resourceful to clean up those lingering gigatons by 2050. We have no choice but to figure it out.

The Race to Capture Carbon

Our planet needs a portfolio of removal solutions. To have even a chance to close our gap to net zero by 2050, we need to start funding and scaling all of them now.

In both natural and engineered carbon removal, the challenges are complex. Natural solutions are vexed by issues around standards, accounting, verification, and "additionality," a gauge of whether the removal would have happened in any case. While opportunities abound for forest-related investments at affordable prices, a market standard for additionality has eluded us. Moreover, these solutions generally compete with land needed for agriculture and development. Finally, they're hampered by uncertain durability. Trees get burned. Carbon-rich topsoil gets tilled. Stored carbon gets released back into the air.

Objective 6
Remove Carbon

Remove 10 gigatons of carbon dioxide per year.

KR 6.1	**Nature-Based Removal**
	Remove at least 1 gigaton per year by 2025, 3 gigatons by 2030, and 5 gigatons by 2040.
	↓ 5 Gt

KR 6.2	**Engineered Removal**
	Remove at least 1 gigaton per year by 2030, 3 gigatons by 2040, and 5 gigatons by 2050.
	↓ 5 Gt

Remove Carbon

Carbon removal: the many ways to do it

CDR approach	Description
Afforestation & reforestation	Sequestration of CO_2 in newly grown forests (afforestation). Regrowth of degraded or removed forests (reforestation).
Improved forest management	Alteration of forest management practices to increase carbon storage.
Biochar	Incorporation of solid residue from thermal degradation of biomass into soils.
Bioenergy with carbon capture & storage (beccs)	Sequestration of CO_2 in biomass, which is not released upon energy transformation, but rather captured and stored.
Building materials	Concrete curing, integration of mineralized carbon materials and plant fibers.
Carbon mineralization	Reaction of natural or artificial alkaline minerals with CO_2 to form solid carbonate minerals, such as calcite or magnesite.
Direct air capture with carbon storage (daccs)	Chemical separation of CO_2 from ambient air coupled to permanent storage.
Increasing ocean alkalinity	Enhancement of stored dissolved inorganic carbon in the ocean through the addition of alkalinity, generally through mineral dissolution or electrochemistry.
Soil carbon sequestration	Enhancement of carbon in soils by adjusting land management, e.g., reducing tillage or establishing agroforestry.
Coastal blue carbon	Sequestration of CO_2 in additionally grown biomass and soils of restored ecosystems, including peatlands and coasts.
Marine biomass management and cultivation	The cultivation of microalgae or macroalgae in marine ecosystems for the purpose of increasing carbon sequestration in aquatic biomass, and/or enhancing the durability of sequestered carbon through improved management or utilization of biomass.

With engineered carbon removal solutions, durability can stretch to a thousand years or more. But this approach has other issues:

Volume: As the World Resources Institute's Kelly Levin notes, we are banking on "off-the-charts scaling for completely unprecedented technologies." While direct air capture is promising, the technology has so far sequestered just 2,500 tons of carbon worldwide—a tiny fraction of 1 percent of a single gigaton.

If we're counting on engineered solutions to remove half of the 10 billion tons of leftover carbon emissions, current technology offers cold comfort. **We'd need an expanse of solar panels as large as the state of Florida.** The process would soak up close to 7 percent of the world's total energy—more than the consumption of Mexico, the United Kingdom, France, and Brazil combined. **To pump that much CO_2 underground would be the equivalent of running the entire oil industry in reverse.** If carbon removal technologies are to meaningfully address our needs, they must become far more efficient.

Cost: No existing mode of engineered removal is nearly economical enough to capture and store carbon at scale. Markets have barely formed. The current ballpark price for direct air capture is $600 per ton, or $600 billion per gigaton. To handle 5 gigatons would cost $3 trillion each year.

That's why we're tying this key result to our **Carbon Removal KR (9.4)** in the innovation chapter in Part II. With future technological breakthroughs and economies of scale, engineered carbon removal could reach a commercial price of $100 per ton by 2030 and $50 per ton by 2040—95 percent less than its cost today.

Equity: As the Intergovernmental Panel on Climate Change noted, our pathway toward a low-carbon, climate-resilient future is "fraught with moral, practical and political difficulties and inevitable trade-offs." In communities plagued by lethal air pollution, carbon removal is no substitute for emissions cuts. When a coal-burning steel factory in China or the United States pays for carbon removal offsets from a carbon capture company in Iceland, it's still harming local people by the factory.

What About Offsets?

Before we go about scaling carbon removal, we must ask how it will be achieved—and by whom. Offsets are programs that enable companies or individuals to pay for emissions reductions or removal that in theory negate their own emissions.

In climate action circles, *offsets* is a loaded term, both heavily criti-cized and broadly used. **At their worst, offsets are an exercise in greenwashing,** a way to absolve companies or individuals from responsibility for bad behavior. While higher-quality offsets can have a positive climate impact, they're prone to overestimates and even fraud. Often they're used to finance green solutions that would have been deployed without them. We cannot solve the climate crisis with greenwashing. Our carbon budget is too small, our time too short.

Gareth Joyce, Proterra's president and former chief sustainability officer for Delta Airlines, believes that we need a new currency for carbon removal credits. The ideal system would reward investments in future tech solutions and at the same time keep money flowing into the nature-based approaches used today.

Before choosing an offset project, ask these two questions:

Have we done everything possible to decarbonize our operations, our supply chain, and the way our product is used?

Have we optimized all available efficiencies?

If the answer to both questions is yes, offsets can be a worthwhile interim solution, as long as they are:

Additional

Verifiable

Quantifiable

Durable

Socially beneficial

Our Speed & Scale Plan asks countries and companies to first get their own house in order and cut emissions through avoidance and energy efficiency. Only then can we turn to offsets—not as an accounting gambit or public relations strategy, but as a genuine effort to help us heal the planet. When you're on the knife's edge, as WRI's Kelly Levin says, "you need to do everything."

Should We Plant a Trillion Trees?

Planting more trees, an obvious and inexpensive way to pull carbon from the air, needs to be approached systematically. Yes, trees are the best natural mechanisms for absorbing CO_2 and ending the feedback loop of global warming. Proposals to plant a trillion trees have drawn considerable fanfare. But these catchy campaigns often skirt the hard questions of tree planting: How much carbon can be absorbed by a tree, and for how long? How will planting a tree affect the local ecosystem and economy? How much land will a trillion trees require?

To work at scale, tree plantings demand forethought, planning, and regulation. Reforestation is most successful when it adds species that are native or complementary to the local ecosystem, in spots where they can thrive. To restore and expand the carbon sink and meet our **Forests KR (4.2)**, we need trees that will live to see us through years of crisis, to 2050 and beyond.

We also need to be mindful about land requirements. **For trees to remove the emissions of Americans alone, we'd need to dedicate half of the world's landmass.** That's not to say tree planting doesn't have a role to play. Quite the opposite. But as with all other carbon removal solutions, it's important to remember that it's not a panacea. Tree planting is an important strategy that comes second to stopping unchecked deforestation and runaway emissions.

With efforts under way from China to Ethiopia, tree planting is on the upswing. Under this canopy of good intentions, it is crucial that we focus not on arbitrary targets, but rather on lasting impact.

Out of Thin Air

In 2009, as engineering grad students in Switzerland, Christoph Gebald and Jan Wurzbacher founded a direct air capture startup called Climeworks. Eight years later, they'd built a bank of eighteen gigantic fanlike units that could filter carbon dioxide from the air. After a debut demonstration, a neighboring greenhouse bought the carbon dioxide to fertilize its fruits and vegetables and spur their growth. As a side business, Climeworks pumped captured CO_2 into tanks and sold it to a local bottler for the fizz for its Coca-Cola.

Those early deals were a bridge to larger commercial ventures. In collaboration with Reykjavík Energy, an Icelandic utility, Climeworks built a pilot rooftop facility to capture 50 tons of carbon dioxide per year at $1,000 per ton, a significant proof-of-concept test for mechani-

cal removal. The next step is a plant that captures 4,000 tons of CO_2 a year. Instead of selling this captured carbon, Climeworks will mix it into geothermally heated water and inject it into underground reservoirs. Over two years, a gradual chemical reaction will produce a solid mineral, calcium carbonate. Past emissions will be forever sequestered in rocks.

Carbon Engineering, a British Columbia startup, plans to deploy similar technology at a much greater scale. Backed by Bill Gates and Chevron Corporation, the company is building what it says will be the world's largest direct air capture plant, with the capacity to remove 1 million tons of CO_2 per year.

Swiss startup Climeworks is a pioneer in direct air capture, an engineered form of carbon removal and storage.

Direct air capture isn't the only way to get this job done. In 2018, a group of aerospace engineers formed a San Francisco startup called Charm Industrial. "We spent a year of Saturdays looking for ways to sequester carbon dioxide economically," says founder and chief executive Peter Reinhardt. They settled upon a process called "fast pyrolysis," a rapid-fire, high-temperature decomposition of plant material into liquid fuel. Agricultural farm waste, once a source of carbon emissions, became a "bio-oil" pumped into old petroleum wells. "It's no longer a gas, so it sinks and remains in the ground," Reinhardt says. "The carbon is never coming up again." Charm can already sequester CO_2 at $600 per ton and turn a small profit.

These potential solutions will hinge on governments setting a price on carbon, which could then be applied as a payment for each ton sequestered. By 2030, if engineered removal costs drop as we believe they will, a carbon price of $100 per ton could offset the entire cost of the capture process. Additionally, it would spur on larger markets and help lower the green premium. Captured carbon dioxide could be used to make cement, for example, or as a jet fuel. Removal credits could be sold to net out a company's emissions in other areas. "We are not just founding a company," Jan Wurzbacher, Climeworks' CEO, told *The New York Times*. "We're really founding a new industry."

Catalyzing the Market for Carbon Removal

You cannot build a new industry without a market for your product. The practical issue with carbon removal is a lack of any real incentive to pay for it. Why would anyone fork over six hundred dollars or even three hundred dollars to erase a ton of carbon from the air? The price is substantial, and that 1 ton gets you just one billionth of the way to the first gigaton of the five we need to eliminate through engineered removal. If you splurged and bought a million tons, you'd be on the hook for half a billion dollars—and you'd still need a thousand friends to do the same to remove a single gigaton in a single year.

Enter Stripe, the internet payment processing company that opened shop in 2009 in Palo Alto, California. The company's founder, Patrick Collison, had a noteworthy track record in software before he was old enough to vote. At age sixteen, Patrick won Ireland's BT Young Scientist Competition for the creation of Croma, a programming language for artificial intelligence. After enrolling at the Massachusetts Institute of Technology, he soon dropped out to start Stripe with his younger brother, John. These days, the company provides financial services for the likes of Amazon, DoorDash, Salesforce, Shopify, Uber, and Zoom. The once tiny family enterprise is now valued at $95 billion. With a cloud-based software services business operating across 120 countries, Stripe is uniquely positioned to build a large-scale market for carbon removal. In October 2020, under the leadership of Nan Ransohoff, it launched Stripe Climate, which makes it easy for any business to direct a small fraction of its revenue to the cause. As of June 2021, Stripe Climate had more than two thousand businesses purchasing carbon removal at an average price per ton of more than five hundred dollars.

Nan Ransohoff

Stripe's initial foray into carbon removal was in late 2019, when we committed to pay at least $1 million per year for the direct removal of carbon dioxide from the atmosphere and its permanent storage, at any price per ton.

This stemmed largely from a key takeaway from the 2018 IPCC report: that carbon removal, in addition to emission reduction, will be critical to mitigate the worst effects of climate change.

While some of the solutions needed to get there exist today—like planting trees and soil carbon sequestration—it's highly unlikely that these solutions by themselves will get us all the way there.

Our focus at Stripe is on filling that gap. In practice that means we are buying carbon removal from early companies, often as the very first customer, with the theory being that early customers can help accelerate promising new companies down the cost curve and up the volume curve. This is not a new idea; experience with manufacturing learning curves has shown repeatedly that deployment and scale beget improvement, a phenomenon seen across DNA sequencing, hard drive capacity, and solar panels.

We turned theory into practice in spring of 2020 when we made our first carbon removal purchases. In response to this announcement, two things happened. First, the carbon removal community had a surprisingly positive reaction—which was mostly indicative of how starved for capital the field had been. Second, we received a flood of outreach from many Stripe users sharing that they'd wanted to make their own climate commitments but found it challenging to know where to begin—much less to develop their own criteria for evaluating projects.

These two insights led to the genesis of Stripe Climate, which makes it easy for any business to direct a fraction of their revenue to fund frontier carbon removal.

No single business can create enough demand to scale carbon removal on its own. But the millions of businesses running on Stripe, collectively, can help grow and sustain this new industry. Our goal is to create a large market for carbon removal by pooling demand. If successful, this market will accelerate the availability of low-cost, permanent carbon removal technologies, and increase the likelihood that the world has the portfolio of solutions needed to avoid the most catastrophic effects of climate change.

Our goal is to create a large market for carbon removal by pooling demand.

As more customers rise to the moment and make voluntary purchases, they'll enable carbon removal companies to scale their operations and cut prices. The next advance would be to audit the amount of carbon actually captured and put a premium on durability.

One pacesetter in this arena is Microsoft, which recently made an unprecedented commitment to become carbon *negative* by 2030. The tech giant pledged that its total operations—plus supply chain—will remove more carbon than they emit. By 2050, Microsoft aims to eliminate its entire history of emissions, dating back to 1975, when Bill Gates dropped out of Harvard and cofounded the company with Paul Allen.

The company's pledge came attached to a $1 billion investment to accelerate and scale a growing portfolio of carbon removal projects already under way. As Microsoft's team leaders wrote, "Those of us who can afford to move faster and go further should do so." In the spirit of OKR tracking, they promised an annual sustainability report to "detail our carbon impact and reduction journey."

Microsoft issued a request for proposals that elicited 189 ideas from 40 countries. The 15 winning plans would remove a total of 1.3 million metric tons—more than 500 times direct air capture worldwide to date. Climeworks and Charm both made the final cut.

To be sure, 99 percent of the volume of Microsoft's first crop of removal projects resides in natural solutions, mainly forest and soil projects with one hundred years of durability or less. The company plans to create bigger and broader portfolios each year. Over time, it expects a growing percentage of its removal efforts to be longer duration, engineered solutions.

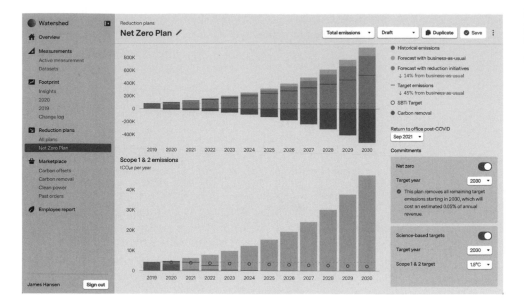

Finding Each Organization's Path to Net Zero

Watershed's carbon accounting platform enables companies to track and reduce their emissions.

Getting a company to net zero is hard work. It means finding ways to cut existing emissions to the bone, maximizing energy efficiency, measuring progress across far-flung supply chains, and calculating how many offsets are needed to cancel out the carbon left over. Not least, a meaningful net-zero effort entails the submission of results to investors with the same rigor and transparency that are expected in corporate financial reports.

Two decades ago, my friend Mike Moritz of Sequoia Capital and I teamed up to back Larry Page and Sergey Brin in launching Google. We joined forces again this year to back three entrepreneurs from Stripe—Christian Anderson, Avi Itskovich, and Taylor Francis—in starting Watershed. We all agreed that a software platform to drive carbon reduction would indeed be a watershed moment in our campaign for net zero.

As we can see from Taylor's experience, **if you don't measure it, you can't manage it—or, even more important, change it.**

Taylor Francis

I remember seeing *An Inconvenient Truth* the summer after eighth grade. I walked out of the movie theater terrified, but also riveted and energized. This climate crisis felt like a generational challenge, something that would affect me and my friends as we grew up. And something we could solve if we got started now.

I wanted to do whatever I could to help. I started cold-emailing Al Gore's office until someone wrote back and said they were training people to give local versions of his slideshow. I went to Nashville for training. (I was fourteen, so my mom had to come to help me get a hotel room.) And then I spent the next four years traveling to high schools from California to China, talking about climate change. I told students that we needed to pressure our parents to act so that we could grow up in a world free from climate catastrophe.

But I struggled to find something that would go beyond talk, that would actually bend the carbon graph. So I put climate on the back burner and worked at Stripe, where I learned a lot about building software products.

In 2019, it felt like it was time to get back into the climate fight. I started working with Christian Anderson and Avi Itskovich to build a startup with the mission of directly eliminating at least 500 million tons of CO_2 per year. Christian had launched Stripe's climate initiative, and we saw that the status quo for corporate climate programs was wholly inadequate. Companies were spending months putting together PDF reports of their carbon footprint that were already stale by the time they were published.

They were buying cheap carbon offsets that didn't actually remove carbon from the atmosphere. That was our *aha* moment: If decarbonization is going to be an economy-wide endeavor, it's going to need software tools that are up to the challenge.

So that's what we're building at Watershed: tools to help companies measure, reduce, remove, and report their carbon emissions. Think of it as a platform for getting to true net zero. We want to enable businesses to embed carbon math into the decisions they make every day.

We know it's possible. Leading companies like Apple, Google, Microsoft, Patagonia, and even Walmart have found that they can cut emissions dramatically while growing their businesses. It's good for their bottom line.

A whole wave of companies is using Watershed to manage carbon just like they manage every other part of their business. Square is sourcing low-carbon materials for its hardware and driving clean power adoption with blockchain miners. Sweetgreen designs its menu by counting the carbon impact of each item at the same time it counts calories. Airbnb, Shopify, and DoorDash have a chance to reinvent hospitality, e-commerce, and logistics in a zero-carbon way.

This is all about bending the carbon graph. The world's total emissions are the sum of billions of business decisions—about how to power buildings, manufacture products, and get them to customers. We'll only get to zero if every business integrates carbon into those decisions.

We'll only get to zero if every business integrates carbon into decisions.

From the Sobering to the Inspirational

In the first six chapters of this book, we have attempted to convey our sense of the scale of the challenges before us, the six essentials that together can turn back the clock on carbon emissions. If you're feeling a little disheartened, I understand; I often feel the same way. Climate change is the product of huge interlocking forces—of biology and physics, government and commerce. The problem is so complex that it tests our ability to comprehend it, much less solve it. There's so much for us to do to have a credible chance to dodge a climate catastrophe. There is so little time. Most of all, the stakes are so very, very high.

But the rewards are high as well. Once we're on the path to net zero, extra benefits will accrue. Once we succeed in cutting and removing carbon emissions, we'll restore nature's healing powers. We'll help the planet absorb more carbon—the ultimate virtuous circle.

Like our OKRs, this book comes in two parts. I've shared with you *what* we need to do to solve the climate crisis before it gets beyond us. That's the hard and sobering part. Now let's address *how* we can get there by our 2050 deadline. We'll examine four sharp-edged tools we can use to work wonders. I call them *accelerants*: policy, movements, innovation, and investment.

I won't suggest that Part II is the "easy" part of the book. But for me, it's the uplifting part. The inspirational part. And while I'm not banking on hope to pull us through, it's the more hopeful part too. It's about the things that can happen within our communities and governments, our companies and nonprofits—the places where we can exert some control.

When all is said and done, we got ourselves into this fix. It's up to us—with all our human frailties, but also our pooled ingenuity—to get ourselves out. Let's consider next how we can do just that.

Speed & Scale: Countdown to net zero

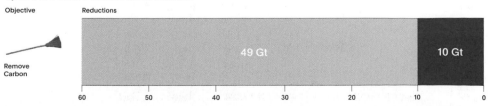

Win Politics and Policy

Chapter 7

Win Politics and Policy

In January 2009, after two years of fighting the good climate fight in California, I entered a bigger arena. I raised my right hand and testified at a U.S. Senate hearing on climate change and energy policy. Speaking before a committee chaired by my home state senator, Barbara Boxer, I warned that America was in danger of being left behind in technology for solar, wind, and advanced batteries. I said that critical solutions would come if we did a better job of funding American entrepreneurs, even if some of them would surely fail along the way. And I argued that putting a price on greenhouse gas emissions—a carbon price—was the one policy that overrode everything else, the one that mattered most. Beyond encouraging emissions cuts, a carbon price would level the playing field between renewable energy and fossil fuels. It could change everything.

"Forgive me for being blunt," I told the senators, "but what we've been doing is not enough. We must act now, with speed and scale."

I'd chosen my words carefully. If we were going to begin to reverse more than a century of climate abuse, our efforts needed orders of magnitude more speed and exponentially greater scale. As a nation with unmatched capacity for innovation, the United States needed to lead the push to check global warming. As the nation most at fault for this predicament, we were obliged to do more than others to solve it.

If one could pinpoint the moment when climate change was widely recognized as a mortal threat to humanity, it might have been in 1992, in Rio de Janeiro, at the United Nations Conference on Environment and Development—better known as the Earth Summit. Scientists, diplomats, and policy makers from 178 nations, including 117 heads of state—convened for twelve days in June. They put their distinguished heads together to begin to figure out just how we might save our planet.

The agenda covered endangered tropical rainforests, looming water shortages, suffocating urban sprawl, and toxins, toxins everywhere, from leaded gasoline to nuclear waste. But one topic rose above the rest. The scientific evidence of mounting CO_2 and other greenhouse gases in the atmosphere, and their link to climate change, cried out for a response.

The Earth Summit became a clarion call for sustainable development, for growth that preserved our ecosystems. The delegates adopted an ambitious-sounding plan with a $600 billion annual price tag. The moment amplified awareness of an emerging political issue and of an ominous term: *global warming*.

And yet the summit was hobbled from the start. President George H. W. Bush, who'd once vowed to be "the environmental president," was up for reelection and loath to offend the fossil fuel industry. Citing potential harm to the U.S. economy, Bush threatened to boycott the convention if it moved toward specific emissions targets. In the end, the United States joined 153 other countries in signing an agreement that fell significantly short of what we needed. That moment would set a pattern for decades to come. Again and again, international climate agreements would be watered down out of deference to U.S. politics and the fossil fuel industry.

Two years later, in 1994, the United Nations Framework Convention on Climate Change called upon wealthier nations to cut emissions while subsidizing poorer ones to protect natural resources. The geopolitics weren't pretty. "Rich countries and poor countries bickered endlessly in Rio over who should pay for various environmental protections," noted *The Washington Post*. "Yet, in the end, the world's nations agreed to keep thrashing these issues out in future U.N. forums. . . . The hope was that lasting solutions would emerge over time."

The signatories clung to that hope with the Kyoto Protocol of 1997, the first international pact to specify greenhouse gas emissions reductions. But in the United States, the Senate voted to block ratification. Climate action stalled for nearly two decades, until the breakthrough Paris Agreement in 2015, joined by the United States by way of President Obama's executive order. One hundred and ninety-five countries called for limiting the rise in average global temperatures to "well below" 2 degrees Celsius above preindustrial levels and for "pursuing efforts" to cap the increase at 1.5 degrees Celsius, "recognizing that this would significantly reduce the risks and impacts of climate change." When negotiators set aside binding targets and timetables, their new tack unlocked greater ambition. For the first time, every country in the world committed—at least on paper—to a shared objective of keeping emissions in check, with efforts to be ratcheted up in the future.

One year later, newly elected president Donald Trump vowed to pull the United States out of the agreement. Four years after that, President Biden rejoined the Paris community. Partisan gyrations aside, here is the harsh truth: in our do-or-die campaign to zero out greenhouse emissions in time, the Paris Agreement—though an important first step—won't get us there. As U.S. climate envoy John Kerry noted, "If you did everything that we laid out that we needed to do in Paris, we would still be rising 3.7 degrees centigrade. That's catastrophic. But we're not doing everything we laid out in Paris, so we're . . . heading to 4.1 or 4.5 degrees—a formula for environmental Armageddon."

Christiana Figueres, a prime architect of the Paris Agreement, stressed that it was designed as a framework for nations to set their own plans and make them more ambitious over time toward the net-zero goal by 2050. Nations' initial commitments, Figueres pointed out, were only "the starting point" for a long process of continual improvement. Based on Paris, governments must reconvene every five years to report on efforts and take the next collective step in emission reductions. With compounding decarbonization and "always rising ambition" over the next three decades, Figueres says, "we would get to net zero by 2050."

The Policies We Need

As always, we aim to be ambitious in designing our objectives and key results, or OKRs. In Part I, we set quantitative goals to cut greenhouse gas emissions. Now, in Part II, we'll address the essential levers to accelerate our transition to zero: policy and politics, for starters, but also movements, innovation, and investment.

The next set of global policy actions must accelerate the world's transition to net zero while defining, with transparency and precision, how each country will tackle the challenge. This unprecedented threat is also an unparalleled opportunity. With the United States back on the job, we have the broadest global consensus for climate action in history. The significance of this opening became evident in April 2021, when President Biden convened a virtual climate summit with forty world leaders around Earth Day.

In the world of policy, it's imperative to know how to win. But what of the policies themselves? As with any set of goals, we must focus on the essential ones. We've distilled dozens of possibilities into nine that matter most.

s Unies

ements Climatiques 2015

CMP11

France

PRESIDENT   SECRETAIRE

The Paris Agreement provided a framework for nations to set more ambitious targets for emissions reductions over time.

Objective 7
Win Politics and Policy

(We will track this objective by country for
the top five global emitters.)

KR 7.1   **Commitments**

Each country enacts a national commitment to
reach net-zero emissions by 2050, and gets at
least halfway there by 2030.*

KR 7.1.1   **Power**

Set an electricity sector requirement to cut
emissions 50% by 2025, 80% by 2030, 90% by
2035, and 100% by 2040.

KR 7.1.2   **Transportation**

Decarbonize all new cars, buses, and trucks by
2035; freight ships by 2030; semi trucks by 2045;
and make 40% of flights carbon neutral by 2040.

KR 7.1.3   **Buildings**

Enforce zero-emissions buildings standards for
new residential by 2025, commercial by 2030, and
prohibit sales of nonelectric equipment by 2030.

KR 7.1.4   **Industry**

Phase out fossil fuel use for industrial processes
at least halfway by 2040, and completely by 2050.

* This is the timeline for developed countries. For developing countries,
 this key result is expected to take more time (5–10 years).

KR 7.1.5	Carbon Labeling	Require emissions-footprint labels on all goods.
KR 7.1.6	Leaks	Control flaring, prohibit venting, and mandate prompt capping of methane leaks.
KR 7.2	Subsidies	End direct and indirect subsidies to fossil fuel companies and for harmful agricultural practices.
KR 7.3	Price on Carbon	Set national prices on greenhouse gases at a minimum of $55 per ton, rising 5% annually.
KR 7.4	Global Bans	Prohibit HFCs as refrigerants and ban single-use plastics for all nonmedical purposes.
KR 7.5	Government R&D	Double (at minimum) public investment into research and development; increase it fivefold in the United States.

The **Commitments KR (7.1)** requires firm national commitments to carbon net zero by 2050, with internally enforceable action plans aligned with ambitious 2030 reduction targets.

Our **Power KR (7.1.1)** tracks national targets for zero-emissions electricity. Escalating targets—50 percent by 2025 and 80 percent by 2030—are powerful market signals that prod utility companies to make the transition on schedule. They also steer governments to invest in critical infrastructure for clean energy.

Our **Transportation KR (7.1.2)** measures national incentives for electric vehicle purchases. Tax credits and rebates are popular across the United States, Asia, and Europe. Even if EVs are cheaper to operate than combustion vehicles, these "money off" deals can offset higher initial purchase prices.

There is no shortage of ideas for how to ramp up EV miles and ramp down combustion miles. In Norway, the government waived import duties for EVs and offered owners tax breaks and discounts on toll roads and public parking. In the United States, a short-lived "cash for clunkers" trade-in program in 2009 showed the potential of paying owners to get old cars off the road. National auto mileage standards are reliable tools for improving fuel efficiency. Raising caps on tax credits can further encourage purchases of EVs. We'll need just a few smart policies—and the political will to fund them—to move to an all-electric global fleet.

The **Buildings KR (7.1.3)** calls for a zero-emissions standard for all new residential buildings by 2025 and for new commercial buildings by 2030. It would mean turning from oil- and gas-fired furnaces and stoves to electricity for all heating and cooking. In addition, it incorporates efficiency goals for both new and existing buildings. The global model is California's green building codes, which have helped state residents save more than $100 billion since the 1970s. Their average annual household electricity bill is seven hundred dollars lower than the average bill in Texas. How was this achieved? With insulation and appliance standards, improved building design, and far more efficient light bulbs. Most importantly, California's requirements escalate up over time.

The **Carbon Accounting KR (7.1.4)** proposes carbon emissions labels on all consumer goods, including food, furniture, and apparel. The aim is to enable shoppers to make lower-emissions choices by disclosing all products' carbon footprint.

Our **Leaks KR (7.1.6)** specifies that nations pass regulations that control flaring, prohibit venting, and mandate prompt repair of leaks at oil and gas drilling sites. Due to lax regulation and enforcement, these "fugitive emissions," mostly methane, exceed 2 gigatons of CO_2 equivalents. Methane is a "short-term climate forcer," a substance that stays in the atmosphere for far less time than carbon dioxide, but creates more warming in the short term. Every nation must squarely address the emergency of preventable methane pollution.

Our **Subsidies KR (7.2)** eliminates government subsidies that effectively fund carbon emissions. It repurposes the money into energy efficiency and our clean energy transition. Fossil fuels receive $296 billion in annual direct subsidies and $5.2 trillion in indirect subsidies, nearly 6.5 percent of global GDP. (The larger figure includes corollary factors, like health-care costs from air pollution.) In addition, the United States alone spends $81 billion on military and security outlays to protect oil and gas sites and global transportation routes. This key result would also end agricultural subsidies for high-emissions practices and substitute incentives for regenerative agriculture and other climate-friendly measures.

Our **Carbon Price KR (7.3)** places a fee on greenhouse gas emissions. While implementation would vary from country to country, the basic idea is simple: greenhouse gas pollution comes with a price tag attached, a rising penalty for emissions of carbon dioxide, methane, and other climate-warming gases. A carbon price would make fossil fuel energy more expensive and less competitive, discouraging its use. It would send a strong signal to markets to accelerate adoption of cleaner, more efficient alternatives.

Our **Global Bans KR (7.4)** calls for universal adoption of the Kigali Amendment to the Montreal Protocol. This international treaty prohibits the use of all hydrofluorocarbons (HFCs), the heat-trapping refrigerants that are thousands of times more potent, pound for pound, than CO_2. More than 120 nations have ratified the Kigali

Amendment to phase out HFCs—but they've yet to be joined, as of this writing, by three top emitters: China, the United States, and India. Shortly after President Biden's inauguration, he submitted the amendment to the U.S. Senate, where it is expected to be ratified. In addition, the Environmental Protection Agency is formulating a rule to curb greenhouse gas refrigerants. This key result also proposes a global ban on single-use plastics for nonmedical purposes, including a rapid phaseout for grocery bags and single-serving beverage containers.

Our **Government R&D KR (7.5)** funds the discovery of breakthrough technologies, which in turn drive down costs for adoption of clean technologies. It prescribes a worldwide doubling of government-funded energy research and development, with a minimum five-times jump by the United States, to $40 billion per year. The additional money would cover basic and applied research, including early trials. Even small government grants can make a big difference for clean-technology startups. They can also pay big dividends to a nation's economy down the road.

We'll Always Have Paris

More than five years after Christiana Figueres took a lead role in delivering the Paris Agreement, I asked her about the upcoming United Nations climate conference in Glasgow, set for November 2021. She said she expected it to build on the Paris Agreement by asking countries to submit a second set of emissions reduction efforts through 2030. And she hoped the Glasgow delegates would agree on a global price on carbon, a mighty leap toward the goal of net zero.

Christiana Figueres

The Paris Agreement of 2015 was the first legally binding treaty adopted unanimously; all 195 UNFCCC member nations signed. The United States subsequently took a "little vacation" from the Paris Agreement and then returned. In the meantime, thanks to speedy ratification by signatories, the Agreement went into force in record time.

Paris is unique in mandating a process of constant improvement, taking into account the realities of each country as baseline starting points. The Agreement also establishes an end point for our targets: net zero by 2050. That was the most difficult piece to achieve consensus on.

We knew from the start that there would be many different paths to get to net zero, a different one for each country. Allowing for what we called "nationally determined contributions" made the agreement more flexible. It wasn't punitive. It was based on countries' enlightened self-interest, a powerful force for change.

Whatever your emissions are, you have those emissions. There's no pointing fingers and no blaming. There is a starting point, a common direction, and a shared outcome. As long as we are at net zero globally by 2050, each country can determine its own path.

There's also what we call a ratchet mechanism, a series of check-points. Every five years, countries must come together and report on what they have done to reduce emissions. And they must outline their next-level ambition for emissions reductions. These plans are necessarily fluid. They're based on rapidly evolving solutions and technologies, quickly shifting financial consider-ations, and the landscape for the most effective policies.

As for top priorities for Glasgow, I would love to see what we wanted in Paris to finally be made a reality: a global price on emissions, because a carbon price is so critical to decarbonizing entire economies and to fighting deforestation.

Right now, we have sixty jurisdictions with a price on carbon, but it's ludicrously low—generally between two dollars and ten dollars per ton. To make a real difference, that price needs to go up to one hundred dollars a ton over time. With rigorous cross-border standardization and methodologies for quantification, I'm convinced that a global carbon price would be absolutely transformational.

The piece I am now most concerned about is nature: our lands and oceans. We are much better at transforming energy and transport and finance than restoring the nature that surrounds us. We haven't really incorporated how we'll regenerate soils or reforest degraded lands or protect standing forests. Very few standing forests are left. We won't have a business model to sustain our lands and oceans until we have that price of carbon that rises over time. That's how we'll be able to tie all the mitigation measures together. And that's what keeps me up at night.

As long as we are at net zero globally by 2050, each country can determine its own path.

Over two thirds of emissions come from just five countries
Emissions share 2010–2019 (%/yr)

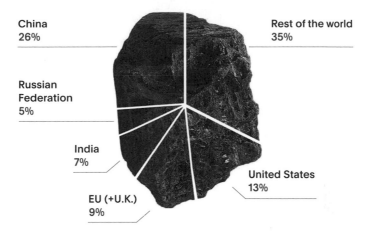

China
26%

Rest of the world
35%

Russian
Federation
5%

India
7%

United States
13%

EU (+U.K.)
9%

Adapted from data by UN Emission Gap Report 2020

I wholeheartedly agree with Christiana that it's best to look forward. Previous international climate treaties have set the stage for where we are today. The stakes for Glasgow and beyond are very, very high.

Focusing on the Big Five

Five entities account for nearly two thirds of total global greenhouse gas pollution: China (26 percent), the United States (13 percent), the European Union and the United Kingdom (9 percent), India (7 percent), and Russia (5 percent).

To further sharpen our focus, we target the leading sources of emissions in each of the Big Five. In China and India, it's coal power. In Russia, it's drilling for oil and gas and mining for coal. Once we factor in fugitive emissions and end-use combustion, the energy industry accounts for 80 percent of Russia's greenhouse gas pollution. In the United States and Europe, where emissions from transportation rose in both 2018 and 2019, it's the reliance on gasoline and diesel fuels. To get to a net-zero world in the little time left, we must continuously measure the Big Five's real progress—both according to their latest official goals and against our own plan's ambitious key results.

Where do the big emitters stand? Which policies will be most pivotal? In the chart below, we show where the Big Five have met one or more of our KRs. (An expanded policy tracking board with additional countries is regularly updated at speedandscale.com.) As you can see, we have miles to go.

Win Politics and Policy

The varying decarbonization policies of the top 5 emitters

Meets/exceeds | Directionally meaningful | **Insufficient**

KR Policy & Policy Objective	China	U.S.	EU + U.K.	India	Russia
7.1 Commitments: Each country enacts a national commitment to net-zero emissions by 2050, and at least halfway by 2030*	Net zero by 2060	Cut emissions in half by 2030	**Net zero by 2050**	**No commitment**	**No commitment**
7.1.1 Power: Set an electricity sector requirement that cuts emissions 50% by 2025, 80% by 2030, 90% by 2035, and 100% by 2040	**China has committed to strictly control coal and peak emissions before 2030.**	10 states, DC, and Puerto Rico have binding legislation requiring 100% clean or net zero power by 2050.	The EU has committed to a target for renewable energy that is at least 32% by 2030.	India has committed to at least 40% of country's electricity will be generated from non-fossil sources by 2030.	**No targets**
7.1.2 Transportation: Use subsidies and mandates to accelerate turnover of cars, buses, and light/medium trucks by 2035 and heavy duty trucks by 2045	**China issued a New Energy Vehicle Industry Development Plan (2021–2035) issued by State Council in October 2020, 20% of sales by 2025, EV becomes mainstream of sales by 2035 (means >50%).**	U.S. offers a federal tax credit up to $7,500; credit ends once manufacturer sells 200,000 units of a given EV model. Some states offer additional incentives (e.g., California, Colorado, Delaware).	The EU has proposed ratcheting down of vehicle CO_2 standards from 37.5% below 2021 emissions levels to 50% below by 2030, and an effective ban on internal combustion engine autos by 2035.	India passed FAME II, effective from April 2019, that has an outlay of INR 100 billion (USD 1.4 billion) to be used for upfront incentives on the purchase of EVs and for supporting the deployment of charging infrastructure.	Russia has waived import taxes for electric cars (through end of 2021).
7.1.3 Buildings: Enforce net zero emissions buildings standards for new residential by 2025 and commercial by 2030 and prohibit sales of non electric equipment by 2030	By 2022, China's Action Plan of Green Building Construction requires 70% of new construction to meet green building standards of China's 3-Star Rating System.	**No federal requirement for constructing net-zero energy buildings. California, Colorado, and Massachusetts have requirements.**	In the EU, new construction is to be "nearly" zero beginning in 2021. EU and various member countries are considering restrictions on the sale of fossil-fueled appliances in new and existing buildings.	**No net-zero building requirements. No restrictions on the sale of fossil-fueled appliances.**	**No net-zero building requirements. No restrictions on the sale of fossil-fueled appliances.**
7.1.4 Industry: Phase out fossil fuel use for industrial processes by 2050, and at least half way by 2040	**No policy**	**No policy**	Work is underway to translate industry strategy from European Comission into robust legislative and financial instruments.	**No policy**	**No policy**

KR Policy & Policy Objective	China	U.S.	EU + U.K.	India	Russia
7.1.5 Carbon Labeling: Require emissions footprint labels on all goods	No carbon labeling	No carbon labeling	No carbon labeling, except a pilot program in Denmark.	No carbon labeling	No carbon labeling
7.1.6 Leaks: Control flaring, prohibit venting, and mandate prompt capping of methane leaks	No laws	Laws under review	Laws under review	No laws	No laws
7.2 Subsidies: End direct and indirect subsidies to fossil fuel companies and for harmful agricultural practices	$1,432 billion	$649 billion	$289 billion	$209 billion	$551 billion
7.3 Price on Carbon: Set national prices on greenhouse gases at a minimum of $55/ton, rising 5% annually*	In China, a national carbon market was launched in Shanghai in July 2021.	The U.S. has no national price. Twelve states have active carbon-pricing programs.	The European Trading Scheme focuses on the power sector. As of May 2021, the price is around US$50/ton. Individual member states have carbon taxes that range from less than $1/ton to more than $100/ton.	No price	No price
7.4 Global Bans: Prohibit HFCs as refrigerants and ban single-use plastics for all non medical purposes	President Xi accepted the Kigali Amendment in April 2021, peak production and consumption of HFCs by 2024. NDC: HCFC-22, reduce 35% by 2020, 68% by 2025.	In May 2021, the EPA proposed its first rule under the American Innovation and Manufacturing Act of 2020 to move forward with the phase down of HFCs. This is pending.	The EU has had an F-gas Regulation since January 2015. The Commission is in the process of reviewing the current F-gas Regulation and strengthening the previous measures.	Proposals, no outright ban	No ban
7.5 Government R&D: Double (at minimum) public investment into research and development, 5x in the U.S.	$7.9 billion	$8.8 billion	$8.4 billion	$110 million	Little to no allocation

* Chart as of July 2021
* Source: See Endnotes

A Sea Change in China

In 2006, China surpassed the United States as the world's largest emitter, and the gap has widened ever since. In 2019, China put more than 14 gigatons of greenhouse gases into the atmosphere—roughly double the emissions of the second-worst offender, the United States. Paradoxically, China has also invested in more clean energy than any other country in the world.

China's most consequential decisions on energy and the environment are made through its famous five-year plans, as formulated by the Central Committee of the Communist Party. Though the process is opaque, these plans are not empty promises or shiny public relations campaigns. Once in place, they have teeth. In September 2020, President Xi Jinping surprised the UN General Assembly by announcing that China would aim to reach carbon net zero by 2060. It was a goal without precedent for the world's most populous nation, and certainly a step in the right direction. But it's still ten years off the target set by the Intergovernmental Panel on Climate Change.

The biggest obstacle to China's reaching net zero? Finding jobs for more than 2 million miners in a nation that now burns half the world's coal and still relies on it for 60 percent of its electricity. On the optimistic side, China's leaders know they must change. According to Christiana Figueres, a shift away from coal is increasingly aligned with China's goals to improve public health and lead the way to a greener global economy. "It is not in China's self-interest to continue to be stuck with obsolete twentieth-century technologies," she says. In 2020, China accounted for half the world's new renewable generating capacity. It matched the rest of the world's wind power the year before.

Most questions of *how* China will approach a coal phaseout remain unanswered, though some clues were dropped at the April 2021 climate summit. President Xi announced plans to "strictly limit" increased coal consumption through 2025, and to "phase it down" between 2026 and 2030, when China's emissions are expected to peak. While the country's energy strategies will likely vary from province to province, these national goals are paramount.

A dose of healthy skepticism is in order; China's near-term actions will speak louder than any declarations from on high. Over the first six months of 2020, as other economic powers shifted from fossil fuels to renewables, China issued permits for more new coal plant capacity than for 2018 and 2019 combined. We also need to keep a close eye on China's financing of fossil fuel projects in Africa,

Ultra-high voltage lines enable China to tap into cleaner power sources far from urban centers.

Eurasia, South Asia, and Latin America. As of December 2020, as part of the nation's Belt and Road Initiative, Chinese enterprises were financing seven coal-fired power plants in Africa alone, with thirteen more in the pipeline.

At the same time, there are signs of a significant shift in China's stance toward climate. Some of the boldest ideas are coming out of Beijing's elite Tsinghua University, set on the site of the great Qing dynasty's royal gardens. Tsinghua hosts the Institute of Climate Change and Sustainable Development. Through much of 2019, Chinese climate scientists huddled there to crunch their way through models for getting to net zero.

As head of the institute and China's special climate envoy, seventy-one-year-old Xie Zhenhua is the most powerful voice on this subject in China. He was tasked with presenting Tsinghua's data to the Central Committee of the Communist Party of China, the nation's top leaders. At the climate assemblies in Copenhagen and Paris, Xie had argued that developing nations, including China, had a moral right to uncontrolled carbon emissions. But by 2017, he'd come around to see the overwhelming benefits of a commitment to net zero. He became a convert. "When you first start, it's just a job," Xie told *Bloomberg Green*. "But after some time, when you see the impact you could bring to the country, the people, and the world, it is no longer just a job. It has become a cause, a higher calling."

Before 2019, "China was reluctant to talk about concepts like net-zero emissions or carbon neutrality," said Li Shuo, an environmentalist who'd spent years lobbying Xie for more aggressive policies. "Xie has helped bridge that gap."

In the sphere of international negotiations, Xie takes a collaborative tack. As he told *Bloomberg Green*, "A climate negotiator has rivals and friends, he has no enemies." As a caveat, we should acknowledge that China and the United States are embroiled in a sharpening global competition over new and emerging technologies, from 5G wireless networks to robotics and artificial intelligence. Cleantech will be no exception. And as Xie concedes, China's stance on "mitigation, adaptation, financing, and technology" still needs to be hammered in Glasgow. For the global push for net zero to move forward, the two largest carbon-emitting countries must find a way to cooperate. Each brings expertise to the table that is critical to their mutual interests in emissions reductions and removal.

Xie maintains that China's 2060 net-zero target will accelerate change in the country's quasi-capitalist markets. "This sends out a clear signal," he says. "We have to transform fast and innovate big." Investments in coal will come to be seen as risky, he says. Markets will adjust. Renewables will ascend and ultimately dominate.

While China's leaders are committed to rapid growth, they are also sensitive to the climate crisis and its threat to their economy. Though the nation has partially abated its horrific air pollution, toxic smog killed an estimated 49,000 people in the first half of 2020 in Beijing and Shanghai alone. Record floods that year afflicted 70 million people and accounted for $33 billion in losses. As Xie says, "The damage done by climate change is not in the future, but right here, right now."

The United States: Back in Business

The United States is the largest *cumulative* contributor by far to the climate crisis, with more than 400 gigatons of carbon—and counting— poured into the atmosphere to date. Over the past two decades, the U.S. stance on climate change has swung like a pendulum, depending on who's living in the White House. President George W. Bush took his cues from the fossil fuel industry. He backed the construction of more coal-burning power plants and refused to implement the Kyoto Protocol.

President Barack Obama, for all his success on health-care reform, was unable to get climate change legislation passed. Even so, Obama understood that investments in clean energy would create jobs in the wake of the Great Recession. His American Recovery and Reinvestment Act channeled more than $90 billion into clean-power initiatives: wind farms, solar panel innovations, advanced battery programs. Obama's administration used the Clean Air Act to boost fuel efficiency mandates in new cars and light trucks by 29 percent between 2009 and 2016 and to set a historic goal of 54.5 miles per gallon by 2025.

Much of this progress was reversed after Donald Trump ordered a wholesale rollback of environmental protections. But shortly after Joe Biden's inauguration, he reversed several Trump orders and put forward a landmark climate plan. It commits the United States to 100 percent clean electricity by 2035 and to net-zero emissions by 2050. The Biden Plan offers the boldest vision to date for American leadership on climate action. **It's more than a fundamental reversal of the prior administration's policies—it is truly a leap forward.**

As I write this, however, the pendulum politics continue. The U.S. Congress is considering a much smaller infrastructure package that may exclude critical elements of Biden's plan. Much hangs in the balance. Failure on climate is not an option, but at times it is a choice—one our children can't afford.

The surest way for the United States and the other Big Five emitters to get to net zero by 2050 is to adopt *all* of our policy KRs, including a national carbon price. But where the United States can lead best is in research and development, a traditional American strength that cries out for renewed commitment. For nearly two decades, federal spending for energy research and development has remained stuck below the 1980 level of $8 billion per year, adjusted for inflation. That's less than what Americans spend in a week on gasoline. In fact, it's less than we spend annually on potato chips. To achieve needed breakthroughs like cheaper, lighter batteries or the scaling of green hydrogen, the United States must step up its public-sector research and development by a factor of five, to $40 billion per year. In other words, we're proposing that the U.S. government match what it now allocates to the National Institutes of Health: around $40 billion per year. With properly funded public R&D and a price on carbon, the United States can go a long way toward reducing the green premium, to the entire world's benefit.

Europe: Leading, but Not Fast Enough

Nearly two decades ago, the European Union established what is now the world's largest cap-and-trade system to impose a price on carbon. Flash forward to 2019, when the U.K. became the first big emitter to legislate a net-zero goal for 2050. The following year, the EU set its own 2050 net-zero target and called for a minimum 55 percent reduction by 2030. While these actions might seem impressive at first glance, climate activists argue that EU members aren't moving fast enough to build clean transportation infrastructure and cut emissions in line with the Paris Agreement.

One problem, says Patrick Graichen, executive director of Germany's preeminent energy think tank, Agora Energiewende, is a gap between understanding and action. "If you ask a politician what's most important, they'd say it's a coal phaseout. But you can't phase out coal without replacing it with wind and solar. That's not well understood, much less being acted upon with appropriate urgency."

As Europe's largest economy, Germany and its energy policy are of special importance. In response to a recent constitutional court ruling, Berlin pledged carbon neutrality by 2045, with a 65 percent reduction by 2030. German companies are leading the ramp-up of green hydrogen fuel production for fabricating clean cement and steel. Perhaps most encouraging, the nation has set a carbon price on fuels for building and transport. But Germany is almost certain to fall short of its 2030 emissions targets unless it accelerates a shutdown of the country's coal plants, a plan that now extends to 2038. As Germany looks toward its September 2021 elections and a new chancellor for the first time in sixteen years, it must lift its ambition.

As the world's epicenter of climate action, Europe has a lot going for it: strong public support, technological momentum, and climate-friendly national courts. With aggressive commitments already in place, the EU and its member states must now work to build essential clean energy infrastructure while reducing dependence on fossil fuels—and all at a record pace.

India: The Challenge of Growth

The Indian subcontinent offers a sobering preview of the potential climate catastrophe in store for us all. In recent years, tropical cyclones, rising sea levels, and killer droughts have all intensified, with a corresponding toll on human life and food production. India has promised to keep its per capita emissions no higher than those of more developed nations. But by 2050, the country's population is projected to rise by nearly 20 percent, to 1.6 billion, the largest in the world. Coupled with a poverty rate of more than 60 percent, India's explosive growth makes a net-zero goal especially challenging.

"As a developing nation, India has valid reasons to defer setting an economy-wide deadline for net-zero emissions," says Christiana Figueres. Shutting down existing energy sources too soon, she points out, would throw even more people into poverty: "India has been consistent for years in saying they would get to net zero through sectoral targets while protecting their biodiversity. And they are meeting their own Paris targets, sector by sector."

For India, the surest route to net zero lies in transforming its power sector and electrifying its transportation fleet. In an attempt to accelerate the transition to a zero-emissions future, Prime Minister Narendra Modi announced a national moonshot—a truly monumental effort—to reach 450 gigawatts of renewable energy by 2030. But despite areas of progress, the Modi administration remains uneven in navigating the nation's transition and steering away from coal.

All the while, India has pointed to the disparity between its progress and the mediocre records of more developed countries. As Environment Minister Prakash Javadekar said in 2020, "We've gone well beyond. Why don't you ask the countries lecturing us to mend their own ways instead? None of the developed countries are Paris Agreement–compliant."

When we look at cumulative carbon dioxide emissions, says Indian climate policy expert Anumita Roy Chowdhury, "India is asking, 'How do you divide the carbon pie and share responsibility?'" Historically, the United States has emitted 25 percent of the world's CO_2, Europe 22 percent, China 13 percent, Russia 6 percent, and Japan 4 percent. India? Only 3 percent. As Javadekar and Roy Chowdhury point out, the global transition to clean energy must be fair and just. It must reflect countries' proportionate contributions to the emissions emergency.

At the same time, India has a historic opportunity to leapfrog past dirty fossil fuels like natural gas. By bypassing investment in obsolete infrastructure, it can reduce pollution-linked mortality and stake out a position of global leadership, economically and environmentally. To be sure, this won't come cheap. Meeting India's renewable goals alone will require an investment of at least $20 billion per year.

Despite its impediments, India needs to do more if we're going to reach net zero. Though its per capita usage is less than half the world average, India is now the globe's third-largest energy consumer and fourth-largest emitter. It is adding the equivalent of Los Angeles to its urban population each year. Untold millions will be buying new appliances, air-conditioning units, and many, many cars and trucks. As demand for building materials and electricity explodes, it will be essential to meet it with more zero-emissions energy and a stronger emphasis on energy efficiency. Because of India's sheer size, any climate actions the country takes today will be deeply felt around the globe for generations. If it expands and accelerates its efforts to decarbonize, India just might save the world.

National moonshot: India aims to reach 450 gigawatts of renewable energy by 2030.

Will Russia Rise to the Challenge?

The world's fifth-largest emitter dumped 2.5 gigatons into the atmosphere in 2019, a number that has trended upward each year for the last two decades. For pessimists, Russia is Exhibit A for why we won't solve the climate crisis. Their concern is twofold: an absence of any long-term commitment to net zero, and the nation's extremely modest short-term targets set at Paris—and later pushed downward.

Under president-for-life Vladimir Putin, Figueres says, "Russia's autocratic political system does not allow for bottom-up, transparent, objective analysis to make its way to the decision table." Putin has alternated between openly doubting the science of climate change and suggesting that the warming might be beneficial for Russia. In a warming world, huge tracts of uninhabitable Siberian tundra might become farmable or at least more exploitable for the drilling of oil and gas—a potential windfall for Putin's inner-circle oligarchs. Sadly, Figueres adds, "If Arctic ice sheets disappear in the summer, it opens up a new maritime route for international oil shipping that benefits Russia."

The tragedy is that Putin may get his wish. Russian lands are warming more than twice as fast as the rest of the world. Siberian permafrost is thawing, releasing carbon dioxide and methane that had been frozen for millennia. All told, the Arctic permafrost stores 1,400 gigatons of carbon. Every bit that escapes works against our plan.

Russia's Energy Strategy 2035 is a big step backward. It calls for boosting oil and gas production while expanding petroleum exports. Solar and wind have no place in the portfolio. Russia's own 2050 projections presume that greenhouse gas emissions will be even higher than their current egregious levels.

How might this rogue actor be induced to change? The most obvious levers are market forces. One strategy would pressure Russia by imposing a carbon price on its sales of oil or gas, placing the country's core exports at a competitive disadvantage. Even without active pressure, Russia is swimming against the renewable energy tide. As China and Europe move to decarbonize, their imports of Russian fossil fuels may become a thing of the past.

With its vast expanse of territory, nearly twice the size of the United States or China, Russia has tremendous untapped potential for renewable energy and regenerative agriculture. It could be a major player in the net-zero economy—if it embraces the challenge.

For now, though, prospects are bleak. The Kremlin is actively contesting international penalties against climate laggards. But as everyone knows, it's hard to argue against penalties when you're already sitting in the penalty box. If Russia continues to isolate itself by opting out of the net-zero economy, it faces a grim future.

One potential path forward for Russia, Figueres suggests, would be to follow the lead of the United Arab Emirates, an oil-rich Persian Gulf state that is moving to diversify its economy into renewables. Oil kingpin Saudi Arabia is doing the same. The problem with Russia, Figueres says, is that it "doesn't have a plan."

The Gravity of Glasgow

Where do we go from here? With uneven national commitments and inconsistent progress toward meeting them, it's easy to lose hope that the world's nations will ever come together for meaningful climate action. To better see how we might collectively move forward, we turned to the individual tasked with executing the United States' climate plan, the special presidential envoy for climate and former secretary of state. As John Kerry reminded me, he was present at the first Earth Summit in 1992, and at nearly every major climate convention since.

John Kerry

In Paris, countries did what they wanted to do. The difference now, going into Glasgow, is that it's what we have to do. And that is a very different, much tougher exercise.

The reality today is if we do not reduce emissions sufficiently between 2020 and 2030, we cannot hold [the increase over preindustrial temperatures to] 1.5 degrees [Celsius]. We will have ceded that for all generations. We will have given up on it, with great consequence.

We set out earlier this year to make it clear to nations that we were going to press hard to adopt and hold on to the 1.5-degree target. At the Leaders Summit on Climate, the United States announced its nationally determined contribution, a 50 to 52 percent reduction between now and 2030.

If we don't achieve what we need to between 2020 and 2030, we can't achieve net zero by 2050. We cannot afford to simply wait to discover something. That would be the height of irresponsibility and recklessness.

We've got to take the technologies we have today and put them to use as much as possible. We're also not doing enough to discover new technologies. We talk about how this is an existential threat, but it's not being treated as if it is truly existential. We're certainly not behaving like they did in World War II, when they knew they had to gain control of the oceans and the air and learn how to bust the defenses that Hitler had built.

What makes today different is that the task is getting harder. It demands a greater response than is being given, even as we are making some rather remarkable progress.

In economic terms, nations representing 55 percent of global gross domestic product are now committed to 1.5 degrees. Can we bring the other 45 percent of nations—or at least the critical mass of them—to the table? India, Brazil, China, Australia, South Africa, and Indonesia all need to be brought into the fold.

You can't run around the world pointing fingers at people and saying, "You've got to do this and you've got to do that," without putting some money on the line to help them do it. We have to get the developing world to develop, but develop smart, develop without making the mistakes we made. In most cases, the developed world will need to help the developing world. As of yet, there isn't a sufficient plan on the table to do that.

What gives me hope—and I do have a lot of hope—is that when we put our minds to things, we get them done. We weren't sure how we were going to go to the moon. We went to the moon. We invented the COVID vaccine in record time. In my lifetime alone, the number of people living in severe poverty has dropped from 50 percent to 10 percent.

This is a matter of organizing ourselves. We know what has to be done, and now we have to go do it. Glasgow is the most immediate moment where the world can come together and address this crisis.

We know what has to be done, and now we have to go do it.

My First Climate Fight

I will admit that global warming wasn't front and center on my political radar through much of the 1990s. But in the year 2000, I became an active supporter of Al Gore's bid for the presidency. The climate crisis was on the cusp of becoming front-page news. That December came the heartbreak: *Bush v. Gore*, when the Supreme Court voted five to four to halt the Florida recount. George W. Bush's standing margin? A miniscule 537 votes. (Don't let anyone say your vote doesn't matter!)

The consequences of that ruling cannot be overstated. The fight against climate change lost twenty years. A President Gore would have made it a priority before it became the drastic crisis it is today.

By 2006, after the screening of *An Inconvenient Truth* and our fateful dinner roundtable, I was all in on climate action. There's a saying, "As California goes, so goes the nation." My attention turned to my home state.

In 2008, some of the globe's smartest leaders on climate policy joined forces around Assembly Bill 32 (popularly known as AB32), California's watershed cap-and-trade bill. It was the nation's most ambitious program to put a price on carbon pollution by imposing a fee on the biggest emitters. After a full-court press in Sacramento by California business executives, we eventually got the bill passed and signed into law by Republican governor Arnold Schwarzenegger. Though the fossil fuel lobby prevailed in limiting the scope of AB32 to oil and coal, excluding natural gas, it nonetheless became an international model, from Canada to China. California would ultimately succeed in cutting greenhouse gas emissions below their level of 1990—four years ahead of schedule. And the program contained a strong equity component, with around half the carbon fee proceeds channeled into cutting air pollution and funding housing retrofits in poor communities.

AB32 proved that the doom-and-gloom forecasts by fossil fuel interests were dead wrong. California showed that you could cut emissions and still promote a thriving economy. Indeed, the two worked hand in hand as California proceeded to outpace the nation in economic growth.

On a personal note, I learned invaluable lessons from AB32, mostly about what's needed to prevail in politics: broad and bipartisan coalitions, strong campaign leadership, clear messaging, aggressive media outreach, and committed allies. By that point I'd evolved from

Silicon Valley libertarian to big government reformer. Indeed, I'd come to view government as an essential partner in getting things accomplished at scale.

Which brings us to 2009, the year I testified before the U.S. Senate, the year of the Waxman-Markey climate change bill that would have put a national price on carbon. We almost did it. We almost got a rising schedule of greenhouse gas fees assessed to the fossil fuel companies. In June 2009, after some intensive wrangling by Speaker Pelosi, Waxman-Markey squeaked through the Democratic-controlled House of Representatives, 219 to 212. Negotiations in the Senate were epic, with competing bills and clashing interests. In the end, the climate bill was strangled by special interests and a lack of coordinated leadership. Once it became clear that we couldn't get sixty votes in the Senate, it was over. The bill died without a vote on the Senate floor.

The following year, the Democrats lost control of Congress, and you know the rest. As of mid-2021, the U.S. Senate has yet to vote on a major climate bill.

California kept moving forward, though. In 2015, we finally brought natural gas into California's cap-and-trade system. All told, the program reduced California's greenhouse gas emissions by as much as 15 percent.

In the world of public policy, the political calculus is always changing. Even so, I've found four rules to be of consistent value:

1. Go for the gigatons: To reach net zero, we'll need to focus on the Big Five emitters and target solutions to the sectors that matter most, the ones accountable for the most greenhouse gas pollution. We'll need action on all major greenhouse gases: CO_2, methane, nitrous oxide, and fluorinated gases.

2. Find out how—and where—decisions are made: National legislation is just one piece of the puzzle. Activists aiming to make change need to know the venues at every level. Building codes, for example, are set by cities, with wide-open opportunities for input at public meetings. Decisions on efficiency or electrification of heating and cooking carry ramifications for years to come. Often the public doesn't bother to show up at these meetings. But the company that sells gas boilers will be there—and will be heard.

Similarly, in molding U.S. energy policy, one powerful platform is often overlooked: the public utility commissions in each state. Whether elected or (more commonly) appointed, these commissioners are policy gatekeepers. They set the all-important renewable portfolio standards that determine future targets for the grid. Typically, there are five commissioners per state. Say we decided to focus on the thirty highest-emissions states and aim for a simple commission majority in each one. You are down to ninety individuals who control almost half the emissions in the United States. Pressing these handful of state officials can make all the difference.

Before bringing pressure to bear, it's important to learn how decisions are made. A sense of urgency is certainly necessary, but it's not sufficient. As Hal says, "If our concern about climate change is not properly aimed, it just dissipates." Are there movements to leverage? Could a massive public rally carry the day? Or might the balance be swung by a pointed economic analysis, or getting the right person elected? Is there a legal angle? How can issues of equity, jobs, and health be driven home?

No tool is more powerful or accessible than civic involvement at the local level. Across the country, people are becoming community activists to demand that their public transportation move off fossil fuels. In June 2021, after persistent advocacy from the local community, the public school district in Montgomery County, Maryland, announced a transition of its fleet of 1,400 buses to electric over the next decade. "We were seeing lots of interest and pressure from all around," said Todd Watkins, the district's transportation director. "I was hearing from lots of environmental groups, elected leaders, board members, student groups, wondering when we were going to go electric."

In Phoenix, Arizona, a group of cross country runners from South Mountain High School persuaded their district to purchase its first-ever electric school bus. Moved to act one day after feeling the effects of the area's dirty air, they corralled their coach and a local advocacy group named Chispa to push for change.

The particulars—and pace—of the clean energy transition will vary from place to place. But few forces are more powerful than a personal motivation to act, coupled with an understanding of how and where decisions are made. In tandem, they will be essential to usher in better, healthier outcomes for the future.

3. Focus on real-life benefits: In working for climate action, we need to get our facts straight and the science right. If we're trying to pass a law or elect a candidate, it's our job to persuade others by conveying technical issues in an approachable way. "People don't know about kilowatt hours," Hal says. "But people do care about energy that's affordable, reliable, safe, and clean."

What else do they care about? Jobs and the economy; their health and their children's well-being. Effective leaders tell stories that connect these concerns to public policy. Al Gore's Climate Reality Project has trained 55,000 climate action leaders to build narratives tied to shared values and real-life benefits. (I invite you to join them at www.climaterealityproject.org.)

4. Fight for equity: Equity matters, both as a moral imperative and a practical necessity. Politically speaking, we need to build coalitions among new voters, new leaders, and new legislators from previously marginalized groups. We need to enlist people who have never before been active in politics.

It's one thing to craft a bold and imaginative policy. It's far more demanding to guarantee that its implementation is fair and just. The 50,000-mile Interstate Highway System, launched by the Eisenhower administration in the 1950s, is widely hailed as a triumph of big government. But rarely is it acknowledged how many poor Black neighborhoods got paved over and deliberately destroyed, communities like Paradise Valley in Detroit or Treme in New Orleans. Around the globe, the climate crisis is devastating those least responsible for it. Our campaign to reach carbon net zero must protect the health and livelihoods of lower-income communities and Indigenous peoples.

Models Matter

To ensure that our climate policies have the desired impact, we need more than good intent. Shouldn't every policy be scored for its climate impact? The intrepid analysts at Energy Innovation have built a dynamic energy modeling tool that forecasts emissions impact in real time. EI's policy design experts, Megan Mahajan and Robbie Orvis, make a compelling case for the use of these models in crafting any net-zero plan.

Megan Mahajan & Robbie Orvis

Emissions are rooted in the physical world. Cutting emissions means changing the efficiency, energy consumption, and outputs of the things we use. If you don't have a strong sense of how policy will affect those factors and how they add up over time, you can't design sound policy.

So how do you model what different policies can achieve? We were asked precisely that question by Chinese policymakers in 2012 in the context of peaking China's emissions by 2030. There are rigorous models out there that take technology choices as the input, but we wanted to use policy as the starting point. Our lead model developer, Jeff Rissman, created a model that could do just that, and the Energy Policy Simulator was born.

The simulator takes policy as the input and estimates how any modeled scenario affects emissions, costs, jobs, and health outcomes. It accounts for interactions among policies, allowing us to identify which policies leverage one another and are most cost-effective. The simulator is updated regularly and incorporates the latest costs of technology, such as falling prices for solar, wind, and batteries.

https:// energypolicy .solutions

Our model is open source, which means that all of our data is publicly available. Anyone can download the tool and dig into our assumptions. This is vital for building trust and getting buy-in, especially in places outside the United States.

This work gives countries and states a realistic understanding of the results that will be driven by specific policies, and allows them to separate the wheat from the chaff. Ultimately, what the model highlights is that a small number of big policies will make a huge difference.

The Powers Behind the Politics

Policy and politics are deeply entwined; any headway in the first is contingent on the other. There's a natural link between directives and how decisions get made. But the two are also in tension. At best, politics is the art of the possible; at worst, it's where great ideas go to die. From my own experience, I can attest that good policy must run a political gauntlet to get passed. Bills get bottled up in committee. Votes get blocked or vetoed. Treaties don't get ratified. The most earnest policy efforts can fail repeatedly—for years, even decades. **You may think you have a great policy idea, but until you get the little *p* of policy past the big *P* of Politics, you have nothing.**

The greatest obstacles to effective climate policy aren't bad ideas or even backward politicians. They're the gallery of "incumbents" whose future is tied to greenhouse gases. Historically, entrenched fossil fuel interests in the United States have a high success rate in quashing progress on climate. They funnel money to politicians on both sides of the aisle to fight progressive policy, stall it, or ignore it. They fund deliberate efforts to cloud people's understanding of the harms from fossil fuels—of late through Facebook and Twitter, which have poisoned public discourse across the world.

Public interest groups have documented disinformation campaigns by the likes of ExxonMobil and the Koch family. More insidious still are the false headlines or misleading videos on social media from Russia-funded propaganda cells or other hazy sources. More mainstream entities—including Fortune 500 companies—enlist top-shelf ad agencies to promote climate change denial. In 2019, *The Washington Post* found that incumbents undermined climate science and public consensus with two parallel strategies: "First, they target media outlets to get them to report more on the 'uncertainties' in climate science. . . . Second, they target conservatives with the message that climate change is a liberal hoax and paint anyone who takes the issue seriously as 'out of touch with reality.' "

These efforts have not been for nothing. In the wake of the 1992 Earth Summit, 80 percent of Americans agreed that something must be done about climate change. Strong majorities of both Democrats and Republicans shared this view. But by 2008, the Gallup poll found evidence of deep polarization and a wide partisan divide on the subject. By 2010, nearly half the American public (48 percent) believed the threat of climate change was exaggerated.

Our hope is that the tide will change with the next generations. In a 2020 survey by the Pew Research Center, nearly two thirds of Republicans between eighteen and thirty-nine agreed that climate change is driven by human activity, and that the federal government is doing too little to stop it. According to Kiera O'Brien, founder of Young Conservatives for Carbon Dividends, young Republicans are "light years ahead of their elder counterparts on this issue."

For the open minded, one strong political selling point for the transition to a net-zero economy is that it will create millions of well-paying jobs—up to 25 million worldwide, according to the International Energy Agency. Beyond solar installers and wind-farm technicians, two of the fastest-growing job categories, millions of workers will be needed for building retrofits or power-grid upgrades.

Ultimately, passing smart policy will hinge on our ability to overcome the incumbents. Well financed, politically connected, and quite often nefarious, they're a formidable foe. We cannot defeat them with politics as usual. To win, we'll need an even more powerful force.

A force like movements.

Turn Movements into Action

Chapter 8

Turn Movements into Action

For Greta Thunberg, it began with getting mad. And the more the Swedish teenager learned about the climate emergency, the madder she got. For every fraction of a degree of global warming, storms and floods and wildfires would be hitting us that much harder. By 2030, at the rate things were going, another 120 million people would be pushed into extreme poverty. Whole cities—including Thunberg's native Stockholm—could be underwater by the end of this century.

Thunberg wasn't the only student who took in the grim reports and grasped their implications. She wasn't the only young person with raging climate anxiety. But she didn't get discouraged; she grew defiant. At fifteen, she started skipping school. In 2018, she camped out before the Swedish Parliament, holding a white sign with bold black lettering: SKOLSTREJK FÖR KLIMATET, or "School Strike for the Climate." At first hers was a protest of one. But then another teen joined in, and another, and before long it became a movement. All of this from an adolescent who shied away from crowds and hated the idea of being famous.

In January 2019, Thunberg was invited to address the World Economic Forum in Davos, Switzerland. "I often hear adults say, 'We need to give the next generation hope,'" she told the assembled chief executives and world leaders. "But I don't want your hope. I want you to panic. I want you to feel the fear I do. Every day. And I want you to act. I want you to behave like our house is on fire. Because it is."

Amplified by social media, Thunberg's words inspired thousands of young people to lead their own climate strikes wherever they were. On September 20, 2019, 4 million people across the world joined in the largest environmental demonstration ever. Then Thunberg spoke to another roomful of adults, this time at the United Nations: "You have stolen my dreams and my childhood with your empty words. And yet I'm one of the lucky ones. People are suffering. People are dying. Entire ecosystems are collapsing. We are in the beginning of a mass extinction, and all you can talk about is money and fairy tales of eternal economic growth. How dare you!"

Then, not long after Thunberg addressed the British Parliament, it passed a law to eliminate its carbon footprint by 2050, the first major nation to do so. As the teenager spoke to more world leaders, including the pope, she could see that her movement was beginning to create real change. She herself began to change in kind, from anger to cautious optimism. Back at school, she told her classmates, "We can't just continue living as if there was no tomorrow, because there is a tomorrow."

For becoming "the most compelling voice on the most important issue facing the planet," Thunberg was named *Time* magazine's Person of the Year for 2019. Her organization Fridays for the Future has reached every corner of the globe. People in very high places took her message to heart. "When you are a leader and every week you have young

Greta Thunberg's School Strike for Climate started small but soon commanded the attention of global leaders.

people demonstrating with such a message, you cannot remain neutral," French president Emmanuel Macron told *Time*. "They helped me change." Leaders respond to pressure. Pressure is created by movements. Movements are built by many thousands of individuals.

But sometimes they start with just one.

What Makes a Movement Matter?

When an issue matters to people—when it *really* matters—things begin to happen. Legislation gets introduced. Counterlegislation is proposed.

Turn Movements into Action

There's conversation, debate, media attention. Ultimately, the issue becomes catalytic and brings voters to the polls. When an issue rises to the top of the agenda, it gains what the political world calls "high saliency." Despite significant progress, the climate crisis has yet to gain global saliency. By and large, it doesn't yet turn out people to vote or guide their choices when they do.

Movements drive saliency. But to succeed, they need to wield two kinds of power. First, there is *people power*, a broad base of supporters, plus a narrower group of activist leaders and participants. Second, there is *political power*, when allies in public office are enlisted to introduce, champion, and defend legislation. The goal of a movement may be a political realignment, a fundamental reset of public senti-ment, a new set of leaders, or all of the above. In any case, movements give policy makers cover for political courage.

Political realignments are game changers, though they don't come around too often. In the United States, Franklin Roosevelt's New Deal was significantly rooted in his ties to the organized labor movement, which supported his first presidential run in 1932. In the depths of the Great Depression, people clamored for a social safety net and job security. In 1935, at FDR's urging, Congress passed the National Labor Relations Act, which laid down guidelines for collective bargaining. The labor movement suddenly wielded political power. Politics realigned.

The movement that helped give rise to the New Deal leveraged both types of people power: a mass of voters and less active supporters, and a smaller, deeply engaged cohort of active supporters to protest, strike, litigate, and raise others' awareness. Between 1900 and 2006, according to a Harvard University study, every political movement that gained the active and sustained participation of at least 3.5 percent of the population wound up succeeding. In the United States today, that's fewer than 12 million people!

At their best, movements forge a new awareness that results in clear action and lasting change. India's nonviolent revolution for indepen-dence is one legendary example. The U.S. Civil Rights Movement of the 1950s and 1960s is another. The impact of movements on policy and culture cannot be overstated.

As we push for the climate problem to attain salience as a political issue, we must also insist on fairness. The climate crisis exacts a devastating toll on human health in poor communities. It widens economic disparities and intensifies racial injustice. The crisis cannot be solved without tackling these inequities.

Our Movements OKR relies on traction with three vital constituencies: voters, government representatives, and corporations.

Objective 8
Turn Movements into Action

KR 8.1

Voters

The climate crisis is a top-two voting issue
in the twenty top-emitting countries by 2025.

KR 8.2

Government

A majority of government officials—elected or
appointed—will support the drive to net zero.

KR 8.3

Business

100% of Fortune Global 500 companies commit
immediately to reach net zero by 2040.

KR 8.3.1

Transparency

100% of these companies publish transparency
reports of their emissions by 2022.

KR 8.3.2

Operations

100% of these companies achieve net zero
in their operations (electricity, vehicles, and
buildings) by 2030.

KR 8.4

Education Equity

The world achieves universal primary and
secondary education by 2040.

KR 8.5

Health Equity

Eliminate the gaps among racial and socioeconomic groups in greenhouse gas-related
mortality rates by 2040.

KR 8.6

Economic Equity

The global clean energy transition creates 65
million new jobs, equitably distributed and
outpacing the loss of fossil-fuel jobs.

Do Voters Care?

Our **Voters KR (8.1)** measures the subject's importance to the electorate. Despite recent headway, the climate crisis has yet to place among
the top-two issues in elections or opinion polls in most top-emitting
nations. It regularly takes a back seat to immigration, taxes, and
health care. To build the climate movement we need, we must inspire
more urgency.

Let's look at issue priority data across the top five emitters. In the
run-up to the 2020 presidential election in the United States, according to a Gallup poll, only 3 percent of voters rated the climate crisis as
the top problem facing the nation, behind COVID-19, the economy,
poor leadership, and race relations. Even before the crisis-ridden year
of 2020, climate and the environment rarely ranked among the
electorate's top-ten issues.

In Europe, public sentiment is shifting more quickly. In a spring 2018
Eurobarometer survey, just before Greta Thunberg launched her youth
climate strike, voters in the European Union's twenty-eight member
countries ranked climate and the environment seventh overall,
behind immigration, terrorism, the economy, public finances, unemployment, and the EU's influence in the world. In the fall of 2019,
when Thunberg became internationally known, the issue leapt to
number two, trailing only immigration.

For citizens in China, India, and Russia, the perceived importance
of climate change is murky at best. In China, people's immediate
concern is air pollution. Since 2000, a rising urban citizen movement,
working through local politics and the Chinese legal system, has
pressed demands for cleaner air. In 2013, the central government
declared war on pollution with the National Clean Air Quality Action
Plan. Over the next five years, China reduced smog in its big cities
by up to 39 percent. In a 2017 national survey, "Climate Change
in the Chinese Mind," 90 percent supported implementation of the
Paris Agreement.

India's government has yet to make an economy-wide net-zero pledge,
focusing instead on commitments for individual sectors. Among
voters, top concerns in 2019 were inadequate government support for
farmers, rural poverty, unemployment, and the water crisis. Regardless of whether people were making the connection, all of them were
aggravated by climate shocks. Though youth-led protests have taken
hold across India, climate change has yet to break into the ranks of
the most salient issues.

In Russia, public interest in the climate crisis is growing slowly from a low baseline. Of those polled in 2019, 10 percent cited it as a major concern. Even after dozens were killed in the Siberian wildfires that year, the issue ranked fifteenth among voters, well behind corruption, high prices, and poverty.

Russian activists are often openly criticized by Putin and run the risk of imprisonment or worse. A one-day climate strike in 2019 in Moscow and dozens of other cities drew seven hundred peaceful protestors. For the most part, however, the grassroots climate movement is limited in its breadth and impact.

Electing Pro-Climate Officials

Movements must aim for tangible results. Where people power is a matter of activists galvanizing the public, political power centers on the role of elected and appointed officials. Our **Government KR (8.2)** tracks the stance of political leaders worldwide, at every level. To enact aggressive policy measures, we need a working majority of these officials to strongly favor climate action.

Many people are skeptical about the impact of movements. I, too, have wondered why so many of them have failed—and how we arrived at this point of climate desperation, despite decades of activists sounding the alarm. The truth is this: when well organized, movements can be remarkably effective in shaping policy. The question then becomes: *What does it take for a movement to succeed*?

Catalyzing Debate and Action: The Impact of the Sunrise Movement

Varshini Prakash traces her passion for climate activism back to 2004, when she was in sixth grade. That's when the Indian Ocean tsunami hit Chennai, India, home to her grandmother. With the phone lines down, Prakash anxiously watched the news from her home in the tranquil town of Acton, Massachusetts, and gathered cans of food for the Red Cross. Though she was relieved to find out that her grandmother was safe, the crisis left a lasting impression. Eager to learn more about natural disasters and their origins, Prakash read about the growing wave of warming-related events around the world. She felt overwhelmed. So she focused on small things, like recycling.

By the time Prakash arrived as an undergraduate at the University of Massachusetts Amherst, she felt angry and frustrated. After joining a campaign to pressure the university to divest its investments in fossil fuels, she spoke at a climate action rally. As she told *Sierra* magazine, "I just fell in love with organizing in a way that I had never imagined."

In December 2015, another big flood struck India, this time in the state where Prakash's father was born. Scrolling through images of the disaster on her computer screen, Prakash recognized streets where she'd walked or ridden with her grandparents—except now they were packed with women and children straining through chest-deep water to find sanctuary. Though her grandparents were out of town at the time, hundreds of others had died. Thousands were left homeless. "That was a big wake-up call to me that the climate crisis was right now," Prakash told *Sierra*. "We didn't have time to waste."

Within weeks, Prakash and a friend cofounded what later became the Sunrise Movement with a dozen other young activists. They created a blueprint for a youth-led, decentralized, grassroots effort to stop climate change and promote economic justice. A pivotal moment came shortly after the 2018 U.S. midterm elections. The group sought to turn the Democrats' newly won control of the House of Representatives into a mandate for climate action. They set up camp outside congressional offices and staged a series of sit-ins.

By then, Sunrise had learned how to get attention. The fledgling movement was armed with facts and compelling narratives. The youngest woman ever to win a seat in Congress, Alexandria Ocasio-Cortez, brought along some of her freshman colleagues, later known as "the Squad." And they listened.

"We didn't just deliver a petition with a bunch of numbers about parts per million or 2°C," as Prakash recalled in *Sierra*. "We shared stories about what we had lost because of the climate crisis, or what we were afraid of losing. We told stories about what we hoped for our future."

In 2018, the Sunrise Movement set up camp in the halls of Congress to press for climate action.

After that electric moment, Prakash and other Sunrise activists held higher-profile protests across the country to move climate policies to the top of the Democratic Party's agenda. They helped build enthusiasm for the Green New Deal, a legislative proposal introduced in 2019 by Ocasio-Cortez. They dove into Democratic primaries and pushed candidates to forgo donations from fossil fuel companies. Their biggest win was to help climate stalwart Ed Markey fend off a challenge for his U.S. Senate seat in Massachusetts. Some Democrats may have disagreed with their splashy approach, but everybody was paying attention.

In the 2020 presidential primaries, Sunrise gained the support of U.S. senator Bernie Sanders. As the movement gathered momentum among younger people, it sharpened the clash between the fervent minority of Democrats supporting the Green New Deal and the majority who had reservations about it. An ideal wedge for Fox News, the issue threatened to rupture the party.

For Prakash and other Sunrise leaders, the last thing they wanted on the debate stage was for centrist Democrats to attack Sanders for backing pro-climate legislation. There was nothing more vital than stopping top Democrats from airing their disagreement and undercutting climate action in the process.

Evan Weber, the movement's cofounder and political director, worked the phones. He got through to several presidential candidates: Kamala Harris, Pete Buttigieg, Joe Biden. "We said, 'Hey, we know you guys have your own plan, but it's really not helpful for you to be shitting on the Green New Deal,'" Weber recalled.

The appeals worked. Though the other Democrats wouldn't endorse the Green New Deal, they stuck to their point of unity: 100 percent clean electricity.

In March 2020, after Biden sewed up the nomination, Weber prodded the presumptive nominee's campaign to refer to the Green New Deal as a "useful framework" for boosting the economy, fighting for environmental justice, and addressing the climate crisis. In August, when it came time to craft the party's platform at the Democratic Convention, key planks from the Green New Deal were rejected. Because of Biden's need to carry Pennsylvania in the general election, there would be no proposed ban on fracking, a significant source of methane emissions. Nor would the candidate clamp down on dairy emissions—he also needed Wisconsin. Even so, Biden baked several Sunrise Movement proposals into his Build Back Better plan, including a 40 percent allocation of infrastructure funding to disadvantaged communities.

Throughout the fall, Sunrise and the Biden campaign kept lines of communication open. In the end, their mutual accommodations amounted to smart politics. When the votes were counted in November, Biden won Pennsylvania by 1.2 percent and Wisconsin by 0.7 percent. The result: a 2020 victory, and a 2021 White House ready and willing to lead strong climate action.

For Sunrise, politics is a perpetual balancing act. As CNN noted, the organization was striving to "keep one foot inside the halls of power, and another with its activist ranks on the streets." For Sunrise, that's a point of pride. The movement's young leaders have learned the importance of cultivating the "grass tops," their direct ties with decision makers, as well as the grass roots. In politics, that's nothing new. It's a model well understood by the Sierra Club, the seminal environmental organization in the movement-building business for well over a century.

Lessons from Beyond Coal

In 2005, just days after Hurricane Katrina tore across the Gulf Coast and deluged New Orleans, the Sierra Club was preparing for its first-ever climate action conference. Founded by the naturalist John Muir in 1892, the organization was formed to protect forests and other wildlands, an inherently defensive strategy. Now it was moving beyond conservation to play offense against carbon emissions. Five thousand climate activists nationwide helped shape the agenda for a convention in San Francisco, the organization's home city. Al Gore showed up with a slide show that would evolve into *An Inconvenient Truth*.

"We were about to find ourselves in a different business," recalls Carl Pope, the group's executive director at the time. From that conference emerged a surprising new number-one goal: to halt the planned construction of 150 coal-fired plants. If unimpeded, by Pope's estimate, the plants would add 750 million tons of carbon to the atmosphere each year, making it mathematically impossible to tame the global-warming monster. To prevail, the Sierra Club would use any legal means necessary and all the public pressure it could muster.

Spearheaded by Bruce Nilles and Mary Anne Hitt, the Beyond Coal campaign didn't set out to change national policy. It was attempting something even more difficult: to put boots on the ground to rally hundreds of communities, organize local protests, and win court injunctions.

Bruce Nilles

In 1990, as a geography and environmental science student at the University of Wisconsin, I remember well my first class on climate change. Each time I'd head over to the geophysics building, where I was becoming more and more worried about rising CO_2 levels, I'd pass the piles of coal being fed into the very old boilers that were powering the campus. I was struck by the disconnect. I wrote my senior thesis urging the campus to phase out its coal plant—and learned that bringing about such change would require more, much more.

After a year of temping in San Francisco during the doldrums of the dotcom bust, I returned to Madison for law school. This is where I learned about many of the great social struggles in U.S. history, and the role of lawyers in bringing about social change as a part of broader social movements. I learned about legal rights and how to enforce contracts, and got to practice on my overzealous landlord.

Fresh out of law school, I had a remarkable four-year stint at the U.S. Justice Department's Environment and Natural Resources Division in the Clinton administration. Before long, I volunteered to be on point to help implement the Department's obligations under Clinton's executive orders on environmental justice and children's health. I got to investigate and prosecute the first cases to enforce a new federal statute that protected kids from the dangers of lead paint. I was awestruck when Attorney General Janet Reno, HUD Secretary Andrew Cuomo, and EPA Administrator Carol Browner

all showed up at a press conference to announce three settlements that I had negotiated. It was an insight into how government works that has served me well ever since.

Armed with this experience, I joined the Sierra Club to launch a campaign to clean up the air in greater Chicago, an area that was home to 9 million residents and where the air was regularly unfit to breathe. This is where I learned about the power of grassroots organizing and how to organize when up against powerful interests.

Initially, I immersed myself in the data and regulatory issues to understand what was going on. I saw where the regulators were failing, in spite of the 1970 Clean Air Act's promise of healthy air for all Americans. I discovered medical waste incinerators behind hospitals that were regularly violating their permits in the middle of residential neighborhoods. I met residents whose complaints about the pollution had for years fallen on the deaf ears of profit-blinded hospital executives and timid regulators.

With a small band of relentless volunteers, we picked a particularly egregious incinerator target in Evanston, Illinois. With growing crowds behind us, we held up business at the City Council until they took up the issue and ordered the hospital to close its dioxin-spewing incinerator. The hospital tried every dirty trick, including

Beyond Coal activists put boots on the ground to roll back plans to build new U.S. coal plants.

a threat to close. Then one night, long after midnight, I witnessed over two hundred residents cheering as the council ordered closure of the incinerator. The icing on this story was how our local movement had caught the attention of Rod Blagojevich, the governor at the time. He showed up at one of our rallies to announce he was going to support legislation to close all ten medical waste incinerators in Illinois. People power!

A similar citizen-led fight was going on in a much more difficult venue, in the heart of coal country. After President Bush reversed his commitment to regulate carbon dioxide, Peabody Energy, the nation's largest coal producer, decided to get in the business of building coal plants to expand the market for its filthy product. One of its proposed plants was in Muhlenberg County, Kentucky, where Peabody thought it would have a smooth path to build the massive 1600-megawatt thoroughbred coal plant. How wrong they were.

Led by the local chapter of the Sierra Club and funded by bake sales, local activists fought the project at every turn. Most amazing, they found experts and lawyers who provided testimony and evidence for why the state should not grant Peabody a construction permit. After a record sixty-three days of administrative hearings, they won.

It turned out that Peabody's three proposed coal plants were a tiny tip of the iceberg—three of more than two hundred on the drawing board. With an oilman in the White House, the company saw an opportunity to win quick approval and lock the United States into another fifty years of coal burning. But inspired by the Kentucky activists, I set about with a small group to oppose the first of seventeen proposed plants in Illinois. Neighboring activists in surrounding states were soon reaching out to compare strategies and build a Midwest network of volunteers and a few staff to "leave no coal plant unopposed." We started winning, and then winning some more. Our campaign expanded south to Texas, and within three years we launched Beyond Coal—a nationally coordinated, locally led campaign of dozens of organizations collaborating to do what most experts said was impossible.

I saw firsthand how people who had never met were sharing a common bond and purpose. They were connecting across the web and through conference calls, and they were united in their fight to protect their communities from coal. When activists defeated a coal plant in Florida, there were celebrations all over the United States among people who were working to stop their local coal plant.

Turn Movements into Action

Beyond Coal played a lead role in stopping the rush to build nearly two hundred proposed coal-fired power plants, a stunning achievement. To be fair, the campaign enjoyed some strong tailwinds: new clean energy policies that drove massive investments in wind power, plus the boom in shale gas from hydraulic fracturing, or fracking. As more and more coal plants were defeated, wind and natural gas became the leading replacements—a mixed bag for the climate.

In 2008, with the election of Barack Obama, the Sierra Club suddenly had the Environmental Protection Agency on its side. Building on Beyond Coal's initial success, Bruce Nilles envisioned a second phase for the campaign: to shut down all *existing* U.S. coal plants, a group of more than five hundred polluters that belched two gigatons of carbon dioxide into the atmosphere each year. The aim was to replace them with solar and wind power. It was an effort that would require a lot of political clout and a lot of money.

An influential ally stepped in: Michael Bloomberg, mayor of New York. Elected in the wake of the 9/11 terrorist attacks, Bloomberg had gained a reputation as a climate crusader. His strategic plan for the city included more than one hundred initiatives for cleaner air and quality of life, notably "congestion pricing" tolls to reduce traffic, pollution, and emissions. In 2007, joining forces with California's Governor Schwarzenegger, Bloomberg formed the C40 Climate Leadership Group. It brought together the mayors of dozens of global cities, from London to Rio de Janeiro.

In 2007, California governor Arnold Schwarzeneger and New York City mayor Michael Bloomberg brought together an international coalition of big-city mayors to take climate leadership.

Now the billionaire mayor wanted to see if a targeted investment could make a difference for Beyond Coal. After talks with Carl Pope and Bruce Nilles, Bloomberg was ready to commit $50 million. The goal was to shut down one of every three existing coal plants by 2020. It was a limited but realistic objective, which made it appealing to Bloomberg. "I like to fight battles that are winnable," he says.

Bruce Nilles

It became a question of showing how investment could translate into results. Mike Bloomberg said, "Great, I'll give you fifty million dollars, I'll raise another fifty million dollars from others, and you raise forty-seven million on your own." We hit 95 percent of that goal, raising $143 million. We were able to expand from fifteen states to forty-five and to fund the development of much better data and analytics.

We launched dozens of lawsuits, forcing the oldest plants to retire. We won with both top-down leadership and a bottom-up, grass-roots campaign. We killed one plant on a Navajo reservation and replaced it with renewables.

It was emotional for me when we sued my old school, the University of Wisconsin at Madison, to finally shut down the coal plant that I'd passed every day as a student—and we won.

We ended up retiring more coal plants under Trump than we did under Obama. Of 530 existing plants, we shut down 313 of them. We need to close all of them, of course. But already coal has dropped from providing 52 percent of America's electricity in 2005 to 17 percent in 2020.

Clean electricity makes everything else possible. Now our focus is on what's happening with building codes for homes, offices, and stores. We need to get the oil and gas out of buildings. We need to stop it through rules for new construction: No more gas appliances. This isn't hard to do. You can go to Home Depot and get four kinds of electric-powered appliances: a water heater, a heating furnace, a washer/dryer, and a cooking stove. All electric.

If we hit the 2030 target—to make 75 percent of electricity zero-emissions worldwide—we'll have a chance to eliminate carbon emissions from the entire power sector.

The Corporate Transformation Movement

Movements consist of more than citizens and consumers. For maximum impact, they also enlist corporations and shareholders. Of late, pressure has mounted for stronger corporate commitments to decarbonize. The world's largest companies bear a heavy responsibility to cut emissions and scale net-zero solutions. According to an oft-cited *Guardian* report, just one hundred companies account for 71 percent of global greenhouse gases. While we know that markets are driven more by consumption than production, decisions by leading companies can make a difference.

A corporate sustainability movement has percolated for some time now. Walmart has set new standards for retail energy efficiency, installing solar power for stores in 12 states. In 2016, the last year of the Obama White House, Walmart became one of 154 companies to sign the American Business Act on Climate, a pledge to uphold the Paris Agreement.

The technology sector has taken the lead in scaling renewable energy for operations and data warehouses. For four years in a row, Google has matched 100 percent of its global electricity use with purchases of renewable energy. Since April 2020, Apple has been carbon neutral across all its corporate operations. The company's goal is to zero out their products' carbon footprint by 2030.

The beauty of this phenomenon lies in its ripple effect. When corporations make pro-climate commitments, suppliers tend to fall in line. The pace of change accelerates. To create products with net-zero impact, Apple is moving to enlist its suppliers to commit to plans of their own. What we're seeing is an active shift from "carbon neutrality" pledges to "net-zero emissions" pledges, a company's promise to balance any residual emissions of *all* greenhouse gases (not just carbon dioxide) with corresponding removals in that calendar year.

In 2021, after announcing Apple's business would be carbon neutral by 2030, Apple CEO Tim Cook said, "The planet we share can't wait, and we want to be a ripple in the pond that creates a much larger change."

Our **Corporations KR (8.3)** tracks the global business community's stated commitments to net zero by 2040. Our key result is to have 100 percent of the Fortune Global 500 in the fold. How will we get there? In the business world, pressure is most effective when it comes from industry leaders. A new standard has been set by Amazon founder Jeff Bezos, whom I first met in 1996. Five years later, Jeff sent me a memorable present—a wooden paddle—inscribed to "the man who always has a spare paddle when you find yourself up a creek." Recently we were talking about the climate emergency when I pulled out his gift. With his signature honking laugh, Jeff exclaimed, "John, it looks like we're going to need a lot of extra paddles!"

What I've always admired about Jeff is his ability to identify giant opportunities, plot a course of action, and execute with relentless precision. (One original name for Amazon.com was Relentless.com.) Once Jeff decides to do something new, he moves with speed and at scale.

For Jeff, the climate crisis became one of those opportunities. Historically, Amazon had trained exclusively on its customers. Now its mission would expand to include climate action, a decision with a sense of urgency baked in.

Amazon built a team of sustainability experts from rival businesses, academia, and across the company. The seed for a net-zero target was planted during an operations meeting in 2016. As the sustainability team grew from fifty to two hundred, it gained the capacity to quantify carbon emissions across the business, from delivery trucks to warehouses. With this research in hand, Amazon had an audacious goal to share. In September 2019, Jeff put forward a plan for Amazon to reach net zero by 2040. His announcement had a ripple effect on the company's vast networks and connections for the good of the planet. Amazon wouldn't be satisfied merely to decarbonize; it would actively recruit others to do the same.

Jeff Bezos

Amazon makes an ideal role model for climate action because people know how difficult the challenge is for us. We don't just move bits and bytes around. While data centers are heavy users of electricity, it's relatively easy to take what's already electrical and switch it over to sustainable energy. In 2019, we said we'd power our operations with 100 percent renewable energy by 2030.

Now we are going to do it by 2025. We're five years ahead of schedule, so that is going really well.

But net zero is hard for Amazon in particular because we move physical packages around. We deliver 10 billion items a year, and air transport and delivery vehicles play a gigantic part in what we do. That's deep, large physical infrastructure at real scale.

It's hard to electrify an entire delivery fleet, but we've made a good start there too. We invested in a company called Rivian and bought one hundred thousand electric delivery vans from them, with the first ten thousand coming online by the end of 2022. We've got that part of the plan rolling.

But Jeff was by no means finished. To expand on Amazon's 2040 commitment, he cofounded a corporate movement called the Climate Pledge. It calls on all signatory companies to follow Amazon's lead and reach net-zero emissions by 2040, meeting the Paris target ten years early. The ramifications are hard to overstate.

When Colgate-Palmolive signed, it made an added guarantee to move to fully recyclable toothpaste tubes by 2025, and to adhere to steep plastic and water reduction targets. When PepsiCo signed, it announced a comprehensive menu of clean energy solutions, from wind-powered Tropicana orange juice plants to electrified Doritos delivery trucks. The company mandated regenerative agricultural practices for its food suppliers across 7 million acres of farmland in sixty countries—all by 2030.

Jeff's vision is for entire supply chains and value chains to become climate action movements unto themselves. Now that Amazon and its suppliers are in the thick of taking on this colossal challenge, Jeff stresses the difficulty—and the urgency—of the work.

Jeff Bezos

It is daunting. It is very hard, and it should be hard, and if you don't start off expecting it to be hard, you'll get disappointed and quit. But we can make the argument—and we plan to do so passionately—that if Amazon can do this, anyone can do this. It's going to be challenging, without a doubt. But we know we can do it. Even more important, we know we have to do it.

We have to act now, and I believe that there is a collective energy to act now. We're at a tipping point where Fortune 500 companies are getting very serious about the climate crisis. Governments are getting serious about it too. For the first time, the people in charge are willing to make it a priority.

With the Climate Pledge, we're seeing organizations commit to reach net zero in their operations by 2040. It's a very unifying idea that large companies can get behind.

We now have more than one hundred signatory companies with $1.4 trillion in annual revenue and more than 5 million employees around the world. You can't get to net zero on your own. It can only be done in collaboration with other large companies, because we're all part of each other's supply chains. To make the kind of change we're talking about, you have to get these supply chains to move together. We all depend on one another.

You can't do this on your own.

To make corporate leadership real, notes Kara Hurst, Amazon's head of worldwide sustainability, the Climate Pledge requires regular self-reporting on greenhouse gas emissions. "We don't prescribe what companies should do, just that they should do it," Kara says. "It's not reporting for reporting's sake. It's a mechanism to share learnings: What could we do differently going forward?" By measuring, tracking, and sharing their progress toward net zero, Climate Pledge signatories are paving the way for others to do the same.

For corporate climate activists, momentum is building. In August 2019, the Business Roundtable, the de facto steering committee for the corporate world, made a historic pivot with its "Statement on the Purpose of a Corporation." Since its founding in 1972 by the heads of major U.S.-headquartered companies, the Roundtable had affirmed that a corporation's cardinal purpose was to seek out the highest rates of return for invested capital—first, last, always. "Corporations exist principally to serve their shareholders," the group's charter avowed. While sustainability was all well and good, it was never viewed as a governing corporate principle.

But times were changing. As more and more CEOs broadened their mission, the Business Roundtable responded in kind. The new statement emphasized the importance of serving customers; of building a workforce of diversity, inclusion, and respect; of protecting the environment with sustainable practices. With the planet in such peril, the Roundtable's new tack couldn't have come at a better time.

How Walmart Leads

As the Business Roundtable charted a new path forward, it was chaired by Doug McMillon, Walmart's chief executive and a longtime champion of customers and employees. Doug learned the business from the ground up, unloading Walmart trucks as a teenager for an hourly wage. He rose through the ranks to become chief executive officer of Sam's Club, the company's warehouse membership division, and then chief executive of Walmart International, before being named CEO of the whole company in 2014. In talking to Doug, I was struck by how much leadership matters in creating a movement—and that real action begins when leaders make a conscious choice to break with the status quo. Doug candidly discussed how and why Walmart embraced sustainability and set its goal of carbon net zero by 2040.

Doug McMillon

Sam Walton founded Walmart in 1962. Like any good entrepreneur, he was really focused on his customers and associates from the start. And he would say if we serve those two stakeholders well, our financial investors will do well.

Fast-forward to the company growing, achieving scale and getting into groceries in the 1990s and early 2000s. We were facing a lot of societal criticism and pressure over a variety of issues. And we didn't respond to it as well as we could have in the beginning. We didn't really understand it.

Lee Scott, our CEO at the time, made an important choice. Rather than defending ourselves and responding with our version of the facts, he led us to listen and learn from our critics. As we listened to thought leaders like Peter Seligmann, Paul Hawken, Sr. Barbara Aires, Amory Lovins, and Jib Ellison, our mindset shifted. We began to see that we could do more and it would be good for our business.

Then in 2005, Hurricane Katrina hit. The levees broke and New Orleans flooded. People were dying. Families were on their rooftops hoping to be saved. The federal government was slow to respond. Our leadership team in Bentonville was on a conference call over the long weekend, working to help our associates and customers in the market, watching it all play out, painfully, on TV.

People needed help, and they weren't getting an adequate response. Lee told the team to unleash everything we had to help. He said we would add up the costs later and if we missed the quarter, we missed the quarter.

We wound up sending 1,500 truckloads of food and supplies. We brought in our people from all over the country, our store managers, our market managers. Many of them worked there for weeks. They slept inside our stores and warehouse clubs at night because there was no safer place to stay. We had associates help guide rescue helicopters as they used our parking lots. One of our officers administered CPR to a customer in one of our stores. There was story after story of our brave associates making a difference.

Many in the country saw what we were doing, and we were proud. The learning journey we'd been on before Katrina prepared us for that moment. Lee seized it, and then he said, "What would it take for us to be that company and feel that way every day?" Under Lee's leadership, we moved quickly to lay out some big goals on social and environmental sustainability. We set goals to create zero waste, to shift to renewable energy, to sell sustainable products.

We were now on the path to becoming systems thinkers, working to design our entire business to benefit all our stakeholders and to strengthen communities and the planet.

Walmart's new sustainability goals rippled out from its headquarters to 6,000-plus facilities, stores, and clubs around the world, and to more than 1.6 million global employees at the time. What made these new goals even more significant was their impact on suppliers, big players in their own right in clothing and apparel, food and agriculture, and industrial materials.

Doug McMillon

Our initial math showed that somewhere between 8 and 10 percent of our carbon footprint was driven by our own assets—our trucks, our stores, the things we owned. The other 90 to 92 percent was driven by our supply chain. So we could not achieve our goals without embracing the issue and engaging our supply chain pretty aggressively.

And that's what we did—not only with our big suppliers and brands, but with factories all over the world. In the United States, about two thirds of what we sell is made, grown, or assembled here, while the other third comes from China, India, Mexico, and Canada, along with components from all over the world. We developed a plan that invited all our suppliers to join in our goals. We formed what we called Sustainable Value Networks.

Suppliers engaged in topics that were relevant to them, like: How do we reduce the carbon footprint of our transportation fleet? How do we remove unwanted chemicals from our products? How do we improve our packaging? We invited our suppliers to come in and help us think about these issues and create policy. We also invited universities, nongovernmental organizations, and other thought leaders. We basically created a collective that was broader than business to help us use science to make smart policy choices. And then we acted on them.

We've found that our suppliers are very like-minded; this wasn't a situation where we had to force anybody to do anything. Rather, it was an open door and an educational experience, and they came willingly.

Near the end of 2020, we set the next set of goals as a result of two developments. One is our own maturation process. We've made progress in renewable energy, eliminating waste, selling sustainable products, and doing whatever else we can that relates to environmental and social sustainability. So we're ready for the next phase of goals.

The other thing that's happening is that the world is not in good shape. Our sense of urgency has to be higher and our goals need to aim higher. We set a goal in 2019 to achieve zero emissions in our own operations, without offsets, by 2040.

At the same time, we need to not only slow the harm and get to carbon neutral, but to find ways to add back. Some experts estimate that nature itself can provide up to a third of the solution to climate change. So while we're trying to get to renewable energy and eliminate waste, etc., we're also going to protect at least 50 million acres of land and 1 million square miles of ocean. We're out to become a regenerative company.

OUR SUSTAINABILITY GOALS

CLIMATE

Target **zero emissions** in our own operations by 2040.

Reach **100% renewable energy** by 2035.

Work with suppliers to avoid **1 gigaton of greenhouse gas emissions** from the global value chain by 2030.

NATURE

With the Walmart Foundation, help protect, manage or restore at least **50 million acres of land** and **1 million square miles of ocean** by 2030.

Source at least **20 commodities** more sustainably by 2025.

WASTE

Achieve **zero waste** in our operations in the U.S. and Canada by 2025.

Reach **100% recyclable, reusable, or industrially compostable** private-brand packaging by 2025.

PEOPLE

Make **responsible recruitment** a standard business practice by 2026 to promote **human dignity.**

Walmart's drive to net zero is a multifaceted effort that enlists a broad set of stakeholders.

Investments that support employees, communities, and the planet are in the absolute best interests of customers and shareholders.

Walmart is now an undisputed leader in the climate action movement. The company is constantly looking for ways to become even more energy-efficient and sustainable, and to bring still more urgency to the broader problem at hand. Walmart's climate leadership serves Sam Walton's original mission: to help people save money and live better lives. By making its truck fleet more efficient, for example, it avoided more than 80,000 metric tons of carbon emissions. As an associated benefit, cost savings are passed on to customers.

That's merely one example of the company's core conviction: investments that support their employees, the communities they serve, and the planet are in the absolute best interests of customers and shareholders. As Walmart has found, a multistakeholder approach, over time, is the best way—perhaps the only way—to maximize value for those that own the business.

The Risk of Not Joining the Business Movement

While Amazon and Walmart have emerged as exemplary climate action models, what about everybody else? For many companies, taking initiative on climate boils down to risk. A failure to meet emissions goals can bring unsavory consequences, from shareholder lawsuits to depressed market value. Among those sounding the alarm is the world's largest investment manager with $8.7 trillion under management. According to BlackRock, a "climate-aware portfolio" is no longer a choice—it's an imperative.

In his 2021 open letter to the heads of the companies BlackRock invests in, chief executive Larry Fink declared that his industry was on "the cusp of a transformation." As more investors tilted their portfolios toward sustainability, Fink noted, "the tectonic shift we are seeing will accelerate further." Companies that fail to prepare for the transition to a net-zero economy, Fink warned, will see their businesses and valuations suffer. He issued a broad and persuasive challenge for investors and corporate leaders to seize this double opportunity—to deliver long-term returns while building a brighter, more prosperous future for the world.

Larry Fink

Five years ago, I began writing letters in support of the corporate sustainability movement. Most of the response I got from the 2020 letter was positive. About 40 percent were highly negative, and half of that came from environmentalists who said we weren't doing enough. I'll acknowledge that the investment community is not perfect. We have failed to address some of the underserved parts of society that are often left behind.

The other half of the criticism came from the far right. Some conservative-leaning newspapers ran a cartoon of me hugging a tree. But make no mistake. While I consider myself to be an environmentalist, I wrote that letter as a capitalist, as a fiduciary to our clients.

I do believe that BlackRock should have a voice on important issues that can affect the value of our clients' assets. Over the years, my letter to CEOs has increasingly focused on the responsibility companies need to shoulder to solve the climate crisis. And it's had impact. When the Business Roundtable decided to broaden its view on the role of corporations to be more inclusive of all their stakeholders, I believe it was in part a reaction to my 2018 letter that focused on companies having a purpose.

In the year leading up to my 2020 letter, I witnessed a bleached Great Barrier Reef. I witnessed wildfires in South America and a drought in Botswana. It was devastating, both for the climate and for business. Sustainability became a topic of every conversation everywhere I went. I came to see more clearly that climate risk is investment risk.

Awareness is rapidly rising. I believe we are on the edge of a fundamental reshaping of finance. The evidence on climate risk is compelling investors to reassess core assumptions. They're thinking twice about investments in impaired companies that refuse to change.

It is our fiduciary responsibility to make sure the companies in which we invest on behalf of our clients are addressing these material issues, both in managing climate risks and capturing opportunities to grow their business. Only then can they generate the long-term financial returns our clients depend on to meet their long-term investing goals.

In 2020, we saw an acceleration in climate-aware investing. And the movement of capital has continued to accelerate in 2021. I became more hopeful in my 2021 letter. Can capitalism shape the curve of climate change? The answer is yes. I believe it can.

But we have a lot more work to do.

The more we understand both the risks and opportunities, the faster we can get to that tectonic shift across all industries. Johnson & Johnson has a higher price-to-earnings ratio today than most of its peers in part because CEO Alex Gorsky focuses on lowering J&J's carbon footprint.

We can show the California Public Employees' Retirement System that an investment fund with higher sustainability scores than the S&P 500 Index can also perform better than the Index. We'd like to give every pension fund the option of not owning the S&P 500 Index if it contains companies dragging their feet.

Thanks to the rise of Tesla and others, you can already see what's happening in the stock market. The price-to-earnings ratios for cleantech companies range from 26 to 36, versus 6 to 10 for hydrocarbon companies.

The greatest danger is that publicly traded fossil fuel companies will divest their hydrocarbon assets to a private company. Divestiture of hydrocarbons can be greenwashing. If energy companies sell their hydrocarbon assets to a private equity firm, for example, nothing has changed. In fact, they've made the problem even worse, because the asset has been moved to less public, less transparent markets.

Climate risk is investment risk.

BlackRock and other large institutional investors are advancing their pro-climate sensibility through both carrots and sticks. As more and more investors insist upon sustainability, corporations that are slow to respond will face higher costs of capital. Those that jump on board will be better positioned to deliver higher shareholder returns, a prime measuring tool for chief executives.

ExxonMobil, the largest oil company in the United States, is a dramatic example of how heightened risk awareness is already forcing change. As global oil prices peaked in 2007, the company's market value topped $500 billion, making it the most valuable—and most profitable—company in the world. But when oil prices tanked and demand flatlined, Exxon's fortunes fell in kind. By the end of 2020, the company's market value had collapsed to $175 billion. Over the past decade, total return has dropped by 20 percent, versus a gain of 277 percent for the S&P 500 Index. Not surprisingly, the people who own Exxon stock are unhappy. Some have become activist shareholders, seeking seats on the board. They aim to compel the company to align its long-term strategy with the transition to renewable energy sources. As one headline read, "Green sharks are circling ExxonMobil."

On December 7, 2020, as part of a campaign called "Reenergize Exxon," the activists dropped a bombshell of a letter. "No company in the history of oil and gas has been more influential than Exxon Mobil," it read. "It is clear, however, that the industry and the world it operates in are changing and that ExxonMobil must change as well." As the activists noted, the current board included no one with a background in renewable energy. In response, Exxon published its first-ever emissions profile, alongside details of its efforts to reduce the company's damaging climate impact.

Unimpressed, shareholder activists pushed on for a radical metamorphosis—for Exxon to steer away from fossil fuels. They pointed to European oil and gas companies that have diversified into biofuels, hydrogen, and offshore wind farms. In May 2021, a small hedge fund named Engine No. 1 led a shareholder revolt that captured three independent board seats—"a landmark moment for Exxon and for the industry," said Andrew Logan of Ceres, a nonprofit investor network. The same day, activist investors rebuffed Chevron's board by voting to cut greenhouse gas emissions from the petroleum giant's products. Almost simultaneously, a court in the Netherlands ruled that Royal Dutch Shell, the world's biggest private oil company, must slash its emissions by 45 percent from 2019 levels by 2030. Oxford University economist Kate Raworth called it "a social tipping point for a fossil-fuel-free future."

When even the mightiest oil companies are being forced to adapt, it's clear that no one in the industry is exempt. Climate leaders like Al Gore have long predicted this day. Citing figures from the Intergovernmental Panel on Climate Change, Gore notes that $28 trillion in fossil fuel companies' carbon assets have yet to be exploited—and that $22 trillion of those assets, more than 75 percent, may be stranded forever in the ground. "Companies are marking down the value of their reserves," Gore says. "Those reserves are toxic, subprime assets which will never see the light of day. It's an utter catastrophe for them."

Fossil fuel company leaders must accept the new reality and dedicate themselves to accelerating the transition to clean energy. It's not enough to ramp up the net-zero economy; we need to shut down the remnants of the old one.

The Move Toward Environmental Justice

As we strive to hold on to a livable world, we also must create a more equitable one. The Greek root of *crisis* is "krisis"—to choose. Solving the climate crisis confronts us with a myriad of choices in redressing social and economic injustice, health disparities, and gender inequality. If we fail in our net-zero ambition, these problems will surely get worse. But here is a more positive outlook: the current emissions emergency is an extraordinary opportunity to address deep inequities that have persisted for generations. More pointedly, accelerating our path to net zero *depends on* our commitment to equity and justice. We cannot do the first without the second.

One leader in this fight is Dr. Margot Brown, head of the Environmental Defense Fund's environmental justice and equity initiatives. In August 2005, while she was immersed in research for her doctoral dissertation at Tulane University, the warning came. Two days before Hurricane Katrina hit, Margot packed her data into her suitcase and left the city. She watched from afar as New Orleans flooded.

Margot Brown

For decades there was a disconnect between the prevailing environmental movement and the environmental justice movement. The first was focused on protecting natural systems and wildlife; the latter was focused on protecting disadvantaged communities from environmental hazards. What we have learned is we must do both and we must do them together. They're both integral components of a larger system.

I am frequently asked how I approach this tension and promote equitable and just solutions that have been too often overlooked. I use a systems approach. That means reconciling nature-focused solutions and environmental justice by making health and welfare of humans a critical part of the evaluation.

By expanding environmental programs to protect people of color and people of lower income, we can ensure that these individuals will no longer be the first to be affected by climate change and the last to be served.

In 2005, a few days before Hurricane Katrina made landfall, I evacuated New Orleans. I watched from afar as the climate crisis punished disadvantaged communities. The devastation to communities of color was so severe that it put their disadvantages into sharp focus for the world to see.

Seven months later, I was back in New Orleans to defend my dissertation. Vast swaths of the city were still closed, and entire neighborhoods were still devastated and in disrepair.

The Uptown Whole Foods, however, looked exactly the same as it did a few days before the storm. Why? Because the store was situated on higher ground in a neighborhood where people had the economic resources to rebuild immediately.

Just a few miles away, in the Lower Ninth Ward, even sixteen years after Katrina, many vacant houses still had faded yellow notices pinned to the front doors. Those who once owned their homes couldn't or didn't come back because they lacked the required resources to rebuild.

This low-income Black community was flooded by an industrial canal that was built in the 1950s to shorten the distance for commercial shipping. The canal's construction wiped out protective natural barriers and exposed residents to hazardous exhaust from industry. The neighborhood was eventually destroyed by the lack of natural protection.

Across the U.S. and around the globe disadvantaged communities are suffering due to environmental and socioeconomic factors. We need to be concerned for these communities' well-being, while also recognizing that the issues they face undermine the health and safety for all.

We need to learn from the factors that led to Katrina's devastating outcomes.

Working people—maids in hotels, janitors, security guards in my building, members of my church—lost the houses they were proud to own, the communities they were part of, the lives they had led, and far too many friends and family.

That's what it means to be the last to be served.

More than any other factor, where you live determines the quantity and quality of education you'll receive, the income you'll earn, the health you'll enjoy, and the number of years you can expect to live.

When people ask me about the impact of environmental injustice, I reply quite simply that the impact is death. The impact is not just loss of homes, but also the destruction of communities and ways of life—entire cultures. We saw that after Katrina in 2005, and we're seeing it again in the way COVID has disproportionately hit minority communities.

A just transition will require a holistic approach. We need an economic shift, and a living-wage shift, and an educational attainment shift, and an economic opportunity shift. We must address every piece of the puzzle.

The impact is death.

With guidance from Margot and other experts, we've grouped core elements of environmental justice into categories we can measure and track: the education gap, the health gap, and the economic gap.

Closing the Education Gap

Climate change is not gender neutral. Because of deep underlying inequalities, women and girls are more vulnerable to its worst effects. They are also invaluable allies in mitigating it. One front-and-center fight is the struggle for equity in girls' education, especially in developing countries in Africa, South Asia, and Latin America. Education, in the words of Project Drawdown, is "the most powerful lever available for breaking the cycle of intergenerational poverty, while mitigating emissions by curbing population growth." Each additional year of secondary school increases a girl's future earnings by 15 to 25 percent. Better-educated women marry later in life and have fewer, healthier children. They have more productive agricultural plots and better-nourished families. Not least, they are better equipped to withstand the effects of climate change. The link between climate impact and girls' education was studied by the Malala Fund, founded by the youngest-ever Nobel Peace Prize laureate, Malala Yousafzai. In 2021, at least 4 million girls in low- and middle-income countries will be unable to complete their education due to climate-related events: drought, food and water scarcity, displacement. By 2025, that number is projected to grow to 12.5 million girls.

In total, 130 million girls are being denied the fundamental right to go to school. The reasons are many: families with limited resources; deep-seated cultural biases; safety concerns within schools and en route to them. An authoritative book on the subject—*What Works in Girls' Education: Evidence for the World's Best Investment*—highlights some promising solutions. Schools must be made affordable, with family stipends to enable parents to keep their daughters in class, even in the face of family emergencies or economic downturns. Girls need access to high-quality schools without onerous travel. They need support to overcome health barriers—with deworming treatments, for example.

There is no shortage of successful efforts to matriculate girls and keep them enrolled. Educate Girls has shown that strong leadership and adequate funding can attract thousands of boots-on-the-ground volunteers, who in turn can get millions more girls inside schools. The group's founder, Safeena Husain, speaks from personal experience.

Safeena Husain

I grew up in Delhi, falling in and out of education, but eventually becoming the first member of my family to study overseas. I graduated from the London School of Economics.

When I returned to India many years later, I was profoundly aware that all the opportunities I had received were only because of my education. But I also knew that millions of other girls in India were being denied that same right and opportunity. Despite its great progress in expanding elementary education, India is still home to over 4 million out-of-school girls.

I started Educate Girls in 2007. We work to bring a positive mindset shift in India's remote and marginalized communities. We get help from village-based volunteers—whom we call Team Balika, after the Hindi word for girl. They are highly motivated individuals, often the most educated young people in their villages. They go door to door to identify every single girl who is not going to school. It's like a census survey, with data recorded on our own Educate Girls mobile phone app.

Using this data, we geotag each village to quickly pinpoint the clusters of out-of-school girls and prioritize the areas of greatest need. Once we know where the girls are, we start bringing them back to school. The journey begins with our community mobilization process: village and neighborhood meetings and individual counseling of parents and families. It can take anywhere from a few weeks to a few months.

Once we bring the girls into the school system, we work with the schools to make sure they are able to retain them and ensure strong attendance. We address some of the safety and hygiene barriers that often force girls to drop out, like the lack of clean drinking water or a separate toilet.

But all of this would be meaningless if our children weren't learning, and so we run a remedial learning program. Most of our children are first-generation learners with no support at home to help them with their homework. Their parents are often illiterate. Our program helps bridge this learning gap.

Educate Girls' audacious plan is to enroll up to 1.5 million out-of-school girls over the five-year period ending in 2024. This would significantly reduce the gender gap in education. We started with fifty schools, went up to five hundred, and then covered an entire district, doubling in size every eighteen months!

In South Asia, the biggest obstacles to getting girls into school are people's mindsets and traditional and discriminatory social practices. Renegotiating those positions is the most difficult part. We use volunteers to try to get over that, but the key is to be in the district for the long term, for six to eight years. When you do that, you create a new normal.

You don't just run a campaign, and everyone goes to school and you're done. The challenge is to sustain it over time. In the time that Educate Girls stays in a region, six to eight years, it will have covered ten cohorts of students, a generation. If we're able to hold a generation of girls in school, their own children will start from a different baseline.

It's about breaking the cycle of girls being denied access. The data says that once you break it, it stays broken, because an educated mother is more than twice as likely to educate her children. That's what we're aiming for.

It is important to link girls with the issue of climate change. It's not just because girls are a key to reducing future emissions by having smaller, healthier families. It's also because poor and vulnerable women and girls are the ones who will pay the highest price of climate change.

For me, education is at the heart of all of this.

Our **Education KR (8.4)** calls for universal primary and secondary education, assuring that all girls—and boys—worldwide stay in school until age eighteen, what should be a basic human right. The means for achieving this KR will vary across urban and rural settings and between developed and developing countries. We'll need to confront the obstacles between girls and their schooling and offer locally relevant solutions to overcome them.

It's a big challenge. But as Malala Yousafzai reminds us, "Getting millions of girls into school in the next fifteen years may seem impossible, but it is not. The world lacks neither the funds nor the know-how to achieve free, safe, high-quality secondary schooling for every girl—and every boy." There's a powerful climate incentive to get this done. According to Project Drawdown, a combination of "voluntary reproductive health resources and universal access to and equal quality of education to boys and girls" would reduce global CO_2e emissions by close to three gigatons per year. The calculus is clear: Girls' education must be universally guaranteed.

Closing the Health Gap

There is more than one way to unlock a clean economy and a world of net-zero emissions. But not all of them will make the transition just and equitable. The Speed & Scale Plan seeks to seize this moment to close the gaps between racial and socioeconomic groups along two crucial axes: health and wealth.

It's well documented that communities of color have borne more than their share of harm from the climate crisis. Let's dive into the health issues linked to greenhouse gas emissions. The most dangerous class of pollutants is known as "fine particulate matter," or "PM 2.5"—the microscopic solid or liquid particles, no more than 2.5 micrometers in diameter, that burrow deep into people's lungs. Mostly generated by gas- or diesel-powered vehicles or by fossil-fueled power plants, they cause an appalling one of five premature deaths worldwide. In 2019, in India alone, toxic air killed more than 1.6 million people. In the United States, it leads to 350,000 premature deaths per year. Black and Hispanic communities are disproportionately affected. The Black community in particular has more than 50 percent more exposure to PM 2.5 than the overall population.

Our **Health KR (8.5)** aims to close the chasm in climate pollution-related mortality rates between racial and economic groups. To gauge our success in forging an equitable transition to net zero, it is critical to measure health outcomes. Our sector KRs—closing coal plants, switching cars and trucks to electric, and electrifying stoves and heat within homes—address this challenge.

The scope of this key result is undeniably aggressive. Mortality rates are stubborn things. Nonetheless, we must be resolute in our ambition. We need to aim for breadth: equity in health outcomes in every corner of the world. And for depth: anything more than a 0 percent difference in mortality rates will be a failure. On my more cynical days, I wonder if achievements like this are possible. In brighter times, I realize that it's less useful to debate what we can or can't do, and more exciting and worthwhile to fight for a better future. That is the work—and promise—of movements.

Widening Opportunity

Of all the potential benefits of pursuing a net-zero world, job creation might draw the most attention in the political world, and for good reason. The economic opportunity of a transition to a clean energy economy has been estimated at $26 trillion. By 2030, as the world retrofits urban centers, scales up renewable energy, develops grid-scale storage, and revamps whole economic sectors, we'll create an estimated 65 million new jobs—and untold wealth.

Our **Economic Equity KR (8.6)** calls for an economically equitable transition, as measured by the distribution of good-paying clean economy jobs. It's essential that the clean energy windfall includes underserved populations. Disadvantaged communities must be prioritized for training programs and clean economy jobs. We cannot leave anyone behind, including former coal miners and petroleum and natural gas workers. Higher-wage jobs in particular must be broadly and inclusively distributed.

As the Climate Justice Alliance wrote, "A transition is inevitable. Justice is not." In putting forward this key result, we make explicit our aims for a just transition and for broad and equitable access to new opportunities. To be sure, this is only the start. True economic justice will require addressing long-standing, deeply entrenched wealth inequities that plague historically disadvantaged communities and divide the developed from the developing worlds.

Considering the magnitude and ambition of the key results in this chapter, some readers may be incredulous—or dismissive. But perhaps more than any other accelerant, movements require bold, imaginative, and unencumbered thinking. By definition, they disrupt the status quo. Most important, they are our best hope for swift, lasting policy change. They usher in new, previously unimagined futures.

A Megaphone for Movements

When I came of age in the 1960s, movements were crystallized in memorable moments and places—the March on Washington, or Bloody Sunday at the Edmund Pettus Bridge in Selma, Alabama. As news director for KTRU at Rice University in Houston, I witnessed firsthand the intensity of campus activism and the movement against the Vietnam War. The protests of that era lived or died by in-person participation at live events and the media attention they drew.

The world has changed. With platforms like Twitter and YouTube, movements no longer require live gatherings. Calls for change can spread at cyberspeed. Advocates and supporters can participate in unprecedented ways and numbers.

Before social media, a megaphone for movements was born in 1984, at a one-off event for leaders to share ideas at the junction of technology, entertainment, and design—hence the acronym TED. In 1990, after some fits and starts, TED became an annual conference and expanded its scope to all fields of innovation and knowledge. In 2006, the first six TED talks went online. The rest, as they say, is history.

Sustainability has long been a central theme for this forward-looking organization. In 2006, after Al Gore delivered a preview for his upcoming *An Inconvenient Truth*, TED's Chris Anderson was stunned by how many people in the audience "changed their life's goals from that moment on."

Urged on by key members of its community, TED's leadership embarked on a bold new approach to address the greatest global challenge in human history. Countdown is a TED-created platform to champion and accelerate solutions to the climate crisis. It brings together numerous groups, amplifies their best ideas, and tries to turn those ideas into action. Launched by Chris Anderson in partnership with Future Stewards, a group led by Lindsay Levin, Countdown brings a diversity of thought and voices to this critical conversation.

In October 2020, the inaugural Countdown event streamed live around the world on YouTube. Seventeen million people tuned in to hear from an array of leaders and luminaries, from Christiana Figueres, the former top UN climate official, to Pope Francis. In the months that followed, as a crop of recorded talks made their way around the globe, the audience grew to 67 million. (You can watch the event in its entirety at https://countdown.ted.com.)

In the weeks that followed, more than six hundred local groups, from Sudan to El Salvador to Indonesia, held their own virtual TEDx Countdown events, interspersing "mainstage" talks with conversations with local leaders and communities. TED had never been so relevant, so personal, so accessible. It began to bring results on the ground.

Philosopher-journalist Roman Krznaric spoke on "How to Be a Good Ancestor," about how the decisions we make today can reverberate across generations. Six months later, citing adverse environmental impact, the Supreme Court of Pakistan ruled against an expansion of a cement manufacturing facility. The justices quoted Roman's TED Talk and linked to the corresponding video, in addition to citing two other Countdown talks on green cement.

As you can see, Countdown's impact runs broad and deep. Local leaders and experts gain access to audiences near and far. Lessons and solutions can be gleaned and streamed to all corners of the world. With bold ideas backed by rich, well-developed storytelling, movements like Countdown are driving change faster—engaging more people, in more places, on more levels. With the power to spread ideas, "people feel ownership," Lindsay says. "They feel like they are stewards of the future."

As I pause to consider so much movement by so many movements, I feel a renewed optimism about the future of our cause. I'm determined to redouble my own efforts to act. America's Youth Poet Laureate, Amanda Gorman, may have captured this feeling best in "Earthrise":

There is no rehearsal. The time is
Now
Now
Now,
Because the reversal of harm,
And protection of a future so universal
Should be anything but controversial.
So, earth, pale blue dot
We will fail you not.

Innovate!

Chapter 9

Innovate!

On October 4, 1957, the Soviet Union launched Sputnik, the first man-made satellite—and Americans grew alarmed over what it could mean to lose the space race. In response, President Eisenhower created the Advanced Research Projects Agency, or ARPA, whose job was to invent the future of our national defense. Congress funded it at a whopping $520 million, or $5 billion in today's dollars. After ARPA's space work was transferred to NASA, its scientists and engineers turned to electronics miniaturization and finding new ways to communicate if the phone lines went down (as they might well have in a nuclear war). Eventually they came up with ARPANET, the 1960s precursor to the internet. ARPA is perhaps the most famous example—but by no means the only one—of how government-backed research can spark innovation and lead to enormous, sometimes unexpected payoffs.

After ARPA was moved to the Defense Department (which tweaked the agency's name to DARPA), its work still aided the space program. The Apollo missions couldn't have happened without DARPA's breakthroughs in transistor-based electronics. DARPA also laid the groundwork for the Global Positioning System. Originally developed for military applications, it became the foundation for satellite navigation in smartphones and cars.

Over the decades that followed, federally funded research and development continued to drive new industries. Today's technology leaders know the legend of ARPA-funded researcher Douglas Engelbart, who created the first graphical interface for computers, with a small device to aid navigation: the mouse. Without taxpayer support for innovation, we might never have had the Macintosh or Microsoft Windows. These early breakout technologies jump-started a global tech sector that now accounts for about 15 percent of the world's gross domestic product.

In 2007, the desire for U.S. energy independence led to ARPA-E, a program within the Department of Energy to spur the development of cleaner energy solutions. But the George W. Bush administration refused to fund it. By 2008, in inflation-adjusted dollars, total U.S. spending on energy R&D was less than in the 1980s, the time when Ronald Reagan removed Jimmy Carter's solar water-heating panels from the White House roof.

Then came the implosion on Wall Street, the Great Recession, and the election of Barack Obama. In February 2009, when Obama signed the American Recovery and Reinvestment Act, he made sure that $25 billion was actually appropriated for energy R&D, efficiency programs, and loan guarantees. A small chunk, $400 million, went to ARPA-E.

Almost overnight, the Department of Energy was inundated with unsolicited ARPA-E project proposals. But there was literally no one there to open the mail, much less get the operation up and running. That's when Duke professor Eric Toone received a phone call. He had no way of knowing the role he was about to play in shaping the nation's energy R&D agenda.

Eric Toone

Unbeknownst to most, I'm from Canada. I got my PhD in organic chemistry from the University of Toronto. In 1990, I joined the faculty at Duke, and now it feels as if I've grown up in North Carolina.

While I was at Duke, Kristina Johnson was dean of the School of Engineering. She was tapped by President Obama to be the under secretary of energy. Kristina called me and said, "Can you come up to Washington and help us out for a couple months?"

I agreed and soon entered the amazing world of Steven Chu, the brilliant Nobel laureate serving as America's twelfth secretary of energy. He selected Arun Majumdar from Berkeley National Labs to be director of ARPA-E.

We got to work opening the mail. I loved hearing from people who had far-out ideas and plans for catching lightning in a bottle. It was a little like that Talking Heads song: How did I get here?

We tried to keep our heads above water. We received 3,700 applications for funding and read all of them. We aimed to pick the best one out of every 100, so we ended up backing 37 projects, from renewable energy to building efficiency to bioengineering.

Innovate!

Things like liquid metal grid-scale batteries. Low-cost crystals for LED lighting. Bacteria that soak up sunshine and spit out hydrocarbon biofuels. And, of course, CO_2 capture in every form, using giant machines or even microscopic artificial enzymes.

One promising idea to make cheaper batteries for EVs came from two researchers at Stanford University, a project called QuantumScape. We granted them $1 million to fund development.

At first, all that mattered was helping to bring innovations to market as quickly as possible. We measured results by tracking licensing deals for the technologies. Pretty soon, however, we realized that what mattered most was whether those innovations could scale up to make a difference in the big energy picture.

Scale is the hardest thing to wrap your head around. Consider ExxonMobil. When they open an oil field, they hire people who could spend thirty years there. Those people might spend their entire working lives stationed at the same set of holes. Yet over all that time, that one oil field will produce only about a week of global oil supply. Imagine your whole career amounting to a week's worth of gas.

Energy technology has mass, which means it doesn't scale like Google or Facebook. Building out capacity can take decades. It's a much bigger effort than we could tackle on our tiny budgets. After four years at ARPA-E, I went back to Duke to lead its new Innovation and Entrepreneurship Initiative.

Scale is the hardest thing to wrap your head around.

Eric wasn't the only one advocating for more congressional support. In 2010, Bill Gates surprised people by giving a TED talk on climate and energy, a topic he'd never addressed before. After stepping down from his active role at Microsoft, Bill turned his focus to philanthropic ventures in public health for 2 billion of the world's poorest people. He came to realize that reducing the price of energy was one of the most powerful factors for lifting people out of poverty. But how could we make energy more affordable and at the same time reduce CO_2 emissions? Bill concluded there was just one way to do it: a massive step-up in spending on research and development.

In 2015, after an intensive study of the climate crisis, Bill proposed a "breakthrough energy effort," essentially a private-sector version of ARPA-E for early-stage cleantech investments. The goal was to invest in the most critical, complex technologies that have yet to scale. It was risky, but when Bill asked me to join the board, I readily agreed. We concurred on the need for more innovations to get to net zero.

And so Breakthrough Energy Ventures was born. Bill gathered a global group of leaders with an abiding interest in solving the climate crisis. Today that roster includes Jeff Bezos, Abigail Johnson, Michael Bloomberg, Richard Branson, John Arnold, Vinod Khosla, Jack Ma from China's Alibaba Group, Mukesh Ambani from India's Reliance Industries, and Masayoshi Son, the chief executive of Japan's Softbank. To date we've committed $2 billion, more than four times ARPA-E's peak annual budget. Perhaps more than anyone else, Bill has helped shape the agenda for climate technology innovation.

Rodi Guidero, Breakthrough's executive director, worked with our board to recruit our technical and business leads, Eric Toone and Carmichael Roberts. Eric built out the technical team so that Breakthrough would be fundamentally science-led. Carmichael helped build the investing team while connecting Breakthrough to academic institutions, corporations, and venture partners. From the start, the goal was to bring a multitude of investors along on the journey and significantly increase overall funding of innovative climate tech. In our first fund, Breakthrough would coinvest with over two hundred partners.

How does Breakthrough choose where to invest? With an eye on the highest-emitting industries, our team vets the science and technology that underpin a company's innovations. The bar is high. To make the short list for investment, a venture must show the potential to reduce greenhouse gases by at least half a gigaton per year, or about 1 percent of annual global emissions.

At both Breakthrough and Kleiner Perkins, our cleantech investing strategies are guided by a set of public and aggressive targets. We don't wait for proposals, though we welcome them. **The science of what *should* be possible guides our search for what we want.** Wherever we find an opportunity for massive impact, especially in difficult areas that have yet to see a breakthrough technology, we roll up our sleeves. We crawl through labs and universities, sponsor challenges, and network like crazy. In the end, we're looking for extraordinary entrepreneurs who can shepherd ideas from basic science to commercial success and even to planetary scale.

Cost is almost invariably critical in these efforts. That's why our Innovate! OKR sets price targets for new technologies we need to accelerate the transition to net zero. Think of them as five leading indicators to signal whether we're on track.

———————————————

Our **Battery KR (9.1)** calls for scaling the production of batteries while lowering their cost from $139 per kilowatt hour to $80. Transitioning all new auto sales to electric—60 million cars per year—will require 10,000 gigawatt hours (GWh) of batteries. We turn out a tiny fraction of that today—and we'll need another 10,000 GWh and more for electricity storage. The world is about to be ravenous for batteries, and scale is hard to attain. To ramp up production by several orders of magnitude, we'll need innovations in both materials and manufacturing.

Our **Electricity KR (9.2)** zeroes in on the cost of delivering energy to the grid. To supplant coal and natural gas, zero-emissions sources must be stable and reliable. Clean energy may come from the sun, wind, or water, from the Earth or the atom. The challenge is to provide steady power during normal times, while ramping up production to meet demand spikes during winter storms or summer heat waves. For any new technology to compete, it must beat the cost of fossil fuel.

The **Hydrogen KR (9.3)** speeds us toward broad adoption of emissions-free green hydrogen. Hitting the target will require huge amounts of clean energy and higher efficiency in converting water into hydrogen

Objective 9
Innovate!

KR 9.1   Batteries

Produce 10,000 GWh of batteries yearly
at less than $80 per kWh by 2035.

KR 9.2   Electricity

Cost of zero-emission baseload power
reaches $0.02 per kWh, with peak-demand
power reaching $0.08 per kWh by 2030.

KR 9.3   Green Hydrogen

Cost of producing hydrogen from zero-
emissions sources drops to $2.0 per kg
by 2030, $1.0 per kg by 2040.

KR 9.4   Carbon Removal

Cost of engineered carbon dioxide
removal falls to $100 per ton by 2030,
$50 per ton by 2040.

KR 9.5   Carbon-Neutral Fuels

Cost of synthetic fuel drops to $2.50
per gallon for jet fuel and $3.50 for
gasoline by 2035.

---

fuel. Low-cost green hydrogen can decarbonize energy-intensive industries that require extreme heat: steel, cement, chemicals.

The **Carbon Removal KR (9.4)** seeks to improve the economics of capturing CO_2 directly from the air and then sequestering it, a technology that has yet to be achieved at scale. We'll also need to find a place to store all that CO_2, which is *really* hard. Scaling carbon removal is a cornerstone of our campaign to get to net zero by 2050. As Bill Gates notes, we've yet to get the cost of direct air capture under $100 a ton. "If somebody comes along who can do it for fifty dollars a ton," Bill says, "that would be really phenomenal. If you got it down to twenty-five dollars, it would be one of the largest single contributions to solving climate change that we have ever seen."

The **Carbon-Neutral Fuels KR (9.5)** provides a path to decarbonize industries that may never entirely electrify, like aviation and cargo shipping. Fleets that cannot run on batteries or hydrogen can be powered by carbon-neutral fuels. The challenge is to find drop-in replacement fuels that can compete on cost with today's fossil equivalent.

We've got many miles to go to hit these five key results.

Mapping Out a New Realm of Innovation

When trying to solve a new problem, it makes sense to be guided by patterns from the past. As Bill Gates said to me the other day, "The technologies that you and I grew up in were magical." Those of us who started out in microchips and software tend to get nostalgic about Moore's Law and the exponential improvements we witnessed over half a century—not only in microchips, but in optic fibers and hard disk storage too. It felt like nothing could stand in the way of rapid progress in technology. But as two alumni of the personal computer industry who are now confronting a very different challenge, Bill and I have found that you can no longer define progress in the same way.

Tech improvement rates still lie at the heart of innovation. But as Bill points out, they are much harder to attain in cleantech. After reading everything he could on climate (including fourteen books by Vaclav Smil, the contrarian Czech Canadian scientist and policy analyst whom we met in chapter 3), Bill developed a sophisticated, multi-faceted approach to achieving critical breakthroughs and meeting this profound challenge.

Bill Gates

You cannot look at energy without looking at civilization too. If we don't confront this problem now, the damage will only get worse over time. By the turn of the next century, large areas of the Earth will be unlivable. We could be looking at extinction as a species.

In the physical economy, as Smil likes to tell us, things are very hard to change. It takes many decades to replace every cement plant in the world, every steel plant in the world. While people are superenthused about electric vehicles, the car industry is so big that electric passenger cars are at only 4 percent of buyers. That other 96 percent has yet to adopt.

But you have to look at it from the point of view of the developing world. The people living in tropical regions did almost nothing to create the historical problem with carbon. They also have the least scientific power to innovate for new approaches. Yet the suffering, the malnutrition and forced mass migration, will fall primarily on them if we don't act now in a pretty urgent way. And so we owe it to them to do just that.

Without dramatic innovation, developing countries aren't going to make the changes the world needs in their physical infrastructure, electricity, transportation, and agriculture. Given the difficult economics, and the imperative of providing basics like shelter and nutrition, it became clear to me that we needed to act. I became impatient with the pace of change. I wanted much higher rates of improvement.

Innovate!

Because any net emissions cause a net temperature increase, and our goal is net zero, we have to get rid of emissions from all sectors. That's why the green premium is so important. The green premium shows you the extra cost of being green in any sector. So if you called India and said, "Hey, make your cement green," India would reply, "What? That costs twice as much." Make your steel green? Not so fast—that costs 50 percent more.

So if you want to get the middle-income countries [like India and Nigeria] to go green, the sum of the green premiums across all sectors has to be brought down over 90 percent. Chipping away at the green premium also serves as a measure for how far along we are—and what rate of improvement we can reach.

We have to focus on where the green premium needs to come down the most sharply—areas like sustainable aviation fuel, clean hydrogen, direct air capture, energy storage, and next-generation nuclear.

The game will be won or lost in the developing countries. To meet a 2050 net-zero target, India will need the green premiums in all of those areas to be very low. So you prioritize the advances that can help bring them down.

The United States has about half of the entire globe's innovation power. We owe it to the world to use that power, to reduce green premiums and enable countries like India to say yes to these solutions.

The game will be won or lost in the developing countries.

At its 2016 public launch, Breakthrough Energy Ventures mapped an initial draft of "technical quests," the innovations we need to help get us to net zero. Each quest targets a scientific pathway where breakthrough technologies could slash greenhouse gas emissions.

Breakthrough Energy mapped the landscape of climate challenges to identify the most promising projects for research and development.

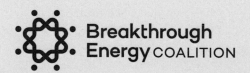

Technical Quests

ELECTRICITY

- Next-Generation Nuclear Fission
- Enhanced Geothermal Systems (EGS)
- Ultra-Low-Cost Wind Power
- Ultra-Low-Cost Solar Power
- Nuclear Fusion
- Ultra-Low-Cost Electricity Storage
- Ultra-Low-Cost Thermal Storage
- Ultra-Low-Cost Transmission

- Low-Cost Ocean Energy
- Next-Generation Ultra-Flexible Grid Management
- Fast-Ramping, Low-GHG Power Plants
- Low-GHG, Reliable, Distributed Power Solutions
- CO_2 Capture
- CO_2 Sequestration and Use

TRANSPORTATION

- Batteries for Gasoline Equivalent EVs
- Lightweight Materials and Structures
- Low-GHG Liquid-Fuels Production—Non-Biomass
- Low-GHG Gaseous Fuels Production—H_2, CH_4
- High-Energy-Density Gaseous Fuel Storage
- High-Efficiency Thermal Engines
- High-Efficiency, Low-Cost Electrochemical Engines

- Low-GHG Liquid Fuels Production—Biomass
- Transportation-System Efficiency Solutions
- Technology Solutions that Eliminate the Need for Travel
- Technology-Enabled Urban Planning and Design
- Low-GHG Air Transport
- Low-GHG Water-Borne-Goods Transportation

AGRICULTURE

- Reducing CH_4 and N_2O Emissions from Agriculture
- Zero-GHG Ammonia Production
- Reducing Methane Emissions from Ruminant Animals
- Developing Low-Cost, Low-GHG New Sources of Protein

- Eliminating Spoilage/Loss in the Food-Delivery Chain
- Soil-Management Solutions for GHG Reduction and CO_2 Storage
- Manure
- Agriculture-Related Deforestation

MANUFACTURING

- Low-GHG Chemicals
- Low-GHG Steel
- Low/Negative-GHG Cement
- Waste Heat Capture/Conversion
- Low-GHG Industrial Thermal Processing
- Low-GHG Paper Production
- Extreme Efficiency in IT/Data Centers

- Fugitive Methane Emissions from Industry
- Extreme Durability for Energy-Intensive Products and Materials
- Transformative Recycling Solutions for Energy-Intensive Products and Materials
- Increasing Biomass Uptake Rate of CO_2
- CO_2 Extraction from the Environment

BUILDINGS

- High-Efficiency, Non-HFC Cooling & Refrigeration
- High-Efficiency Space/Water Heating
- Building-Level Electricity and Thermal Storage
- High-Efficiency Envelope: Windows and Insulation

- High Efficiency Lighting
- High-Efficiency Appliances and Plug-Loads
- Next-Generation Building Management
- Technology-Enabled Design of Efficient Buildings and Communities

Breakthroughs can't be scheduled or dictated; new ideas are unpredictable by their nature. But if we cannot predict which innovation
will be next to bloom, we can seed the soil by funding basic and
applied science. Each technical quest involves chemistry, physics,
biology, material science, or engineering. As new lessons are learned,
we can move them out of the lab and try to push them to global scale.

As Bill says, we're only at the starting line in taking these vital
but expensive technologies to market. It's a circular problem.
To get costs down, you need scale. But to achieve scale, you need
to get costs and prices down.

Our Speed & Scale Plan features a handful of these high-priority
quests. We aim to highlight the obstacles and opportunities
ahead, from better batteries to carbon-neutral fuels. We also
call for innovation that relies less on new choices by society.

In Pursuit of Battery Breakthroughs

For decades, determined scientists and engineers have labored to
make headway on energy storage. Since 1800, when Lord Alessandro
Volta unveiled the first battery, the race has been on for better ones.
Volta's first battery was a set of paper cups filled with electrically
charged fluids connected by wires—not a whole lot of storage capacity
there. But it was enough to draw the attention of Napoleon, who
volunteered to assist with the experiments. In more recent times,
we've seen the transition from bulky, expensive lead-acid batteries to
more efficient nickel hydride varieties, to the lithium-ion models that
now power our computers, phones, and electric cars. Over the past
twenty years, energy density—the energy batteries contain relative to
their weight—has tripled. But that's still not enough.

In 2008, an engineer named Jagdeep Singh set out to make a radically
better battery for electric vehicles. Born in New Delhi, Jagdeep moved
to the United States as a teenager. After earning graduate degrees at
Stanford and Berkeley, he joined Hewlett-Packard before cofounding
his first company. Then he sold it and founded three others, selling
two and taking one public. Only then did he buy his electric dream
car, a Tesla Roadster.

Jagdeep Singh

I was driving my Tesla to work every day. And I thought, there's got to be something better than this battery. That 2008 version contained about as much energy as 8 gallons of gas and accounted for much of the cost of a hundred-thousand-dollar car. That's pretty bad.

It was obvious that the only way for more people to experience EVs would be a drastic drop in the cost of batteries and an equally dramatic increase in the energy density, or range, of the battery.

I was introduced through a classmate to Stanford professor Fritz Prinz and his postdoc Tim Holme. Though he chaired the mechanical engineering department, Prinz was more of a materials scientist. Our original idea was to build a better battery with quantum dots. They're nano particles that we thought could create higher electrical permittivity, increasing the ability to store potential energy in a supercapacitor.

Turns out, quantum dots are pretty hard to get a handle on. But that's how we got the name QuantumScape. After about six months, we concluded it would take longer to commercialize than we wanted.

Instead, we decided the best way to achieve our goals of a disruptively better battery was to make one with a lithium metal anode, which required a solid-state electrolyte between the anode and cathode, instead of the liquid one in a lithium-ion battery. That was the high stakes, bet-the-company pivot. And it was the best decision we ever made.

Innovate!

Singh and Prinz were waist-deep in developing their first battery at Stanford when ARPA-E awarded them $1.5 million. Though they chose to leave the grant at Stanford, ARPA-E's stamp of approval gave other investors more confidence in QuantumScape. Kleiner Perkins joined my friend Vinod Khosla among the startup's earliest backers.

The founders impressed us with a plan to double energy density with a solid-state lithium metal battery. Their engineering team replaced the conventional liquid electrolyte with a custom-built ceramic separator—QuantumScape's secret sauce. The battery packed more energy into a smaller package at a lower cost. And since ceramics are fireproof, they're also safer. Achieving this in the lab would be one thing. Commercializing and scaling it would be another.

Jagdeep Singh

QuantumScape wouldn't have been doable as my first startup. It's too big, too hard. I needed the experience from my first four companies to prepare myself for this venture.

When exploring what to do after my last company, I couldn't think of anything bigger and more important than launching a better battery. That enabled us to build a mission-driven team. Engineers in Silicon Valley tend to move around a lot. But if you have a mission beyond just making money, something to care about, you can retain your people.

Our work caught the attention of the major car companies. Volkswagen signed up with us early and put skin in the game. Once Dieselgate happened in 2015, VW doubled down on electrification. They invested over $300 million in us over the next six years, becoming our largest shareholder and a fantastic partner.

The demand for this kind of innovation is almost infinite. The number of vehicles sold every year is approaching 100 million. Even if we can lower the cost of a better battery pack to $5,000—a lot cheaper than today's packs—that's a $500 billion per year market. We aim to supply 20 percent or more of that demand in the fullness of time.

In 2018, QuantumScape and VW created a joint venture for mass production. The aspirations of a small startup were linked to the ambition and muscle of the world's largest automaker. In 2020, VW committed another $200 million, five months before QuantumScape went public with a special-purpose acquisition company (SPAC). What began as a research project was now worth more than $11 billion.

To electrify vehicles in the developing world, we'll need far denser and cheaper batteries. QuantumScape is at work on a manufacturing line to produce enough solid-state battery cells to test in actual cars. If the company meets its cost and density targets (and beats the competition to the market), it could eliminate the green premium for EVs in places like India or Africa, where new gasoline cars cost less than half as much as in the United States.

Beyond improving energy density, we'll need to scale the battery workforce, factories, and materials well beyond where they are today. Our **Batteries KR (9.1)** tracks both price and volume. To electrify every new car, we'll need to produce 10,000 gigawatt hours' worth of batteries annually, close to twenty times the industry's current capacity.

There's an awesome scale to this undertaking: When completed, Tesla's Gigafactory in Nevada will have the largest building footprint in the world, spanning more than a hundred football fields. It will hire nearly ten thousand people—and produce just 35 GWh of cells per year. To supply a global EV fleet, as Elon Musk acknowledges, we'll need at least a hundred like-size plants. Elon believes we can achieve this goal if leading companies in China, the United States, and Europe collectively "accelerate the transition to sustainable energy."

Even with all hands on deck, the battery industry will face stubborn issues around material scarcity and mining practices. Lithium is reasonably safe to mine, and supply should be able to keep pace with demand. But cobalt, which accounts for up to 20 percent of the material in a lithium-ion cathode, is more problematic. Sixty percent of the world's supply comes from the volatile Democratic Republic of the Congo, where mines are notoriously dangerous and children are forced into labor.

As the world's hunger for battery power keeps growing, we'll need heightened scrutiny on supply chains to ensure that materials are mined responsibly. New versions of cathode chemistries will reduce cobalt content by half. New battery technologies may eliminate cobalt altogether, which would solve one quandary. But given the limited life

span of lithium-ion batteries (typically ten to fifteen years), we'll run
the risk of accumulating a huge waste problem. Fortunately, it makes
more economic sense to recycle batteries than to throw them away.

In 2017, Tesla cofounder J. B. Straubel began recycling used batteries
through his new startup, Redwood Materials. The goal is to cut back
on mining nickel, copper, and cobalt with a closed-loop supply chain.
In the long run, by recycling used-up batteries from electric cars and
the grid, a large-scale battery industry could operate with little or no
new mining.

We'll need many more breakthroughs from battery makers (and
recyclers!) to meet the world's demand for cheaper, environmentally
friendly energy storage. There is room for many winners in this race.

Longer Storage for the Winter

On Valentine's Day 2021, as snow blanketed Texas and temperatures
dropped to single digits, people across the state desperately cranked
up their thermostats. Sixty percent of Texans' homes are heated with
electricity, nearly twice the national average. Since most homes
predate the 1989 state energy code, they're often drafty and poorly
insulated. During the snowstorm, power demand surged. The abnormally cold weather froze natural gas infrastructure and wind turbines, forcing these power sources offline. Millions of homes were left
in the dark amid frigid temperatures, many without water. Over 150
people died.

The Texas failure exposed the grid's vulnerability to extreme weather
events, which have grown increasingly common. It also underscored
the need for robust energy storage and more reliable power systems—
especially during intense storms, when demand spikes. As we saw in
Texas, it can be a matter of life and death.

How can we make variable energy sources like solar and wind
more reliable? And how can we rely upon these zero-emissions
solutions in a pinch? The answers lie in inventing new ways to
store energy *longer*.

Only recently have we scaled up from mere megawatts of grid storage,
reaching 1 gigawatt of storage in 2015. As of 2021, almost 10 gigawatts
have been installed, with another 10 under construction or announced. None of this would have happened without the rise of
electric cars to drive battery costs down.

Storage technologies are defined by their charging and discharge
cycle. Short-duration storage—for phones, laptops, cars, and houses—is

a day-to-day proposition. After capturing energy during periods of excess production, power grids store and then distribute it in times of peak demand. For these short cycles, the popular and cost-effective choice is the lithium-ion battery.

Long-duration energy for our grids must be stored economically for weeks or months at a time, which makes batteries way too expensive. For long-term storage we need more efficient alternatives—like pumped-storage hydroelectricity, which relies on the gravitational power of water. Tucked away in the Appalachian town of Warm Springs, Virginia, the Bath County Pumped Storage Station is now thirty years old. Known as "the largest battery in the world," it provides reliable electricity for 750,000 homes in thirteen states. At night, when demand is low, the station draws cheap power from a nuclear plant, which it uses to pump water from a lower reservoir to a higher one. When electricity is needed, water flows down from the higher reservoir to turn a hydraulic turbine. This technology can be activated far more quickly than natural gas "peaker" plants, the old standard for demand surges.

> The concrete poured to build the storage station was enough to lay 200 miles of interstate highway.

While pumped hydro is well suited for long-duration storage, it's costly to build and won't work on level land. An alternate use of gravity, by a startup called Energy Vault, lifts, drops, and stacks 35-ton composite blocks to store and release energy. A company named Malta stores energy as heat within large tanks of super-hot molten salt. Highview Power and Hydrostor use excess energy to store pressurized air, which is later released to generate electricity. Bloom Energy can use green hydrogen produced and stored on site to power their fuel cells. Finally, Form Energy and others rely on novel chemical reactions.

Next-Generation Nuclear Fission

Nuclear energy is an integral part of our power mix today and will likely remain so in the future. Its drawbacks are well-known; if a plant fails, the consequences can be devastating. In 18,500 cumulative reactor years of nuclear power plant operations in 36 countries, we've experienced three significant reactor accidents: Three Mile Island in 1979, Chernobyl in 1986, and Fukushima in 2011. They remind us of the risks of nuclear power and the need for safer reactor designs.

Can we create safer and cheaper nuclear power through technology breakthroughs? The answer is yes, but only if governments ramp up their funding to improve existing fission technology.

Most reactors today are cooled with ordinary water. To prevent the release of radioactive materials, they feature active safety systems for automatic shutdowns. But as Fukushima revealed, these systems aren't immune to accidents. After a 9.0 magnitude earthquake off Japan's Pacific coast, the plant's six reactors shut down automatically, as designed. But they weren't designed for the 46-foot-high tsunami that breached the 19-foot seawalls, flooded the lower levels, and cut off the reactors' backup diesel generators. When power to the circulating pumps went down, it set off three nuclear reactor meltdowns and hydrogen explosions. A decade later, the water used to cool the plant is still radioactive. The Japanese government is planning to dump it into the sea, a move that some environmental groups fear could harm nearby populations and the area's fishing industry.

While certain safety add-ons to Fukushima-style reactors can stop meltdowns, few reactors have them. The path forward lies in a new breed of advanced reactors, known within the industry as Generation IV. More than fifty labs or startups are going this route, to advance aspects of nuclear power: safety, sustainability, efficiency, and cost.

Nuclear energy comes with significant baggage. Safety and security are legitimate concerns, and poor people get run over when plant sites are chosen. When issues arise, governments rightfully layer on more regulation in service of safety, making an expensive operation even more so. But despite all these impediments, the case for nuclear power is not so hard to make—even beyond the fact that getting to net zero will be exceedingly difficult without it. As Bill Gates noted, "It's the only carbon-free energy source that can reliably deliver power day and night, through every season, almost anywhere on earth, that has been proven to work on a large scale."

Believing that nuclear power is essentially for the huge grids we need to develop, Bill became an early backer of a sodium-cooled nuclear reactor startup called TerraPower. The company's long-standing goal is to build a plant that could provide 24-7 baseload, zero-emissions power to a million homes. Unfortunately, TerraPower has been stymied by runaway nuclear construction costs in the United States and has yet to break ground. After the company reached out to the state-owned China National Nuclear Corporation, in hopes of building an experimental reactor south of Beijing, the deal got snarled by U.S.-China tensions. In February 2021, Bill told *60 Minutes* that convincing people that the reactor should be built would be just as hard as building it. For nuclear power to continue playing its part in decarbonizing our grid, it will need active support and investment by both private and public sectors.

In June 2021, plans were announced for the first TerraPower demonstration plant, to be constructed in Wyoming at the site of a coal-fired power plant slated to be shut down. I asked Bill to assess the company's future.

Bill Gates

TerraPower has the potential to make a meaningful contribution to these gigantic electric grids of the future. It's a very high bar, with four big challenges to overcome: nuclear plant safety, nonproliferation of materials that could be used for nuclear weapons, nuclear waste disposal, and cost.

TerraPower almost died in 2018. If they hadn't won funding for their advanced reactor demonstration, I might have given up. The U.S. government is funding half of the demo plant. I'm orchestrating the private side to fund the other half.

In five years, we may turn to the world and say, "Hey, look, in terms of safety and economics, fourth-generation nuclear power really should be part of the solution." As it is, I'm superthrilled that we're going to get a chance to build a demo plant and prove this technology works.

Fourth-generation nuclear power really should be part of the solution.

The Fusion Moonshot

Scientists have long dreamed of a controlled fusion reactor that actually works. Unlike traditional nuclear fission reactors, which generate energy by splitting atoms, fusion throws off energy by combining them—the same reaction that powers our sun and the stars. It takes absurdly high temperatures and pressure to squeeze the nuclei of separate atoms together. To be practical, fusion reactors must generate more energy than it takes to run them. The first scientist to demonstrate a sustained net energy gain with fusion at scale would have an inside track on a Nobel Prize.

This moonshot of a quest has researchers around the world scrambling to build a reactor that can produce enough heat for a fusion reaction. In healthy competition with an international consortium, Commonwealth Fusion Systems, a spinoff from MIT's fusion science lab, is developing superconducting electromagnets to create a plasma, a super-heated, ionized gas. If they succeed, they'll have the holy grail, a system that makes more energy than it consumes.

Fusion reactors would be fueled by hydrogen, an element that exists in abundance. In theory, you could filter the hydrogen from a gallon of seawater to produce as much energy as three hundred gallons of gasoline. But the technology has yet to be demonstrated. While the components and parts have been researched and tested, we're still waiting on a working prototype.

Some will say we are spending too much money for research on speculative new technologies like fusion when solar and wind are so cheap. But I believe it is critical to fund them, if only to determine whether the science can work at scale. When Bell Labs first demonstrated the solar cell in the 1950s, it was deemed technically brilliant but financially impractical—at the time, it would have cost $1.5 million to power a house. By their nature, innovations may seem impossible at first—even the ones that wind up changing the world.

Carbon-Neutral Fuels

By 2040, there will likely be half a billion electric vehicles on the road, driving 10 trillion miles a year. Assuming the grid is carbon-neutral by that point, they'll be 100 percent emissions-free. But until gasoline and diesel are phased out worldwide, legacy combustion vehicles may continue to log their own 10 trillion miles. Their emissions will continue to pump carbon dioxide into the atmosphere. Setting aside combustion cars and trucks, it's almost certain that long-haul ships and planes will continue to burn liquid fuels for some time to come.

We could reduce transportation emissions by using biofuels created from plants, crops, algae, vegetable oils, grease, and fats. Industrial processes convert these sources of biomass into ethanol, diesel, and jet fuel. When the fuels are burned, their emissions are offset by the atmospheric CO_2 absorbed by the biomass. The offset is less than complete, however. Depending on the process, and the fossil fuel energy it requires, the emissions cut will range from 30 to 80 percent.

As an investor in several biofuel companies, I can tell you that scale is difficult and costs are decisive in determining whether a fuel will be adopted. When crude oil prices are low, the economics for any alternative fuel are more challenging.

One complication that cannot be ignored, however, is the need for biomass. In a perfect world, all biofuels would be sourced from waste streams, like leftover sugarcane, cornstalks, or used cooking oils. But as demand rises, so does the risk that biofuels will compete with food crops or forest preservation. As we scale this industry, we must keep land use issues in mind.

As Timothy Searchinger from the World Resources Institute noted, "In a world that needs a lot more food and a lot more forests, and is clearing forests to produce food, why would anyone instinctively think that the best use of croplands is to produce energy?"

The crux of this dilemma is apparent in Brazil. The sun is full of energy, but Brazilian sugarcane struggles to convert it into something we can use. One acre of solar panels produces as much energy as a hundred acres of sugarcane.

Innovate!

The path forward to net zero needs a synthetic fuel from 100 percent emissions-free energy sources, one that doesn't compete with land or food. A promising approach would be to use solar or wind energy to combine hydrogen from water with CO_2 extracted from the air. Since emissions from these fuels would contain no more CO_2 than had been captured to produce them, they would be carbon-neutral.

If that sounds too good to be true, there's a reason: carbon-neutral fuels are not yet viable economically. For the math to work, the zero-emissions source used to make the fuel would need to be extremely cheap. Or, alternatively, the cost of fossil fuels—with a carbon price—would need to be significantly higher. The good news? Both developments are possibilities. The conditions may soon be ripe for synthetic fuel entrepreneurs to succeed, assuming they have dollars behind them.

Energy Efficiency Breakthroughs

Notwithstanding our huge gains in energy efficiency over the past fifty years, there is potential for much, much more. In the United States, more than two thirds of all energy produced from fossil fuels is wasted—partly in how it's generated, partly in how it's used.

All forms of energy, even solar and wind, consume resources. To carve out greater efficiency gains, we need lighter materials for things that move, and more efficient motors for machinery, heat pumps, water pumps, and fans. We need smarter buildings that use less energy—or none at all—for lighting, heating, and cooling. Supply chains must be reoriented to minimize packaging, minimize material use, and switch to sustainable and recyclable materials. Together, these advances can radically shrink the carbon footprint of our built world.

The BMW i3 EV hatchback, for example, is fabricated from carbon fiber that translates into a smaller battery pack and many more miles of range. Though this ultralight, ultrastrong material costs more per pound than steel, the price gap is offset by fewer batteries and simpler manufacturing. Since lighter cars use less energy to move, even a simple shift from steel to aluminum generates a leap in efficiency. When Ford's popular F-150 pickup switched to aluminum and shed seven hundred pounds, its fuel efficiency rose by 30 percent. A rugged, full-size truck was suddenly getting 26 miles per gallon on the highway.

More than half of the world's electricity passes through motors—in vehicles and appliances, in heating and cooling systems, in industrial machinery. Even when the motors themselves are efficient, poor controls can waste up to half the energy they consume. One novel improvement is a lighter type of motor—a "switched reluctance motor"—that allows for variable speeds and can run forward or backward. In Tesla's Model 3 and Model Y, switched reluctance motors expand driving range while lowering costs. The startup Turntide uses them to improve efficiency in heating, ventilation, and cooling systems.

Bulbs based on light-emitting diodes, or LEDs, show how changing consumer habits at scale can both cut emissions and save money at the same time. By 2018, LEDs accounted for 30 percent of all lighting applications in the United States and were estimated to have saved $15 billion in energy bills and 5 percent of electricity use in buildings. When innovations can simply plug in, screw in, or drop in, adoption rates soar.

In the realm of energy efficiency, seemingly small things can have a big impact. Apple keeps improving its products to minimize energy use, maximize recyclability, and lower its costs at every stage, from production through shipping. The latest iPhone was shipped without power adapters—saving plastic, zinc, and materials. The smaller, lighter package enabled the company to load 70 percent more boxes onto a single pallet. Apple continues to improve the energy efficiency of its products through its new microprocessors and software. That's a win on two fronts: better battery life and a smaller carbon footprint.

Engineering Our Climate

For the sake of discussion, let's say we fail to cut emissions fast enough and fall far short of our net-zero target. We could then be forced into a high-stakes choice. The default would be to live in a world with unchecked global warming and all of its associated human suffering—a nightmare scenario.

Or . . . we might try to alter nature itself.

Humanity has engineered climate adaptations since before recorded history. The first known seawall was erected seven thousand years ago by a Stone Age village off what is now the northern coast of Israel. But geoengineering is something else again. It's not about adapting to climate change; it's the manipulation at scale of nature itself.

One hotly debated notion is to deflect the sun's rays by launching sulfur dioxide particles into the atmosphere. If it worked, it could reduce warming and slow or even stop the melting of the polar ice caps. The empirical case for sulfur dioxide? In 1815, in what is now Indonesia, Mount Tambora gave rise to the most powerful eruption in recorded history. The explosion was heard 1,600 miles away. A column of superheated volcanic ash, saturated with sulfur dioxide, blew more than 70 miles into the upper atmosphere and spread more than 800 miles from the eruption site. Finer particles lingered in the sky for years, blocking significant solar radiation.

The effects were striking. Beyond abnormally brilliant sunsets, 1816 became known as "the year without a summer." It was the second coldest year in the Northern Hemisphere going back at least four centuries. In Albany, New York, it snowed in June. As Tambora's sulfur dioxide gave rise to acid rain and ruined food supplies, tens of thousands of people died from starvation and disease, perhaps as many as from the eruption itself.

Two centuries later, David Keith founded Harvard University's Solar Geoengineering Research Program. It's not work for the faint of heart; Keith has fielded more than a few death threats. Undeterred, he believes that geoengineering research is essential, if only to avoid unintended consequences—to lower the potential risks of this extreme option as much as we can.

Could there be safer tools than sulfur dioxide—limestone particles, perhaps? No one knows for sure. In *Under a White Sky*, Pulitzer Prize-winning author Elizabeth Kolbert describes one unnerving impact anticipated by Keith and other scientists. Were we to launch limestone dust into the atmosphere, it could turn the sky white. We'd have a new kind of cloudiness—all day, every day.

Al Gore argues that once geoengineering moves beyond carbon removal, it's the wrong moral choice—because its impact is unknowable, and because safer, surer options have yet to be tried. Al would say that geoengineering is less a moonshot solution than a Faustian bargain with nature.

Still and all, some top global experts believe we may need this sort of Hail Mary pass—not to substitute for emissions cuts, but as an emergency backup in the event all else fails. In *Under a White Sky*, there's a revealing exchange between two earth scientists, Daniel Schrag and Allison Macfarlane. Schrag suggests that geoengineering may be necessary "because the real world has dealt us a shitty hand."

To which Macfarlane replies, "We dealt it ourselves."

Constructing and Reconstructing Cities

Our climate is bearing the brunt of the most powerful social trend on the planet: urbanization. In 2000, 371 cities worldwide counted more than a million inhabitants. Today, there are 540. By 2030, there will be 700. At present, China pours more cement in two years than the United States used in all of the twentieth century. (At the same time, China has announced ambitious plans for fifty "near-zero" urban carbon zones by 2050.)

As cities around the world chart their future course, three choices define the trajectory of emissions in any urban development:

1. How will we design and construct our buildings?

2. How will people get around?

3. How much green cover can be retained?

For some time now, the prevailing responses have been: with concrete and steel; by automobile; not nearly enough. To achieve and sustain a net-zero world, we need a new set of answers.

How Will We Design and Construct Our Buildings?

In building a new city, urban planners can make zero-emissions choices up front. The critical first step is to lean into maximum efficiency in all aspects of planning. One credible plan for a city of the future is rising in India, where the urban population is projected to double between 2010 and 2030, to 600 million. Currently under construction on the far inland outskirts of coastal Mumbai, Palava City is projected to be home to 2 million people.

The Lodha Group, India's largest real estate developer, is collaborating with the Rocky Mountain Institute to achieve Palava City's net-zero target. Rooftop space will be blanketed by solar panels to power buildings and charge vehicles. Windows and floor plans maximize natural air flow through the apartments, reducing energy needs for heating and cooling. Under the rubric of "deep efficiency," Palava's buildings will use 60 percent less energy than the nation's norm.

Stores, jobs, and apartments will be close enough that people can walk between them. Parks and trees will help absorb CO_2. Rainwater will be captured for use and wastewater recycled. The Palava design will be two thirds more efficient than existing infrastructure.

None of these efficiencies require far-off, unproven technologies. We've had all these solutions for decades; the innovation lies in their integration into a single coherent plan. But to shrink cities' carbon footprint still further, we'll need more breakthroughs in efficiency and in fabrication methods and materials, like lower-carbon cement and steel.

How Will People Get Around?

We'll need to build cities and neighborhoods that make bicycling safe, provide ample public transportation, and reduce the role of cars. Copenhagen has cut emissions by becoming the world's leading bike commuter city. Denmark's capital contains 237 miles of wide, dedicated bike lanes. To keep riders safe, most of them are elevated and buffered by curbs from car traffic. By 2019, more than 60 percent of the city's commuters and students made their daily round trips by bike, up from 36 percent in 2012.

Where urban biking is less popular, surveys say, it's for a lack of protected bike lanes. It's not enough to paint lines on roads shared with cars. During the coronavirus pandemic, many cities in the United States added physically protected lanes. Feeling safer, bicyclists took to the streets in droves.

In Spain, Barcelona is famous for its car-free zones. The city's imaginative urban design has boosted tourism and the local economy. Its "Superblock" model has been copied around the world. In 2020, Mayor Ada Colau announced a $45 million Superblock expansion to create twenty-one pedestrian plazas and sixteen acres of shady new parks. As Colau declared, "Think of the new city for the present and the future—with less pollution, new mobility, and new public space."

In another victory for the net-zero movement, Barcelona has banned gasoline-burning cars bought before 2000 and all diesel vehicles from before 2006. Monitored by video cameras, violators can be fined up to five hundred euros. By doubling down on funding its excellent public transit system, the city plans to push 125,000 cars off the streets by 2024.

Copenhagen and Barcelona, along with Medellín, Paris, and Oslo, are compelling role models in meeting the urban emissions challenge. None of their initiatives lean on national mandates or radical technologies. They show how smart and creative design can go a long way.

How Much Green Cover Can We Retain?

Singapore mandates lots of trees, shrubs, and grass around its buildings, to help cool a hot city. The country originated the green plot ratio, a metric for tracking the proportion of greenery in an urban surface area. High-rise apartment developments can meet the requirements with sky terraces, communal planter boxes, and communal ground gardens. Ground-level greenery reduces peak surface temperatures between 2 and 9 degrees Celsius. Green roofs and walls do even more, lowering surface temperatures by 17 degrees Celsius. They also act as thermal insulation for the buildings themselves.

Over the past two decades, New York City has applied all three of these core urban principles: design, mobility, and green cover. In 2006, the city opened the first section of a raised pedestrian greenway and public park on the abandoned High Line industrial railway. The development was rich with symbolism. An abandoned site was now absorbing CO_2, doing its bit for a net-zero future.

Five years later, in another initiative led by Mayor Michael Bloomberg, New York embarked on a six-year project to turn its famed Times Square into a car-free pedestrian zone. Citywide, Bloomberg's administration built four hundred miles of protected bikeways. Later, under Mayor Bill de Blasio, cars were banned from Fourteenth Street, a major east-west thoroughfare. Within a year, with crosstown bus speeds on the rise, ridership climbed 17 percent. Between 2005 and 2016, even as it added population, New York achieved a 15 percent reduction in CO_2 emissions. The cut totaled 10 million tons per year, a down payment on a plan to reduce the city's emissions by 80 percent by 2050. New York is now a model for the green urban movement. And as Frank Sinatra would tell you, "If you can make it there, you can make it anywhere."

New York City turned an abandoned railway into the High Line pedestrian greenway, a symbol of the transition to clean energy.

Scaling Up R&D Itself

To accelerate our transition to net zero, we must scale existing technologies even as we develop the next generation of breakthroughs. At the same time, we must steer clear of incremental measures that will complicate or delay our transition. For example, we can't settle for natural gas installations just because coal is twice as dirty. There is no acceptable level of greenhouse gas emissions.

When I think about innovation, I'm reminded of the maxim coined by Alan Kay, Apple's first chief scientist: "The best way to predict the future is to invent it." To which I'd add: The second-best way is to fund it. Which brings me full circle, back to where my climate journey began, as I set out to invest in a clean energy future.

Invest!

Chapter 10

Invest!

In 2006, our Kleiner Perkins green team embarked upon our journey into cleantech investing. To start, we put $350 million on the line. Six years in, things weren't looking too rosy—and that's when the sniping began. An article in *Wired*, "Why the Clean Tech Boom Went Bust," called out my emotional TED talk on the climate crisis. It cataloged the Kleiner investments that had failed to pan out in solar, EVs, and biodiesel. To make sure no one missed the point, *Wired* ran an image of a jug of biodiesel fuel going up in flames. Here's how the story closed: "In other words, John Doerr may once again have a good reason to shed a tear."

But that was downright kind compared to what *Fortune* had to say a few years later. Under a headline declaring Kleiner's "downfall," the article lamented that the firm, "once the very embodiment of Silicon Valley venture capital," had taken a "disastrous detour into renewable energy."

It gets to you, I won't deny it. But when you're in the innovation-funding business, zigs and zags are part of the territory. Venture capitalists are prone to waves of false certainty and sieges of doubt. The way forward often gets messy, even treacherous. Most startups fail. Founders like Elon Musk, Lynn Jurich, and Ethan Brown have shared the grit it takes to survive the ups and downs.

Invest!

Over the years, I've found that great ventures are set apart by a handful of factors: technical excellence, an outstanding team, reasonable financing, and laser focus—on either a large, existing market or a rapidly growing new one. Finally, a standout venture needs that paradoxical combination of persistence, patience, and urgency. Few young companies possess all of these qualities, especially at the start. The winners develop them over time.

I look at the risk/reward dynamic in investing this way: You can lose only 1x of your money. But the upside can be many, many times the sum you put in—sometimes 1,000x or more.

Venture capitalists bet on entrepreneurs, those exceptional people who do more with less than anyone thinks possible—and do it faster than anyone thinks possible. Typically we think of internet, biotech, or cleantech entrepreneurs at bleeding-edge startups, but that's by no means the whole picture. Not all entrepreneurs create companies. Their ranks include corporate leaders who incubate new businesses, the *intra*preneurs. There are social action entrepreneurs and policy entrepreneurs, and not-for-profit climate entrepreneurs whose passion and purpose are to stop global warming.

The stages of innovation: from idea to scale

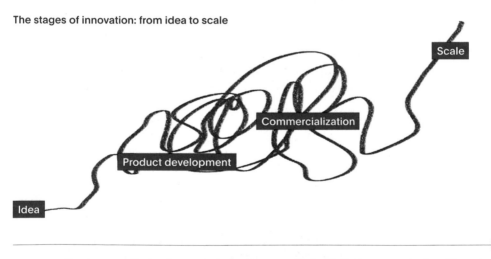

Scale

Commercialization

Product development

Idea

Seed   Series A   Series B   Series C, D, E ...   Project Financing

Steve Jobs toasted them all—to "the misfits, the rebels, the trouble-makers . . . the ones who see things differently . . . They push the human race forward, and while some may see them as the crazy ones, we see genius, because the ones who are crazy enough to think they can change the world, are the ones who do."

Disrupting a huge legacy market—say, the energy market—is a formidable task. In cleantech investing, the outfield walls are high and far and difficult to clear. The wind is in your face. A home run must do more than generate returns to shareholders, though that's what makes our world go round. In cleantech, home runs move us closer to our climate goals. They're all wins for the planet, whether or not Kleiner Perkins or Breakthrough Ventures backed them.

With due respect to *Fortune*, the reports of Kleiner's demise in cleantech investing were premature. One week after *Fortune* published our obituary, the initial public offering of Beyond Meat skyrocketed in value from $1.5 billion to $3.8 billion, validating a new market category. Over the next few months, the company's stock price quadrupled. Enphase Energy, a Kleiner bet that sold solar equipment to homeowners, turned solidly profitable and soared to a $20 billion market value. We'd also gone in early with Proterra, the U.S. electric bus leader, and QuantumScape, the Stanford spinoff that was incubating what might be a landmark breakthrough in batteries.

As of June 30, 2021, Beyond Meat was valued at $9.8 billion.

When it comes to funding the climate challenge, here is the cold, hard fact: We need a quantum leap in speed and scale, unprecedented sums on a very tight deadline. The innovation-funding machine is one of the glories of American capitalism, but we're not investing nearly enough capital to reach our goals. We need more breakouts and more entrepreneurs to lead them, or as Vinod Khosla says, "more shots on goal." Five different classes of funding must fill the gap: government R&D and financial incentives, plus venture, philanthropic, and project financing. While venture capital is where entrepreneurial funding usually begins, it is by no means where a startup's needs end. Much larger infusions come from growth capital and project financing (from banks, companies, or the public sector) at scale.

By our calculations, reaching net zero globally will require as much as $1.7 trillion *each year*—and we'll need to go full throttle for twenty years or more. That's the benchmark we're proposing for this extraordinary effort. Our plan entails five key results, each corresponding to one of the five classes of funding.

Objective 10
Invest!

KR 10.1   **Financial Incentives**

Increase global government subsidies
and support for clean energy from $128
billion to $600 billion.

KR 10.2   **Government R&D**

Increase public-sector funding of energy
R&D from $7.8 billion to $40 billion a year
in the U.S.; other countries should aim to
triple current funding.

KR 10.3   **Venture Capital**

Expand investment of capital into private
companies from $13.6 billion to $50 billion
per year.

KR 10.4   **Project Financing**

Increase zero-emissions project financing
from $300 billion to $1 trillion per year.

KR 10.5   **Philanthropic Investing**

Increase philanthropic dollars from $10
billion to $30 billion per year.

The **Financial Incentives KR (10.1)** consists of programs that governments can use to speed the tempo of change: loan guarantees, tax credits, and grants for zero-emissions technologies. Globally these incentives need to be expanded from the current pittance of $128 billion to $600 billion annually. The money for this key result is in plain sight, though its transmittal would be politically fraught: eliminate financial subsidies for fossil fuels, and the KR can be paid in full.

Our **Government R&D KR (10.2)** tracks public-sector funding for inventing a net-zero future. In the United States, federal funding for basic and applied energy research needs to be expanded by a factor of five. In other words, we're proposing that the U.S. government match what it now allocates to the National Institutes of Health, around $40 billion per year. Other countries should aim to triple their current spending.

Our **Venture Capital KR (10.3)** increases close to fourfold the dollars for building new companies and finding innovative solutions that can scale much faster. This capital is often raised from institutional investors (university endowments, pension funds, governments) and high-net-worth individuals. The dollars are invested in private companies, with checks as small as $250,000 and as large as $250 million.

The **Project Financing KR (10.4)**, our single largest bucket, ties to funding proven technologies. Public and private banks need to lend more money to fortify the deployment of renewable energy, storage, and carbon-reduction projects.

The **Philanthropic Investing KR (10.5)** triples funding for efforts that don't normally generate direct financial returns, such as climate justice or the protection of our lands, forests, and oceans. The nonprofits working in these areas need far more support from foundations that control nearly $1.5 trillion globally—$890 billion in the United States alone.

Turning Government Incentives Around

Governments around the world use favorable tax rates, tax breaks, and military spending to subsidize and safeguard the fossil fuel industry. Meanwhile, the oil, coal, and gas companies are allowed to ignore the destructive effects of their pollution. All in all, we directly subsidize the sector to the tune of $447 billion. Our plan calls for eliminating preferential tax treatment for fossil fuels and redirecting those vast sums to accelerate emissions-free alternatives.

The tax code bestows clear advantages on favored industries. The fossil fuel industry benefits from artificially low prices because it's been free to devastate our environment and collective health at every turn, from extraction to consumption—and without any penalty. Vaclav Smil puts it bluntly: "No fossil fuel has borne the eventual cost of CO_2-driven global warming." If all costs were to be taken into account, from climate change to mortality and disease from air pollution, the sector would owe more than $3 trillion a year.

Our governments possess various tools to speed adoption of clean technologies: grants for specific projects; direct loans, to be repaid with interest; private loan guarantees, where the government assumes all risk of a borrower default; subsidies as incentives to lower the purchase price; and tax credits.

For years, the punching bag of choice for climate action opponents has been the U.S. Department of Energy's Loan Programs Office. Their prime target was Solyndra, a solar startup that received a DOE loan guarantee of $535 million early in the Obama administration. Two years later, overwhelmed by cheap Chinese solar panels, Solyndra went bankrupt. (For the record, Kleiner didn't back Solyndra, but we took our chances with seven other solar PV panel startups. Four of them went under at around the same time.)

Solyndra is a classic example of spin trumping facts. Yes, the company failed and the government lost half a billion dollars. But what the headlines missed was that Solyndra's loan guarantee was but one small part of a broader strategy to keep the United States competitive in cleantech with China and the rest of the world. The overarching goal was to accelerate solar and wind technology and create clean energy jobs in the process. That strategy's success is beyond dispute. Between 2010 and 2019, employment in the U.S. solar industry grew by 167 percent, from around 93,000 jobs to nearly 250,000.

In reality, the Loan Programs Office has returned a tremendous bang for the taxpayers' buck. Whenever you back a portfolio of startups, whether with loans or grants, you expect some to go under. Since its inception, the Loan Programs Office has lent or guaranteed more than $35 billion. Less than 3 percent of the loans have defaulted, with current and future interest payments more than compensating for the losses.

As Jonathan Silver, the office's executive director under Obama, explains, "The role of a federal loan program is to support solutions that are very likely to be both important and commercially viable, but are not yet widely available due to the financial risk inherent in innovation." Government guarantees act as backstops to make private investors and lenders more comfortable in financing these projects. Ideally, he adds, federal backing helps a company develop something new and useful, reach scale in the market, and begin to sustain itself.

As Silver notes, the federal cleantech loans weren't designed to maximize returns. To attract applicants, the interest rate was pegged for the government to break even. A 3 percent default rate for a portfolio of innovative, utility-scale renewable projects would have netted "monstrous profits," Silver points out, had the loans been issued by commercial banks at normal rates.

By way of illustration: In 2010, a $465 million DOE loan went to an early-stage company on life support. Tesla Motors was in crisis. It had committed to manufacturing the capital-intensive Roadster amid the worst economic plunge since the Great Depression. The Loans Programs Office kept the company afloat. In 2013, Elon Musk announced that Tesla was repaying the loan ten years early, with interest—a happy ending for all parties. But lest we forget: no loan, no Tesla.

In the federal fiscal year 2010, the Obama administration spent $400 million on cleantech R&D and issued $70 billion in loan guarantees. It's an impressive-sounding sum, but China is beating us. Between 2012 and 2020, the Chinese government allocated an average of $77 billion *per year* in government support to state-owned or state-sponsored companies making solar panels, electric vehicles, and other cleantech solutions. It was a spectacular jobs program. Every province suddenly had its own solar manufacturing company. If one began failing, the government would typically rescue it.

In a nutshell, that's how solar panels got so cheap and spread so fast—China's gift to the world. It also explains why the five solar companies Kleiner Perkins backed got crushed in the ensuing price war. It wasn't because the United States had spent recklessly or excessively. To the contrary, it was because we had invested so little for so long. As a result, China now owns 70 percent of the international solar manufacturing market.

The Power of Venture Capital

Coming up with a good idea is a far cry from executing at scale. It's not enough to discover something the world needs. For a new company to succeed, it must have something the world is ready to adopt. The critical next steps—building a team, selling, making, and supporting a product—all require money. Enter venture capital. By trading a portion of their company stock for essential funding, founders can push their ideas from the lab into the market. That's the role played by firms like Kleiner Perkins: to find, fund, and accelerate the success of entrepreneurs.

Over the past five years, more than $52 billion in global venture capital has been channeled into cleantech. The first round of funding—the "seed" round—carries the greatest jeopardy, since fledgling companies are more likely to fail and lose their investors' money. To mitigate our risk, Kleiner's approach to cleantech was methodically grounded in science. As I've noted, we identified a small number of climate "grand challenges" that cried out for solutions.

The first decade of early stage cleantech investing: from boom to bust

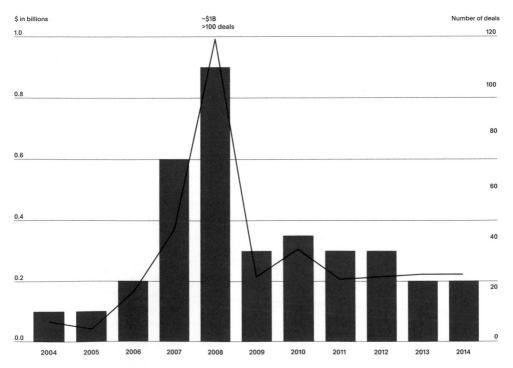

Adapted from data by MIT Energy Initiative

In 2006, not long after my wake-up call from *An Inconvenient Truth,* our team dove in to survey opportunities and meet entrepreneurs. As we reviewed more than three thousand proposals for solar, biofuels, steel, and cement companies, we led a venture capital wave into the search for climate solutions. As of 2001, the VC industry had bankrolled less than $400 million across eighty climate deals. Just seven years later, almost $7 billion had been plowed into four hundred deals.

The timing for this surge of capital turned out to be less than ideal. With the financial crisis of 2008, much of the embryonic cleantech sector fell apart. The implosion was triggered by the fall of oil and gas prices, a resulting credit crunch, and the inability of U.S. companies to keep pace with China's subsidized competition. Some technologies fell short in their leap from lab to commercial market. Others simply didn't work.

In 2009, investing in clean energy nose-dived, with early-stage funding taking a particularly hard hit. Meanwhile, billions flowed into software and biotech, among other industries. By 2012, the year of Kleiner's obituary in *Wired,* most of our cleantech investments had gone broke. It looked like we might lose every penny.

But then—gradually, unexpectedly, even miraculously—a few of our companies crawled out from under the wreckage. Proterra and their electric buses survived. So did ChargePoint, which operates America's largest network of public EV charging stations (112,000 locations and counting) and is now publicly traded on the New York Stock Exchange. Other companies backed by Kleiner were acquired by larger firms. Nest, the digital thermostat startup, was snapped up by Google for $3.2 billion in 2014. Opower, which provides software for utilities, was acquired by Oracle two years after that. A growing sense of opportunity led to the spinout of our cleantech investment team into a new fund, G2 Venture Partners.

The single biggest factor in reviving our cleantech portfolio was the initial public offering of Beyond Meat in May 2019. Kleiner had put $10 million into the company across multiple rounds of investment. Now its stock was listed on the NASDAQ. By January 2021, as Ethan Brown and company raised $240 million to expand the market for their plant-based meat substitutes, the value of Kleiner's shares had grown to $1.4 billion, a standout return of 140 times our original investment. In venture capital, two or three hits, or sometimes just one, can pay for many misses.

Since 2006, Kleiner has invested a total of $1 billion in sixty-six startup cleantech ventures. As of 2021, the value of our stakes had tripled to $3.2 billion, and venture investing in the sector is at an all-time high. We've come away from this roller coaster with a few hard-learned lessons on building a successful climate-forward business:

Be ruthless in identifying the key risk up front—and removing it. Founders and investors must confront technology risk (it won't work), market risk (it won't stand out), consumer risk (it won't sell well), and regulatory risk (it won't get approved). The question becomes: What are the main risks? Can early-stage capital be used to remove them? If not, it will be nearly impossible to raise later-stage capital.

You are always raising money. The message for founders is simple: Be great at fundraising—be better than great. Recruit a range of investors in your funding rounds, especially those who can write large checks. And seek out corporate partners, who can be invaluable.

Costs are king; performance matters. When you're competing in a commodities market like electricity or steel or fuel, unit costs rule. Consumers won't pay more for a lesser product with an ecofriendly badge; they expect something superior, or at least equivalent. Tesla, Beyond Meat, and Nest are three spectacular examples.

Own the relationship with your customer. The companies that fared best in the Great Recession sustained direct relationships with end buyers of their products.

Incumbents will fight. Some will adapt; others will die. But nearly all of them will battle tooth and nail. Their businesses, after all, are built on the premise of free-of-charge carbon pollution.

In absorbing these lessons, Matt Rogers is a quick study who has gone through the gauntlet. Not yet forty years old, Matt has had three successful careers: as a software engineer on the original iPhone; as a climate entrepreneur who cofounded Nest, the energy-saving smart thermostat company; and now as a venture capitalist. In 2017, Matt created the investment fund Incite to back mission-driven cleantech founders who are unafraid to tackle incumbents.

Matt Rogers

I was twenty-six when I left Apple in 2009. I was thinking about the grand challenges of humanity, climate being one of the top ones. At the time we had so much brain power, horsepower, financial resources, and talent going into apps like Angry Birds. But what were we putting into climate?

Together with my cofounder, Tony Fadell, we took an analytical approach to the market. We knew the customer space from building the iPod and iPhone together. We looked at the Department of Energy flow diagrams and looked for what was important and what no one was working on. On a yearly basis, heating and cooling is on the top; it accounts for half the energy a home uses.

At the time I was living in a condo in Silicon Valley that was built in 1973. We put in new floors and countertops, but we still had these beige plastic things controlling the heating and air-conditioning. We'd just built the iPhone 4, which was the sleekest product ever, all brilliant glass and aluminum. But in our condo, the design and technology was from the 1970s. That beige plastic crap was controlling how we spent a thousand dollars a year or more to heat and cool our home.

In the 1980s, they came up with technology to program a thermostat to turn down the heat at night and save energy. But the user interface was so bad that no one bothered to use it. That was the fundamental tenet behind Nest. We could build a beautiful product. But we also needed to build a thermostat that was easy to use and could save energy automatically.

That was our key insight: it was both an energy efficiency issue and a user interface issue. Nest was a mission-first company and a product-first company at the same time.

Invest!

We had no expertise in this area. So we did a lot of research, talked to a lot of experts. We needed to understand how HVAC systems worked and what the Environmental Protection Agency was saying. There was a lot of action at the time in environmental research that hadn't trickled into the consumer market.

It's very hard to create new markets. I like going after established markets. People already had thermostats; we didn't invent them. But when you invade an established market, you have to watch how the incumbents respond to change. Sometimes they will buy companies, either to squash change or incorporate change and grow their business. Sometimes they will sue to scare folks, just because they have market power. And sometimes they will ignore you.

In the case of Nest, the incumbent in the room sued us. Honeywell said we were infringing on patents—for having a circular knob, for example. Four years later, the suit got tossed.

New entrants have something the incumbents don't: agility. It is really hard to make decisions when you have seven layers of management. It's hard for new ideas to percolate up when there are so many people and priorities.

That was our M.O. at Nest: Be fast. Make quick decisions. Evolve more quickly than anyone thinks you can. Early on, we would announce a really cool breakthrough product—but we wouldn't stop there. Three months later, we'd have a new software update. Every year we'd have a new hardware update. We kept rolling as fast as we could. By the time the competition copied our first version, we were already on version three.

We started with an idea: How can we help people save on energy in the home? But the goal was always to bring it to scale, to where there's now tens of millions of thermostats in people's homes, and we're saving tens of millions of megawatt hours of energy every year.

Nest succeeded because we were a team at the right place, at the right time, with the right product.

In the month of January 2021, investments in clean technologies exceeded the total for all of 2015. After a decade-long grind, venture capital cleantech is roaring back. Carmichael Roberts, business lead at Breakthrough Energy Ventures, has overseen investments in over fifty startups at every stage of their development. I asked Carmichael about what it takes to be a successful entrepreneur in this area.

Carmichael Roberts

Successful founders get on their surfboard and paddle out before there's a single wave to be seen. Something in their gut tells them that the water is going to have the most beautiful wave coming and nobody else can see it. They work really hard to prepare themselves so when that wave comes, they can stand up and ride it.

Breakthrough Energy Ventures has thirty full-time, on-the-ground scientists, entrepreneurs, and company builders. No one inside the organization would call themselves purely an investor. We search for groundbreaking climate technologies that we can help shape to be as successful as possible. Sometimes that means paddling out into the water with the founders. Sometimes that means throwing them a life raft.

To be successful, entrepreneurs need to be confident, but also vulnerable and a little bit paranoid. A founder recently came to me and said, "Carmichael, should I be nervous about x, y, and, z?" I said, "Yeah, you should." Then I said, "Now that that's out there, let's deal with it together."

People want to know how the companies we invest in are faring. It's only been four years for BEV, so we're early on in the journey. But here's what no one knows. After every investment decision is made, I sweat. My partner sweats. The entire team sweats. We ask ourselves: Did we make a crazy decision on this thing?

And you know what we do then? For the next several months, we work to make sure the decision was not crazy. We mine our networks to their depths. We scour the landscape for partnerships and bring as many people as we can along on the journey with us.

Invest!

We provide every last bit of technical expertise we have. Our job is to support the entrepreneurs, the ones who are doing the really hard work. If they are successful, our world will be a gigaton lighter in greenhouse gases. And every gigaton counts.

If we are successful as investors, we will be responsible for the 150 companies that are most critical to the outcome in climate. More than that, our collaboration with other venture capitalists and companies will be responsible for the thousand companies that enable us to reach net zero by 2050.

We are late on climate change, there is no denying that. But I do believe that pure human spirit—our imagination and commitment—can save the day. We've seen it before in history and we are seeing it again. You've got to be realistic, no question. But you also have to go for it.

Our job is to support the entrepreneurs, the ones who are doing the really hard work.

This Is Not a Bubble, It's a Boom

Since the dawn of industrial-scale capitalism in the early 1800s, so-called investment bubbles have funded new industries, from the railroads and automobiles to telecommunications and the internet. With each disruptive technology, mounds of money pour in. Much of it is lost. But society gains.

In clean technology, we need to open the floodgates of capital. One trend to watch is the surge of special-purpose acquisition companies, or SPACs. These firms are created to acquire early-stage companies that aren't yet ready for a public offering, usually because they have yet to show a profit. While SPACs are high-risk investments, they will be an important means of funding the technologies we so desperately need. Without them, innovation slows.

ChargePoint, QuantumScape, and Proterra were all acquired by SPACs and converted to companies listed on a stock exchange. Investor enthusiasm is surging, from 46 SPAC deals in 2018 to 248 in 2020. Twenty percent are energy or climate related. It's reached the point where some are warning of a bubble of speculative overfunding.

I'd argue that it isn't a bubble, but a boom. Many SPAC-backed ventures will fail, without question. But SPACs are here to stay. And booms are good. They lead to more investment, full employment, and healthy competition. They spur on complacent incumbents. Through "creative destruction," they transform markets.

A Solar Power Turnaround

Of all the cleantech companies Kleiner backed, Enphase Energy may have taught us the most. My partner Ben Kortlang is probably the world's most experienced solar investor. When Ben led us to Enphase in 2010, the solar tech startup was struggling to scale its inverters, the circuit boxes that link a solar home's rooftop panels to its electrical system. We believed the inverter market was primed to explode, and that Enphase would capture a healthy share of it. But the company's revenue remained stuck at around $20 million. Dozens of other startups had entered the same space. For a time it looked like Enphase might meet the same grim fate as our other failed solar tech investments.

We sought out the advice of T. J. Rodgers, the legendary founding chief executive of Cypress Semiconductors and a board member at Bloom Energy, our first big energy investment. T.J. saw untapped potential in Enphase, which had just shipped its millionth inverter. The missing piece was dynamic leadership that could approach the company's challenges with fresh eyes. T.J. recommended a rising star at Cypress to take the reins as chief executive.

That's how we met Badri Kothandaraman. Born and raised in Chennai, India, Badri launched a twenty-one-year career at Cypress after earning his masters in materials science at the University of California, Berkeley. His deft piloting of Enphase reflects the importance of operating excellence in carving out a new cleantech niche.

Badri Kothandaraman

Every other investor had left the space. They feared that inverters would become a commodity business with zero profit and endless price wars. Their concerns were not unfounded. Enphase was losing money and running out of capital.

I started at Enphase in 2017. For the first two years, my focus as CEO was on operational excellence. We started measuring everything. We set up a war room to manage cash and receivables and payables on a daily basis. We established a pricing team to price products based on the value they generated, as compared to the next best alternative. We said goodbye to price wars and turned down sales that were unprofitable.

We spent a lot of time working on product costs. We created dashboards to measure progress and a quarterly goals system for all employees. Our bonus program paid out depending on the goal-related performance of both the company and the individual employee. No goals, no bonus!

Our strategy with investors was not so different. In June 2017, at our analyst day, we put a stake in the ground. We told investors that it would take six quarters to achieve a 30-20-10 financial model. It was an easy-to-remember way to say that the company was targeting 30 percent in gross margins, 20 percent in operating expenses, and 10 percent in operating income.

Invest!

Our strategy started paying off. We ended 2018 by making our 30-20-10 financial model a reality. From there on, our top line accelerated.

How did we grow? Once we got operations under control, we spent more time on the top line. We focused on product innovation, quality control, and customer service. Instead of running high-DC voltage lines on the roof and having our customers install big inverters in their garage, we made semiconductor-based microinverters small enough to fit below each solar panel on the roof.

If you had twenty panels, you needed twenty microinverters instead of just one, but you gained a significant advantage: safe AC voltage. Our scaled-down inverter family was world class; sleek, high-power, high-efficiency, and easy to install and connect to the cloud.

We relentlessly focused on quality, as measured by customer returns or defects. We gave equal diligence to customer service and started taking calls from homeowners in addition to installers. We staffed up our service centers in the United States, France, Australia, and India. My weekly staff meetings always began with a review of our service dashboard, including our net promoter score, average customer wait time, and first-call resolution rate.

Our net promoter score improved from single digits in 2017 to 60-plus percent in 2020, but we are by no means done. In 2021, we introduced 24-7 service for our customers and created a field service team to help installers be more efficient. We've also added battery storage to our product line and are now on our way to building state-of-the-art home energy management systems that consumers can trust. As with everything else we do, we're measuring customers' energy savings as well. It's the only way we can ensure a great customer experience.

No goals, no bonus!

As you might suspect, Badri's management by metrics is music to my ears. Enphase has justified the confidence of its backers while effectively addressing our emissions crisis. In 2020, ten years after Kleiner's initial investment, it became the world's most valuable solar technology company, with a market cap exceeding $20 billion. In January 2021, Enphase was deemed big and steady enough to join the ranks of the S&P 500 index.

Putting Project Financing to Work

Over the past seventeen years, clean energy project financing for new facilities and retrofits has soared from $33 billion to $524 billion. Most of it is dedicated to solar and wind plants, with growing amounts going toward electrification of heat and transportation. While the trend is promising, those dollars could make a bigger difference if they were channeled toward even newer, critically needed technologies.

Our Project Financing KR (10.4) calls for project financing dollars to reach $1 trillion a year and to be disbursed more quickly. Beyond funding proven technologies, public and private banks need to issue more loans for new energy sources, new types of storage, and new carbon removal projects.

Breakthrough Energy's Catalyst program, created in 2021, is a radical idea to demand more from project financing in reducing the green premium. Catalyst founder Jonah Goldman puts the case bluntly: "The five hundred billion-plus dollars going toward solar and wind is not charity. It's economically profitable—and that's because of fifty years of action by innovators, the climate community, and government." Jonah calls for more "heroic and courageous capital" to create markets for riskier new technologies like emission-free aviation fuels, green cement, and carbon removal.

Invest!

The four indispensable project funders are governments, companies, banks, and philanthropists. If all four commit to pay the green premium and furnish enough money to build out these companies, it will kick-start Wright's Law into action. With larger facilities and greater demand, new technologies can more quickly pare their costs. As Jonah reminds us, "It took fifty years for the solar cost curve to come down. But we don't have fifty years to spare." To accelerate, he says, we'll need to invest significant capital to building the first few demonstration plants of a new technology, to show that it works.

Project financing for clean energy is on the rise

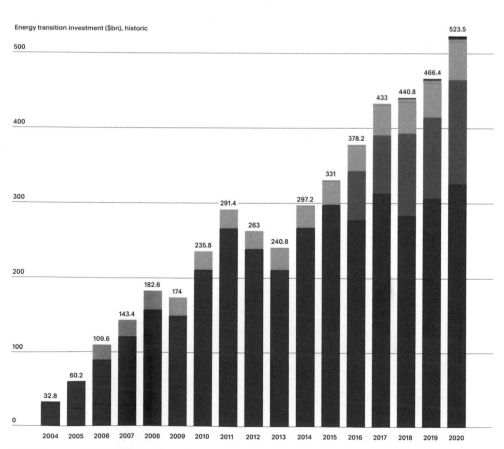

Adapted from data and visuals by BloombergNEF

Project financing naturally gravitates toward proven technologies, like solar deployments and efficiency retrofits. That's a good thing, and we'll need a lot more of it to continue to drive down costs in these areas. But we also need bold moves to buy newer technologies. When a company like Google pledges to purchase energy from a next-generation geothermal company like Fervo, they galvanize the entire market. Just as Stripe has created a market for carbon removal technologies by paying the green premium, project financing dollars can drive down costs by providing demand at scale.

Summoning a New Kind of Capital

In 1998, Kleiner Perkins bet $12 million for a 12 percent stake in a web startup founded by a pair of Stanford grad school dropouts. Sergey Brin and Larry Page entered the search engine business at number six in market share. A year later, hoping that Andy Grove's simple management system might be helpful to them, I arrived at Google's original headquarters to give a lecture on Objectives and Key Results. "We decided to give it a try," Larry said. Since then, thousands of Google employees have embraced OKRs to help inspire them to aim high and go far.

In the quest for net zero, few large companies have moved faster. In 2007, after early purchases of renewable energy and high-quality carbon offsets for any remaining emissions, Google became carbon neutral throughout its operations. In 2012, the company set an even more ambitious goal: to power 100 percent of its operations from renewables like solar and wind by 2020. The company hit that target three years early, in 2017.

Today, the mission for Google and parent company Alphabet is to invest at scale to solve the world's toughest challenges. Both organizations are led by Sundar Pichai, who joined Google in 2004 as a thirty-two-year-old product manager. In 2015, Sundar was named the third CEO in the company's history.

That was also the year that Google recruited Kate Brandt as chief sustainability officer, the same position she'd held for the federal government during the Obama administration. Since then, Kate has set the company's sights far beyond its own footprint—to use Google's technology platform to accelerate global emissions reductions.

Sundar Pichai

Thinking ahead with a multidecade time frame allows you to take moonshots, to be very ambitious. When we bet on wind and solar power, it was considered too expensive, and most people were skeptical it could work at scale. Alphabet is now one of the world's largest purchasers of renewable energy, and that early investment has paid off in helping to drive down costs.

Looking to 2030, we aim to run everything carbon-free, 24-7. That means every query on Google, every Gmail you send, every transaction on the Google Cloud will be done without emissions.

We don't fully know how to get there. We need more innovation. We also need more project financing. That's why we issued the largest sustainability bond in corporate history, $5.75 billion in green project financing.

One such project centers on next-generation geothermal energy. Because of their intermittency, we know wind and solar alone can't run the entire power grid in many places. To make clean power affordable and reliable, we're leveraging geothermal steam heat to drive electricity turbines. The steam is generated by hot water pumped from wells as deep as two miles underground. Starting next year, we'll connect new geothermal sources in Nevada to the grid to power our data centers running Google Cloud. We'll use AI to respond to demand in real time and to achieve always-on, 24/7 power. With our platforms and our scale, we can use the cloud to realize emissions reductions in our operations.

Kate Brandt

We thought about how we could play a role in helping to build this asset class and show the value of sustainability bonds.

We put out a framework that articulated how we would allocate the money. It centered on different categories of our environmental work: our procurement of renewable energy, energy-efficient data centers, and circular materials. Understanding that environmental issues and social issues are so deeply entwined, we also brought in social dimensions like racial equality.

The initiative is something we're really proud of. Our goal is to show that this asset class can bring more capital into sustainability.

It has been great to see other companies following suit, to witness the momentum in funding projects that are both environmentally and socially responsible.

We're passionate about the role of AI to drive deep energy efficiency gains. We've used it to great effect in our own data centers, and now we're trying to take it beyond our walls, so other data center operators and large building operators can realize their own energy efficiency gains.

In Nest, we also have a learning thermostat for the residential market that helps optimize energy use in homes.

With the two together, we see a growing opportunity to use AI in commercial and residential buildings and realize substantial decarbonization.

Invest!

Sundar Pichai

What excites me most about getting the world to net zero is that we'll need end-to-end change—both big and bold moves and small but meaningful changes.

To maximize Google's impact in other ways, we're nudging users to decrease their carbon footprint. For instance, Google Maps now defaults to the most ecofriendly route.

Looking globally, we set a goal to help the world's top five hundred cities drive an added reduction of 1 gigaton by 2030. These cities account for 50 percent of the world's population and 70 percent of emissions. We're doing it using AI, data, and sensors. Cities often don't realize where carbon emissions are coming from. In places like Copenhagen and London, we're working with local leaders to install air quality sensors that detect emissions instantaneously. Having access to this information allows city policy makers to develop a durable blueprint for their emissions reduction programs. We are systemically scaling this program across cities to achieve our one gigaton goal.

I grew up in Chennai, India. My childhood was marked by severe droughts year after year. The scarcity of water meant we had to rely on just a few buckets of water for our daily needs.

In 2015, Chennai got hit by a one-in-a-hundred-year flood. The city had never experienced that kind of rain, and that juxtaposition drove home the impacts of climate change.

In 2020, we were inundated with wildfires here in California. My kids woke me one morning, pointing to the orange skies, looking really concerned. I felt accountability to the next generation in a deep, visceral way.

As a business leader running a company using technology to innovate, I feel a strong sense of responsibility to apply that approach to make progress on the climate crisis. It's one of the biggest innovation opportunities we have.

Larry and Sergey, our founders, were ahead of their time. Google became carbon neutral in 2007. They were talking about sustainability before most companies even had it on their radar. It's an enduring value for the company.

But every company can make sustainability one of their foundational corporate values. It is important for them to do so, because the people who use their products will demand it. So will the best talent.

As a leader, the sooner you embrace the shift to sustainability, the better positioned you are to succeed. It's what your customers and employees will ask for, but it's even bigger than that. It is what's right for your people, your country, and the world.

The climate crisis is one of the biggest innovation opportunities.

How Money Flows

In 2003, David Blood retired from Goldman Sachs and set out to make the case that socially responsible investing would one day outpace all other asset classes. At the time, "green investing" occupied a small niche in the financial world. Substandard returns were viewed as acceptable, if not inevitable. But after David teamed up with Al Gore to cofound the London-based Generation Investment Management, all of that changed. They created nothing less than a new model for money in cleantech.

We realized poverty and climate change were the same issue, just different sides of the same coin.

David Blood

Growing up in Brazil after my dad was transferred there, I was taken aback by the poverty I witnessed. After retiring as Head of Asset Management at Goldman Sachs, I wanted to use the capital markets to help address the challenges of sustainable development.

In October 2003, I met with Al Gore in Boston to talk about sustainable investing. My interest was in poverty and social justice, and Al's, of course, was in climate change. At the first meeting we realized poverty and climate change were the same issue, just different sides of the same coin.

We founded Generation with a dual mission to deliver strong risk-adjusted investment results for our clients and to help mainstream sustainable investing. The investment world at the time did not take sustainability and ESG seriously, so we focused on making the business case.

We see long-term investing as best practice and sustainability as the organizing construct of the global economy. We use environmental, social, and governance (ESG) factors as tools to evaluate the quality of businesses and management teams. We believe this approach reveals important insights that other investment frameworks may leave undiscovered—and that these insights ultimately lead to superior risk-adjusted investment results. To be clear, we are not trading values for value.

And, most importantly, we have happy clients. From zero in 2004 to today, clients have entrusted us with over $33 billion under management.

Invest!

We are pleased with the significant growth of sustainable and ESG investing. We are also encouraged by the important commitments to net zero announced by asset owners, asset managers, banks, and insurance companies. We have indeed made extraordinary progress over the last decade. However, it is not enough. To achieve the objective of limiting global temperature rise to 1.5 degrees Celsius, it will require transformational change.

And make no mistake, we will succeed in addressing the challenges of climate only when we address the impact on people and communities in both the developing and developed world.

At Generation, we believe the next decade will be the most important of our careers. The world needs and deserves leadership from the financial sector. We need to raise ambition. We need to be uncomfortable. We need to change what people think is possible. But most importantly, we need a relentless commitment to action.

We see sustainability as the organizing construct of the global economy.

The Mother of All Markets

In the course of writing this book, I was reminded of a quote that helped inspire the Green Growth Fund. It came from an investment banker in Tom Friedman's *Hot, Flat, and Crowded*, the prophetic 2008 manifesto that called for global action on global warming. "The green economy is poised to be the mother of all markets," said Lois Quam of Piper Jaffray. It's the "economic investment opportunity of a lifetime."

Today, at long last, Quam is being proven right. But keep in mind that cleantech is by no means insulated from the blunt forces that act on any new technology market. "In a real revolution," as Friedman notes, "there are winners and losers."

Nations that lead the way in clean technology will be rewarded with expanded manufacturing, job growth, and in the end, a higher standard of living. Unlike the internet, the clean energy transition will unfold on a local level. It will bring quiet new buses to our communities, solar panels to our roofs, and vast wind farms off our shores. We've debunked the idea of a tradeoff between expanding the world's economy and solving the climate crisis. It's now clear we can have both: profit and planet.

The Growing Need for Giving

Many worthy climate solutions won't come with a ten times return on investment. They're designed for something more than enriching a set of shareholders. All the same, protection of the planet—with a commitment to climate justice—calls for serious capital. For those who care and can afford it, we're asking for their money plus something more valuable still: their time and skill in strategic giving. Some notable individuals and corporations are already answering the call.

Climate action is wildly underfunded. To put things in perspective, philanthropic giving totaled $730 billion in 2019. The portion aimed at the climate crisis was less than 2 percent. Foundations make the bulk of their donations to health care and education. Why? According to Jennifer Kitt, president of the Climate Leadership Initiative, it may seem that climate solutions are less "people-oriented." In fact, the opposite is true. "Many donors have steered

Foundations are rising to the moment to fight climate change

Annual averages 2015–2019

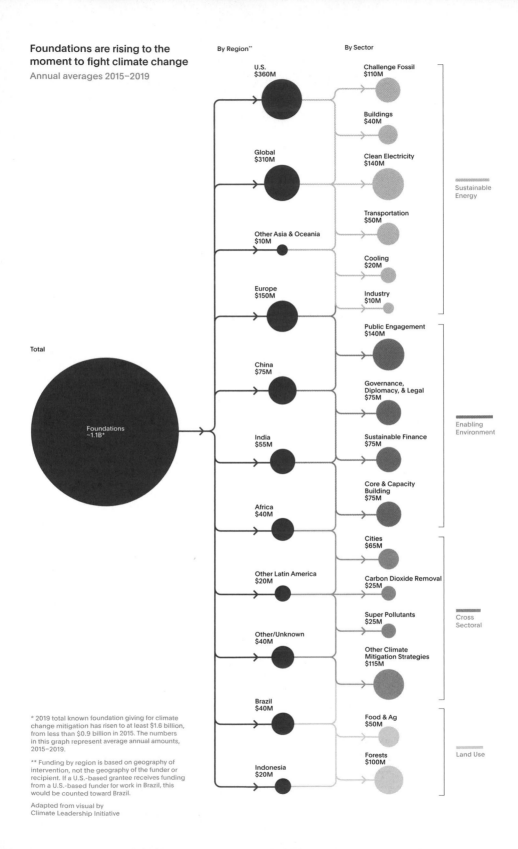

By Region**

- U.S. $360M
- Global $310M
- Other Asia & Oceania $10M
- Europe $150M
- China $75M
- India $55M
- Africa $40M
- Other Latin America $20M
- Other/Unknown $40M
- Brazil $40M
- Indonesia $20M

By Sector

- Challenge Fossil $110M
- Buildings $40M
- Clean Electricity $140M
- Transportation $50M
- Cooling $20M
- Industry $10M

Sustainable Energy

- Public Engagement $140M
- Governance, Diplomacy, & Legal $75M
- Sustainable Finance $75M
- Core & Capacity Building $75M

Enabling Environment

- Cities $65M
- Carbon Dioxide Removal $25M
- Super Pollutants $25M
- Other Climate Mitigation Strategies $115M

Cross Sectoral

- Food & Ag $50M
- Forests $100M

Land Use

Total

Foundations ~1.1B*

* 2019 total known foundation giving for climate change mitigation has risen to at least $1.6 billion, from less than $0.9 billion in 2015. The numbers in this graph represent average annual amounts, 2015–2019.

** Funding by region is based on geography of intervention, not the geography of the funder or recipient. If a U.S.-based grantee receives funding from a U.S.-based funder for work in Brazil, this would be counted toward Brazil.

Adapted from visual by
Climate Leadership Initiative

away from climate because they thought governments or markets would solve this," Jennifer says. But there's a new generation of donors, she adds, that is "awake and terrified, and ready to actually do something." Jennifer believes in the power of philanthropy as a flexible instrument for ambitious projects that can make a real difference. In two years, the initiative has raised over $1.2 billion in new dollars for climate philanthropy.

One front-runner in institutional climate giving is the IKEA Foundation, the philanthropic arm of the umbrella company that owns the Swedish-born retail chain of the same name. With $2 billion available for deployment, the foundation's focus, says chief executive Per Heggenes, is to accelerate the displacement of dirty energy in the Global South by renewable energy. Meanwhile, IKEA's retail business has pledged to go carbon-*negative* by 2030—to reduce more emissions than it emits, including its supply chain.

In February 2020, a new standard-bearer for individual climate philanthropy emerged when Jeff Bezos pledged $10 billion to establish the Bezos Earth Fund. Many expected him to emulate Bill Gates's Breakthrough Energy Ventures and become another early-stage VC firm. Amazon had already backed cleantech startups like Rivian, which is busy building the 100,000 electric delivery vans for the world's largest online retailer. It's an important element of the company's goal to reach net-zero emissions by 2040.

But Bezos had a different idea for the Earth Fund. In announcing the sixteen recipients of its first round of funding, he revealed an approach more akin to highly focused philanthropy. The list of grantees included the World Wildlife Fund, the Nature Conservatory, the Rocky Mountain Institute, the Union of Concerned Scientists, the Environmental Defense Fund, and the Hive Fund for Climate and Gender Justice. None of them plan to turn a profit, issue stock, or go public. Yet in their efforts to protect vital ecosystems and clear the atmosphere of gigatons of emissions, these nonprofits can be as disciplined and determined as any for-profit company.

There are tremendous benefits, both environmental and economic, to be derived from protecting oceans and waterways, preserving rainforests, and fostering regenerative agriculture. As Bezos said, "I've spent the past several months learning from a group of incredibly smart people who've made it their life's work to fight climate change and its impact on communities around the world. I'm inspired by what they're doing, and excited to help them scale." To run his new nonprofit enterprise, Jeff hired Andrew Steer, then chief executive of the World Resources Institute. In speaking with Jeff and Andrew, I could see that they're taking climate philanthropy to a whole new level.

Invest!

Jeff Bezos

This is the decisive decade. If we don't make the right kind of progress by 2030, it will be too late. I believe that it's doable, and that there is reason to be optimistic.

That said, there's not just one thing we have to do, but a whole bunch of things, and it's a daunting problem because of its scale. For a hundred years, we have treated putting carbon in the atmosphere as if it's free. In classic economic terms, that makes it an unpriced externality. It means we've built trillions of dollars of capital infrastructure that makes a false assumption every day. And we're continuing to build infrastructure today that makes that same false assumption. We need to stop doing that and take care of our planet.

The scale of the problem we're talking about requires collective action, and philanthropy can play a very important role to catalyze that. Philanthropists can take risks that governments and companies can't or find difficult to take. Philanthropy can get things started, prove out solutions. And then governments and markets can scale those things up.

The Bezos Earth Fund is purely philanthropic. It isn't going to be funding any for-profit activities. I believe in funding green startup companies with new ways of doing things that have zero carbon, and that's important—but that's not what the Earth Fund is for.

One thing I was surprised to learn is that philanthropy in climate is actually very small, and it was growing just a couple percent annually. That's way too little to fight climate change.

This is the decisive decade.

Andrew Steer

The starting point for the Bezos Earth Fund will be to diagnose the systems changes that will be needed this decade, and to identify where philanthropic funds can unlock and drive change. In each of the major transformations that must occur—energy, transport, industry, food and agriculture, financial systems, etc.—there are several "mini" transformations, although these are themselves quite large. In transport, for example, we need to say goodbye to the internal combustion engine, but we also need to develop hydrogen technologies for shipping and airlines, and we need to radically overhaul public transport and rethink city planning. In food systems we need new climate-smart agricultural technologies, but we also need to reform supply chains, shift diets toward plant-based food, and cut food loss and waste by half this decade.

We now know that these transformations are feasible and are economically, financially, and socially beneficial. But there are all kinds of barriers, knowledge gaps, risk aversion, and path dependencies that prevent them from happening at the pace that is required. That's where the Bezos Earth Fund can come in.

Invest!

Different transitions are different stages, with some well on their way and approaching tipping points, while others are just getting started. Our role will need to be tailored accordingly. In some cases we'll support basic research; in other cases we'll help create markets for new technologies; or de-risk investments. In some cases the need may be for policy change or for information systems and transparency, so we will support advocacy groups and monitoring systems, or convene coalitions of leaders who can work together to create momentum. In all cases we will take social concerns into account, recognizing that issues of environmental justice must be addressed with urgency.

If you look at the first round of Bezos Earth Fund's grants you'll find examples of all of these types of interventions. In everything we do we will seek to accelerate momentum for change—making it irresistible and unstoppable.

Jeff Bezos

We need to inject money in a very careful way, so that if you put the pieces of the jigsaw puzzle together, we're all working together systematically.

We look for where we can engage to get the biggest bang for the buck.

It's not a one-size-fits-all theory of change. One strategy is not going to work for all fifty or so subareas of the total problem. So it's daunting. It's very hard. But it should be hard, and if you don't start off expecting that, you're going to get disappointed and quit.

The tent for this cause is getting larger. The people joining us as allies in this endeavor know that these philanthropic dollars are going to help achieve goals of substantial size.

Let me put it this way: We live better lives than our grandparents did, and they lived better lives than their grandparents did. We cannot be the generation that breaks that cycle. That said, this is not about our future legacy; it's about getting this done today.

The Philanthropic Mission

The growing need for funding has led to the rise of "philanthropic capital," a hybrid category of investing that is being perfected by one of the world's most thoughtful and dynamic givers. In 1989, an entrepreneur named Laurene Powell was getting her master's degree at Stanford Business School when she met Steve Jobs, who came on campus to give a talk. Two years later, they married. In 2004, Laurene Powell Jobs founded the Emerson Collective, a $1.2 billion commitment to education and economic justice in communities like East Palo Alto, California, the low-income city that sits between congested highways and Palo Alto, the wealthy town best known as the home of Stanford University.

The Emerson Collective wasn't set up as a nonprofit. Instead, it uses after-tax money and invests in ventures that could produce a return— you might call it profit-optional. (Sometimes a profit-making business is the best way to get something done.) As more of the Collective's work shifted to climate justice causes, Laurene deepened her involvement. In 2009, she launched a philanthropic venture called Emerson Elemental. It's all part of her plan to invest much of her money into climate action over the next fifteen years.

When I spoke with Laurene, she explained that Elemental is about "taking risks, proving new concepts, and building out demonstration projects." When I asked about her personal story, and what led her to such a visionary effort, I learned that her entire life experience had prepared her to take on this vital work.

Laurene Powell Jobs

I'm from a rural town in northwestern New Jersey, by the ski mountains. Our family's house abutted a watershed area in the back, and across the front yard there was a small lake. My mother was a great believer in fresh air and grew up going to summer camps, and she ran her house like one. My brothers and I swam and sailed on the lake; in the winters, we skated and went skiing. My sense of the world was shaped by the turning of the seasons, by the cadence of creation and destruction, rebirth and renewal.

My father was thirty-five, a Marine pilot, when he died in a plane crash. I was three. My brothers and I understood that life could end early. It may be why I wanted to learn to read very young. I remember a teacher giving me a library card in first grade, when normally that happened in third grade. So my worldview was largely shaped through books, because we didn't travel. Instead, I became a philatelist, a stamp collector. I filled up folios with stamps from every country. It helped build my imagination for what I wanted to see and do in the world.

As I grew up and got closer to my dad's age at his death, it imbued a sense of urgency about my personal vision for my life. Knowing that our time is finite and unpredictable gave me a sense of purpose and passion that might have otherwise eluded me.

I certainly learned that again with the loss of my beloved husband. Steve was fifty-six. We lost him far too soon. Ten years later, it is still so deeply inspiring to see what he manifested in this world. Unlike my father, Steve had time to consider his legacy and his ripple effect, how a life can have lasting meaning. That has also resided in me. What can I do with my time on the planet that is purposeful to me and to others?

Steve used to say: Your work is going to fill a large part of your life, and the only way to be truly satisfied is to do what you believe is great work, and to love what you do. The two of us grew up together during our adult years. I learned an awful lot from him: how to execute exquisitely with a team; how to bring out the best things in individuals they sometimes don't see in themselves.

Thirty years ago, when I was getting my MBA at Stanford Business School, I found out that just a few miles away the city of East Palo Alto was a disposal center for Silicon Valley. A lot of semiconductor debris was dumped there, along with biomedical waste. The city was paid for this disposal, but it was not done properly.

This happens across low-income areas all over the world. There were all sorts of toxicity in the water table, with high levels of arsenic and radon. It gets transmitted into the food that's grown there, it's in the gardens, it's in the drinking water. Since we fund local education through property taxes, the schools in East Palo Alto were far inferior to the ones in West Palo Alto. They don't have a robust tax base. They couldn't afford good roads and sewage systems. They didn't have a grocery store. They didn't have a bank. They didn't have the kind of infrastructure that would yield a healthy community.

In 2004, I started the Emerson Collective on the belief that all the issues we work on, all the systems that touch our lives on the planet, are interlocking.

We started working in education in East Palo Alto. We supported students through college. But the students didn't want to come back there to work because there were no jobs.

Here's the big lesson we learned: You have to address everything at the same time. Environmental injustice presented itself as childhood asthma at five times the national rate. East Palo Alto is a drive-through community. It's bumper-to-bumper traffic for more than five hours a day. The exhaust stays in the community. They collect no revenue from cars going through, and they have deeply negative health issues. When a kid is suffering from asthma, two things happen: 1) they miss a lot of school, and 2) they have lifelong negative health effects.

Invest!

At Emerson, we work in the realm of ideas, design, and action. We know that system redesign often requires local policy redesign, and we've worked on that as well. But I wanted to bring our emerging model to other communities.

We got to know Dawn Lippert, who had founded the first accelerator for deploying climate technology, a nonprofit organization that predated what is now Elemental Excelerator. Dawn had previously led Hawaii's Clean Energy Initiative, where she saw the role that innovation needed to play in transitioning a community off fossil fuels. Elemental was funded by a mix of government and philanthropic capital. It was an experiment to find new ways of combining breakthrough climate innovation with true community voice and leadership for climate solutions.

Elemental has a philosophy that the best climate solutions will also be the most equitable. I was compelled by Dawn's work. Very quickly I said, "How do we supercharge this model?" That's what led to the Elemental Excelerator.

The island state of Hawaii presented a special opportunity for Laurene, Dawn Lippert, and the Elemental Excelerator team. Until around 2008, 90 percent of the state's electricity came from burning oil, the easiest fossil fuel for ships to carry. But the convenience came at the cost of expensive electricity for people who could least afford it, not to mention unhealthy air and significant greenhouse gas emissions.

It was oil's high cost that eased the transition to renewable energy in Hawaii. Solar panels and water heaters made economic sense there years before they would in other places. Hawaii was an ideal canvas to test innovative and equitable technologies for clean energy, water, food, and transportation—solutions that also addressed climate equity and justice.

Dawn Lippert

When I came to Hawaii, I was struck by the many energy-related challenges and got a firsthand look at how the climate crisis would exacerbate so many interrelated problems that communities were already struggling to address. For example, you can't talk about energy for five minutes without bumping into issues about water, or transportation, or education, or the workforce. That's the beauty of working on islands. Because of their size, we can see how the whole system connects.

Our work around the world accelerating climate solutions and creating social equity stemmed from starting to fund clean-technology companies in 2009 in Hawaii. We found a critical gap. We were getting technical results, but not seeing large-scale adoption. We were missing the commercial and community contexts.

Communities are where climate solutions are deployed, and Elemental helps design how technology and people intersect at the very practical, local level. You can have the best technology in the world, but unless people accept it, it won't get to scale. We've seen that while technology may have half the solution, the community has the other half.

One tool we use to help companies bridge the gap with communities—and broader commercialization—is funding demonstration projects. We've deployed more than seventy around the world. For example, the company SOURCE Global makes "hydropanels" that create drinking water from sunlight and air entirely independent of infrastructure. The hydropanels were typically installed on individual homes or schools, but the company wanted to explore a new business model to link up hundreds or even thousands of hydropanels to create a community-scale solution. We used project funding for their first Water Purchase Agreement with an Indigenous-owned community partner in Australia. Successfully showing that this new business model worked at community scale, with real-world data, helped SOURCE secure project financing and enabled deployment in more than fifty countries worldwide.

All told, over the last twelve years we've evaluated over five thousand startups from sixty-six countries and invested in a portfolio of more than one hundred. This robust community of companies now employs more than two thousand people and has leveraged Elemental's dollars eighty times, raising over $4 billion in follow-on funding.

In working with these startups, we've uncovered the art and science of commercializing climate technology and developed new methodologies to accelerate their progress. We love working with entrepreneurs because they are wired to make rapid change and challenge the status quo. And with the right tools and support they're also uniquely positioned to leverage technology to advance social equity—both within their companies and in the broader communities.

Elemental shows how climate solutions can address a community's needs, and how entrepreneurs and investors can weave climate justice into practical, scalable work when they choose to make it a priority. The organization funded a startup whose software can make public transit more equitable. It paid for another startup's energy efficiency retrofits in frontline communities.

In addition, Elemental funds youth internships in its portfolio companies and in other climate-related opportunities. Over the next five years, the venture will launch five hundred new climate careers, mainly for traditionally excluded groups and people of color.

All told, an estimated ten thousand people now work in clean energy in Hawaii. The state leads the nation in both residential energy efficiency and clean electricity, which will soon account for most of its power. In 2020, the state exceeded its 30 percent clean energy goal, and is well on its way toward 70 percent by 2030 and 100 percent by 2045.

As Laurene says, we need to work on the whole system at once. Our energy, food, water, and transportation systems are inextricably tied into our education, housing, criminal justice, and political systems. Entrepreneurs will have a central role to play as they build the big companies of the future.

Laurene Powell Jobs

So, what can philanthropic capital do? It's about taking risks, proving concepts, building out demonstrations. But it should not be taking the place of government funding. We need to prove a concept and then turn it into a business to scale it or hand it over to another business to scale it. Philanthropic capital should not be doing what at-scale capital can do. It's at-risk capital. We're okay if 30 percent of our portfolio fails, as long as we fail fast, learn fast, and move fast.

Another 30 percent can get spun out into companies that are structured in a way that is equitable and generous and can execute. And 30 percent are doing good things but might limp along financially, and that's okay too. We have that 30x30x30 appetite for what success looks like. If we tell ourselves everything needs to be successful, we'll miss out on a lot.

This is a huge opportunity. There are a lot of ideas and smart people out there that are not getting funded. But the question is: Can we make it to net zero? Can we avert a climate catastrophe?

I'm dedicating a significant portion of my resources to the climate crisis. And we're going to spend down those resources over the next fifteen years. We're not moving fast enough in the right direction. And the next ten to fifteen years really matter.

My biggest concern is that when we're looking at all the Speed & Scale OKRs, the degree to which everything needs to change is daunting. It requires all of us. It requires change across sectors and industries across the board. It's not like anything we've ever done before.

But look, we developed, tested, manufactured, and deployed a vaccine against a novel coronavirus—all in less than a year. Taking on the climate crisis is possible. It requires the same level of focus and urgency. Now we have to combat something we can't touch and feel and that we're not feeling the full effects of yet. And as humans we're hardwired to be reactive rather than proactive.

I'm very optimistic when I meet with entrepreneurs. Our ability to innovate, create, and deploy is what our species does best. We need to galvanize and celebrate that ingenuity around this issue.

The great thing is that working to solve the climate crisis will bring us back into harmony with the natural world, with the cadence of the seasons. In the end, it will be a healthy and beautiful thing.

The climate crisis should be looked at as one of the greatest opportunities that has ever been presented to humankind.

Conclusion

Conclusion

At the outset of this book, I promised you an action plan to wipe out 59 gigatons of greenhouse gas emissions and avert a climate catastrophe. I have done my best to identify the goals and levers that can do just that. But for the math to work at this titanic scale, we'll need to get more people in motion and more technologies deployed and more new ones invented than at any time in human history. We'll also need more money and so much more leadership and unity if we're going to save a habitable planet. We have such a long way to go.

I'll be honest: My daughter and Greta Thunberg aren't the only ones who are scared. There are mornings when I wake up terrified we won't make it. It can be alarming out there, in our carbon-heavy air. (There are times when panic is indeed the appropriate response.) If this book scares you into action, if it makes you as fearful as I am, then I have done my job. But for our fear to work for us, it needs to galvanize, not paralyze. To spur us on, it must be tethered to hope.

And so you might be wondering: What gives me hope that we can get to net zero in time? What stops me from waving the white flag? Why not bow to the inevitable and just hold our children close against the coming storm?

My answer begins with the creative genius of humankind—and our knack for collaboration. Our shared saga is of an endless frontier, from fire and the wheel to the internet and the smartphone. While the United States may be justly called the capital of innovation, it recruits genius and inspiration from every corner of the Earth. Now we need to emulate the scale of the transcontinental railroad and the speed of the race to a COVID-19 vaccine—only bigger and faster. It's an unparalleled endeavor, and America is not in it alone; we cannot solve a global problem in isolation.

Do you remember FDR's napkin from the beginning of this book? There was a time when the Axis powers had the Allies on their heels. Hitler's army conquered Denmark, the Netherlands, Belgium, Norway, and France. Imperial Japan rampaged through Southeast Asia. Britain was reeling under the Nazi Blitz. The threat to the free world was absolutely existential.

To turn the tide, it took a united global effort, on a scale never before seen, but one we need to see again. It demanded a wave of new technologies: two-way radio, radar, sonar, more powerful computers— and a speech encryption system, the first of its kind, that allowed FDR and British prime minister Winston Churchill to speak securely across an ocean. The United States and Great Britain and their allies stopped making cars and appliances and turned instead to a historic mobilization of wartime manufacturing: 14,000 ships, 86,000 tanks, 286,000 planes, 2.5 million trucks, 434 million tons of steel, and 41 billion rounds of ammunition.

Beating the climate crisis will require all of that focus and commitment, and more. Greenhouse gas emissions are less visible—and harder to target—than the Luftwaffe. Yet as in World War II, humanity's future is at stake. Like Roosevelt and Churchill, we have no time to waste. We cannot wait for the fossil fuel companies to reinvent themselves to join us. We cannot wait for breakthroughs yet unimagined. We need to move forward with the tools at hand. We must deploy the *now* with no less vigor than we seek the *new*.

Fundamental changes don't happen because they're virtuous. They happen because they make economic sense. We've got to make the right outcome the profitable outcome, and therefore the likely outcome.

To get anywhere close to universal adoption, clean energy must offer a competitive advantage. Entrepreneurs and venture capitalists can't get us there on their own. The most brilliant breakthroughs wither without support. To get the powerful winds of markets at our backs, we need bold national policies. And if we're going to reach net zero by 2050, we need something more: climate equity and fairness. If access to clean technologies is blocked by greed or selfishness or market failures or inept governments, we will fail.

For a cautionary tale, look no further than the COVID pandemic. Not so long ago, many of us were optimistic about the prospects for worldwide herd immunity. Today that future seems remote. It's been compromised by uneven leadership, the vagaries of human behavior, and—most of all—by extreme inequalities in vaccine availability and medical support systems.

On the climate front, wealthier countries—first and foremost the United States, the globe's top historical polluter—must do more. We need a climate Marshall Plan that dwarfs the latest international commitment by the Biden administration. Wealthy countries in

North America, Europe, and Asia must finance and subsidize the transition to green energy in nations not yet able to make it on their own. The shift away from fossil fuels will take hold when renewables become widely reliable and affordable, even in lower-income countries. Then their momentum will be irresistible. Cleantech will become the greatest business opportunity of the twenty-first century.

———————————————

Our net-zero plan runs on the rails of objectives and key results, or OKRs. Since you've stayed with us this far, you've made your way through ten high-level objectives and fifty-five key results—the most instrumental goals, in our view, for the crisis at hand. I'm confident that our OKRs would pass muster with their spiritual father, Andy Grove. Together they cover the *what* and the *how* of this last-ditch proposal to save a habitable Earth. When passionate people use time-tested methods to achieve audacious goals, the results can surpass all expectations.

OKRs foster several virtues: focus, alignment, commitment, ambition. But the most important may be what we call *tracking*, or continuous measurement. It ranks first among equals for this reason: If we fail to measure what matters, there's no sure way to get to where we need to go.

Greenhouse gases are a stubborn and elusive problem, to say the least. To get to net zero in time, we must measure precisely how much carbon the planet is emitting, where it's happening, and who is responsible—all in real time. This requires a kit of precision tools, from mathematical models to artificial intelligence to the latest satellites. We need data we can trust to hold nations and companies to account, and to focus our time and resources where they matter most.

Measurement is the through line of every chapter in this book. It imbues each of our objectives with meaning—it's the universal accelerant. By tracking our way to what seems like an impossible goal, we might even get there.

But while my inner engineer loves the precision of our metrics for CO_2 equivalents and methane concentrations, for degrees Celsius and gigatons of emissions, we also need to be humble about the limits of our knowledge. As some say Einstein sketched on his chalkboard, not everything that counts can be counted. There is no metric, much less a crystal ball, for human ingenuity and inspiration. From now until 2050 is an eon in science and technology. The uncertainty in any thirty-year projection is great indeed. Still, we must do our best to look ahead.

Conclusion

Working against us is the world's steadily growing population. Our 59-gigaton challenge will grow larger before it gets smaller. The "new normal" will soon get worse. Billions more people will require more land, buildings, materials, transportation, food, and energy, including the dirty kinds—unless we can offer them cheaper alternatives.

In our favor: the power of proven, scalable clean technologies, plus the potential of radical innovation. We do not know the ceiling for synthetic fuels, kelp forests, engineered carbon removal, green hydrogen, or nuclear fusion reactors. What seems like science fiction today may be standard practice the day after tomorrow. One or more of these solutions could save an Earth we'd still want to call home. There's a metric ton of hope right there.

Some might call this a leap of faith. I see it as our healthy biological response to mortal threats: fight or flight. But here flight is not an option; we cannot outrun global warming. We're going to have to fight this one out, with every weapon we can muster.

There is no shortage of shining lights in this fight. We have climate warriors who have been at it for thirty years and more. We have younger voices and entrepreneurs who are leading with a fresh perspective. We need to get them the help they need to do even more, to go even faster.

In April 2021, Germany's highest court paid due respect to the *now*. In response to a complaint by young environmental activists, the court ordered the federal government to enforce "more urgent and shorter-term measures" to reach emissions targets for 2030. The judges declared that young people's "fundamental rights to a human future" would be jeopardized if global warming went much past 1.5 degrees Celsius.

International climate conferences in Glasgow and beyond must be approached the same way. Voluntary pledges and promises won't suffice, not anymore. Countries must set goals and execute. They must hold themselves responsible for replacing fossil fuels with green alternatives, and for removing emissions we cannot avoid. There is a rising vision for a world powered by solar and wind and other emerging sources of clean energy. We are overdue to make it a reality.

These pages are a call to arms—and an invitation to you to join the ranks of our planet-saving army. Mostly we have focused on the obligations of the world's prime movers: governments, movements, nonprofits, businesses, investors. But every individual has a role to play. (More than switching to LED light bulbs, you may have to switch out your lawmakers.)

How do you become a leader on climate? First, you figure out *what* must be done, through learning, dialogue, and debate. Second, you get other people to *want* to do it. Third, you move others in your own way, with your own voice.

The blueprint in this book is one earnest attempt to come to grips with the task before us. It's a starting point, no more and no less; it will improve with our diligent attention and joint contributions. It needs you—and you can get involved at speedandscale.com. We look forward to the discussion and debate and criticism. While none of us have all the answers, together we may find a solution.

———————————

I bring climate consciousness to every facet of my life—as a father, an investor, an advocate, and a philanthropist. Then there's the book you are holding in your hands. It's been one of the most rewarding, enlightening, and exhausting efforts of my lifetime—a labor of love, to be sure, but a labor nonetheless. There were moments when I wondered if I'd taken on more than I could handle. (Some critics said as much.) But if ever there was a time to make a wholehearted commitment, come what may, this is it. Because no one can say, "Climate change is not my problem."

Because we're all in this together.

Like other baby boomers, I'm not likely to be around in 2050. My generation came of age in a postwar prosperity that was fueled by fossils from 300 million years past—and the illusion that greenhouse gases had no consequences. I grew up after World War II, when the good life was embodied by the backyard barbecue. Friends would gather from miles around in their gas-guzzling cars. Assembled on a concrete patio, around the shrine of a steel grill, we'd squirt a petroleum-based lighter fluid to set the charcoal ablaze and get the steaks sizzling. The carbon-rich smoke stung the eyes, but we all had a great time. There was next to no data on global warming. None of us gave a passing thought to the implications of our emissions binge.

Of all the obstacles to net zero, the nostalgic picture I'm painting might be the toughest of all. It's human nature to cling to the life we know. It's not an easy thing to relinquish. But again, we have no choice. We are past time for a new, carbon-free paradigm of the good life.

———————————

Conclusion

What do I say to Mary today? What is my message to all of our daughters and sons? First, I own up to the failings of my generation. I pledge to keep doing my part to address this grave emergency. And then I turn Mary's question—*What are you going to do about it?*—around. Because if we're going to save this planet, we'll need her generation, with all its impatience, to seize the reins.

Today's young adults have come of age in a world of climate crisis. Their birthright is a just and livable world through 2050 and far beyond. With help from courageous leaders and activists, farseeing investors, catalyzed corporations, enlightened philanthropists, and—most of all—brilliant innovators, they just might get us to net zero. If we pool our energy and talent and influence, the multiplier effect could move mountains—or, at the least, save our oceans and forests.

In the face of such a wickedly hard problem, against all odds and expectations, these impassioned young people give me hope—and inspiration—most of all.

While none
of us have all
the answers,
together we may
find a solution.

Acknowledgments

Acknowledgments

Winston Churchill wrote, "Writing a book is an adventure. To begin with it is a toy and an amusement. Then it becomes a mistress, then it becomes a master, then it becomes a tyrant. The last phase is that just as you are about to be reconciled to your servitude, you kill the monster and fling him to the public."

As I fling this monster of a book into your life, dear reader, I feel an overwhelming sense of gratitude. First, that I was so lucky to be heir to Andy Grove's system of OKRs to solve big problems and amplify human potential. Second, I'm thankful for my country and, indeed, for all the world's institutions that reward and honor risk-taking. Because we need risk-takers now more than ever.

I offer my undying gratitude to my wife, Ann, and daughters, Mary and Esther, whose patience, encouragement, and love kept me going through this long and challenging project.

I thank you in advance, my readers, for your feedback, engagement, and personal leadership on the climate crisis. I trust that your commitment and cunning will compel others to "want to do what must be done."

And I hope you'll write to me about that at john@speedandscale.com.

The Team: Ryan, Alix, Anjali, Evan, Jeffrey, Justin, and Quinn

The creation of *Speed & Scale* confirms my mantra that it takes a team to win. My partner, Ryan Panchadsaram, is the cocreator of this book. From original concept to OKR essentials, our Speed & Scale Plan—and the book and website it inspired—would not exist without Ryan's orchestration, drive, and superb judgment.

The book and I were both blessed by the good grace of a gifted squad of cowriters. Jeffrey Coplon and Anjali Grover are notably survivors of my first book, *Measure What Matters*. As an engineer, I sometimes struggle with writing and words. Jeffrey artfully smooths my rough edges without losing my voice, which is no small feat. Then there is Anjali, the master of logic, through lines, and intellectual rigor. The book's clarity and grit are a credit to Anjali Grover.

Evan Schwartz is an imaginative, seasoned environmental storyteller and documentary film writer. (And kelp enthusiast.) Alix Burns is a wise, passionate, and powerful partner on all matters of policy and politics. Justin Gillis is a diamond, a tough and brilliant former lead science and climate writer for *The New York Times*. Rooting out fuzzy thinking with surgical precision, he's a fanatic for facts and clarity—acronyms beware and begone!

The mighty and meticulous Quinn Marvin led our research and data team (including Heiker Medina and Julian Khanna), wrangling nearly one thousand datums into more than five hundred dutifully sourced endnotes.

Call Me Al, Call Me Hal

How can I best honor the seminal thinker and activist who inspires me? Well, in 2007 the Nobel Committee singled him out (with the IPCC) for "their efforts to build up and disseminate greater knowledge about man-made climate change, and . . . for the measures needed to counteract such change." Through fifteen years of our weekly calls, Al Gore has been optimistic, resolute, and selfless in his life well spent on this existential crisis. Al's team at Climate Reality Project (Lisa Berg, Brad Hall, Beth Prichard Geer, and Brandon Smith) is awesome. I recommend that you join me and fifty thousand other trained volunteers in Al's Climate Reality Leadership Corps. I'm proud to be his partner and friend.

Learn more at climatereality project.org

Hal Harvey is a humble, low-key climate crusader and Stanford-trained engineer. Policies shaped by Hal have locked in more than two hundred laws and standards that directly cut emissions. But his drive for more remains unabated. Hal is universally trusted, from Washington to Brussels to Beijing. He is a highly effective, intensely focused, data-driven climate advocate. Hal's team at Energy Innovation includes Bruce Nilles, Minshu Deng, Robbie Orvis, and Megan Mahajan, who all contributed mightily to our network, stories, and climate modeling.

The Founders: Jeff Bezos, Bill Gates, Laurene Powell Jobs

Jeff Bezos and his hardy crew of Amazonians are committed and welcome leaders in our multifront climate campaign. Amazon's instinct to "Go Big, Fast" can be seen in their global operations, logistics, supply chains, hundred thousand Rivian e-vehicles, and the AWS "net-zero cloud." Beyond their own operations, Amazon is rallying other enterprises to the Climate Pledge. Then there's the Bezos $10 billion Earth Fund. Thanks to Jeff, Kara Hurst, Andrew

Steer, Jay Carney, Drew Herdener, Allison Leader, Luis Davilla, and Fiona McRaith for shaping our story.

Bill Gates and I first met working in the magical world of microprocessors, Moore's Law, and software. That set the stage for an energetic collaboration on education, global poverty, philanthropy, and the climate crisis. Thanks, Bill, to you and your expert team, including Larry Cohen, Jonah Goldman, Rodi Guidero, Eric Toone, Carmichael Roberts, Eric Trusiewicz, and all the great folks at Gates Ventures and Breakthrough Energy.

Laurene Powell Jobs is the visionary founder of the Emerson Collective. We also profile Dawn Lippert at Elemental Accelerator, one of Emerson's climate initiatives. The Emerson team, including Ross Jensen, is outstanding. Laurene, thank you for your majestic close to our book: "The climate crisis should be . . . one of the greatest opportunities that has ever been presented to humankind."

Global Policymakers: Christiana Figueres, John Kerry

Words can't capture the intensity and tireless, urgent global leadership of Christiana Figueres and John Kerry. Christiana was a prime architect of the Paris Agreement, the first legally binding, unanimously adopted climate treaty. As Jeff Bezos says, "Christiana is fantastic, a force of nature."

John Kerry was U.S. secretary of state and the nation's lead delegate on the Paris Agreement. He is President Biden's brilliant choice as special envoy for climate. John's remit is to get the world to reduce carbon emissions by 50 percent by 2030 (and get to net zero by 2050). He is an eloquent, elegant street fighter who has made a powerful contribution to solving the climate crisis—and to our book.

Global CEOs: Mary Barra, Doug McMillon, Sundar Pichai, Henrik Poulsen

In researching *Speed & Scale*, I've been excited by the power, progress, and promise of global enterprises. General Motors, WalMart, Alphabet/Google, and Ørsted are exemplary world leaders in transportation, commerce, technology, and renewable energy.

Acknowledgments

Mary Barra, chief executive of General Motors, opens our stories with the bold commitment to end her company's production of combustion vehicles by 2035. Mary brings a compelling combination of innovation, execution, customer focus, and urgency to GM.

Doug McMillon is the chief executive of Walmart, and also chairs the powerful Business Roundtable. Doug candidly discusses why and how Walmart will achieve net zero (without offsets) by 2040 and become a "regenerative company" protecting 50 million acres of land and 1 million square miles of ocean. Walmart has engaged its massive supply chain in creating Sustainable Value Networks.

Sundar Pichai is the chief executive of Alphabet, the parent company of Google and one of the largest private buyers of renewable energy. Sundar credits cofounders Larry Page and Sergey Brin, along with Kate Brandt, Ruth Porat, Eric Schmidt, Susan Wojcicki, and Nick Zakrasek, for the company's bold program of investment, industry standards, procurement, AI, and advocacy. Thanks to Tom Oliveri, Beth Dowd, and the great Alphabet team.

Henrik Poulsen, former chief executive of Ørsted (and innovator from Lego) shares the drama of turning Denmark's state-owned fossil fuel company into a world-leading developer of offshore wind.

Thought Leaders: Jim Collins, Tom Friedman, Bill Joy

While I'm quick to say "ideas are easy . . . execution is everything," the truth is I worship at the altar of ideas. I am awed by the genius of these thought leaders.

First there is author, researcher, and ex-Stanford business school professor/rebel Jim Collins. For decades I've been frustrated by well-meaning but fuzzy definitions of leadership. In Jim's recently revised book, *BE 2.0*, he nails it with a quote from Dwight Eisenhower: "Leadership is the *art* of getting others *to want to do* what *must* be done." The Jedi master of Socratic questioning, Jim demanded rigor and clarity in the writing of this book. He asked (and helped answer) the right questions—not just *what* and *how* to proceed, but importantly *why*.

In his books *Hot, Flat, and Crowded*, and *Thank You for Being Late*, columnist Tom Friedman of *the New York Times* brilliantly synthesized the growing tectonic stress between markets (globalization), Moore's Law (the internet), and Mother Nature. (Hint: Mother Nature always wins.) Tom, thank you for connecting the dots, pointing the way, and being you.

Bill Joy is the Edison of the internet and a brilliant engineer. He's a true futurist, seeing things well before the rest of us. Bill has guided Kleiner Perkins' science-based framework of grand challenges to find and develop the clean technologies we need.

Entrepreneurs

To borrow from Margaret Mead, "Never underestimate the power of a small group of *entrepreneurs* to change the world. It's the only thing that ever has." This book and our world have benefited enormously from the stories, struggles, and successes of entrepreneurs. Among the essential contributors to the Speed & Scale Plan are Ethan Brown (Beyond Meat), Amol Deshpande (Farmers Business Network), Taylor Francis, Christian Anderson, and Avi Itskovich (Watershed), Lynn Jurich (Sunrun), Badri Kothandaraman (Enphase), Nan Ransohoff (Stripe), Peter Reinhardt (Charm Industrial), Jagdeep Singh (QuantumScape), KR Sridhar (Bloom Energy) and J. B. Straubel (Redwood Materials). Thanks to you, your teams, and innovators around the world.

Investors

David Blood is chief executive of Generation Investment, a family of sustainability funds, cofounded with Al Gore. A big thank you to David, and to Larry Fink, chief executive of BlackRock, the world's largest investment manager, and widely regarded as the dean of capital markets.

Ira Ehrenpreis, Vinod Khosla, Matt Rogers, and Jan Van Dokkum are friends and standout venture investors. Ira brilliantly backed Elon Musk with Tesla and SpaceX—and is excited about the next generation of Elon Musks. Vinod is a bold and fearless investor, always pushing for more aggressive venture "shots on goal." Matt built the software team for ten generations of the iPod, then five generations of the iPhone. He cofounded Nest, and now leads Incite Ventures, an early-stage investor. Jan Van Dokkum is a gifted operating executive and investor at Imperative Ventures.

And then there's Jonathan Silver, who surely issued more climate loans and guarantees than anyone else as he headed Obama's Loan Programs Office for the Department of Energy.

To those, and so many others, thanks for your courage and insights. And for the investments. We need more!

Acknowledgments

Scientists and Activists

I'm acknowledging the contributions of climate scientists and climate activists together; they combine the best of both worlds. Chris Anderson and the amazing Lindsay Levin have built the TED Countdown platform for a whole new generation of climate voices. Safeena Husain leads Educate Girls, possibly the highest-impact climate program.

World Resources Institute (WRI) is a global nonprofit research organization with deep data and experience in climate systems and an unmatched commitment to precision. Thanks to interim president Manish Bapna, with a special shout-out to Kelly Levin and the superlative team at WRI for their wisdom, clarity, and unfailing collaboration.

Brian Von Herzen is chief executive of the Climate Foundation at Woods Hole Institute and an expert in kelp permaculture. Robert Jackson is a Stanford University authority on carbon.

Fred Krupp has been president of EDF, the respected Environmental Defense Fund, since 1984. Fred and his team—including Steve Hamburg, Amanda Leland, Nat Keohane, and Margot Brown—made keen contributions regarding the methane emergency, satellite surveillance, and, not least, climate justice.

Amory Lovins is chairman/chief scientist of the Rocky Mountain Institute and an advocate for greater efficiency in energy systems. Tensie Whelan built the Rainforest Alliance and now runs NYU's Stern Center for Sustainable Business.

Patrick Graichen is executive director of Germany's preeminent energy think tank, Agora Energiewende. Anumita Roy Chowdhury is executive director of Center for Science and the Environment in India. Bob Epstein (E2) and Bill Weihl (Climate Voice) are effective activists from the tech community.

James Wakibia is a Kenyan photographer and environmental activist. Nigel Topping and Alex Joss are brilliant UN-affiliated climate champions for COP26, ably joined in their work by Kelly Levin from WRI.

Advocates, Philanthropists, and Partners

I am inspired by the work of an amazing group of philanthropic world changers. In addition to those already acknowledged, they include

John Arnold, Josh and Anita Bekenstein, Mike Bloomberg, Richard Branson, Sergey Brin, Matt Cohler, Mark Heising and Liz Simons, Chris Hohn, Larry Kramer, Nat Simons and Laura Baxtor-Simons, Tom Steyer, and Sam Walton.

Jennifer Kitt is the dynamic president of Climate Leadership Initiative, nurturing new climate philanthropists.

At Kleiner Perkins, my partners' commitment to our climate and entrepreneurs lifts me up every day. I sincerely thank you for our journey together: Sue Biglieri, Brook Byers, Annie Case, Josh Coyne, Monica Desai Weiss, Eric Feng, Ilya Fushman, Bing Gordon, Mamoon Hamid, Wen Hsieh, and Haomiao Huang. Also, Noah Knauf, Randy Komisar, Ray Lane, Mary Meeker, Bucky Moore, Mood Rowghani, Ted Schlein, and David Wells.

Ben Kortlang, Brook Porter, David Mount, Dan Oros, Ryan Popple, and Zach Barasz are outstanding Kleiner Perkins cleantech alumni who formed G2VP and raised two funds to focus on sustainable investing.

The Manuscript

For those friends and partners who reviewed the manuscript, my deep gratitude. May you soon have a weekend respite from OKRs! Rae Nell Rhodes, Allie Cefalo, Cindy Chang, Sophia Cheng, Jini Kim, Glafira Marcon, Lisa Shufro, Igor Kofman, Debbie Lai, Leslie Schrock, Sanjey Sivanesan, and John Strackhouse, thank you.

From inception to finished product, I thank the Portfolio/Penguin team that made this book possible: my publisher, Adrian Zackheim, who foresaw its potential, and my superlative editor, Trish Daly, who went so many extra miles and somehow kept her good humor. And Jessica Regione, Megan Gerrity, Katie Hurley, Jane Cavolina, Megan McCormack, Jen Heuer, Tom Dussel, Tara Gilbride, and Amanda Lang. I am grateful as well to my agent, Myrsini Stephanides, and my attorney, Peter Moldave.

Order has turned this tome of numbers, facts, and figures into a work of art. I am grateful to Jesse Reed, Megan Nardini, and Emily Klaebe for their unflappable collaboration and amazing work. Rodrigo Corral Design has hand-drawn beautiful portraits of our collaborators.

As these acknowledgments make clear, I have tapped into a broad spectrum of experts and leaders. Each of them cast new light on the solutions the Earth keenly needs. Their insights touch every page; the errors, though, are all my own.

How the Plan Adds Up

Resource 1 # How the Plan Adds Up

Our Emissions Baseline

Speed & Scale uses greenhouse gas emissions numbers from the United Nations. Specifically, the UNEP Emissions Gap Report in 2020, which details 2019 emissions by sector. According to the report:

Global GHG emissions continued to grow for the third consecutive year in 2019, reaching a record high of 52.4 $GtCO_2e$ (range: ±5.2) without land-use change (LUC) emissions and 59.1 $GtCO_2e$ (range: ±5.9) when including LUC.

We use 59 gigatons of carbon dioxide equivalent as our present-day emissions.

Sector by Sector

The UN report defines the overall emissions trend from 1990 to 2019 and each sector's percentage contribution.

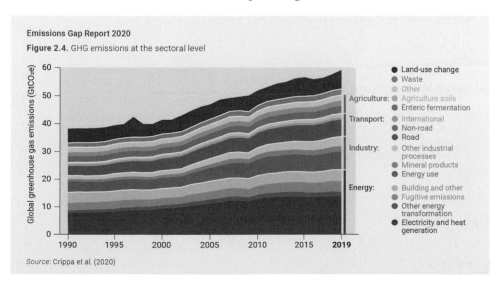

Emissions Gap Report 2020

Figure 2.4. GHG emissions at the sectoral level

Source: Crippa *et al.* (2020)

We have combined and aggregated this data into five main sectors: Transportation, Energy, Agriculture, Nature, and Industry. Since the data is approximate, we round to the nearest whole number.

GHG Emissions by Sector in 2019

Sector	% of $GtCO_2e$	$GtCO_2e$
Transportation	14%	8
Energy	41%	24
Agriculture	15%	9
Nature	10%	6
Industry	20%	12
Total	100%	59

These five sectors represent the first five chapters of *Speed & Scale* and illustrate the extent of human-caused emissions.

Projections for 2050

While *Speed & Scale* focuses on present-day emissions, we keep in mind what emissions would look like in 2050 if business as usual prevails, the world's population grows as projected, and industrialization continues at its current pace.

Projections of GHG Emissions by Sector in 2050

Sector	% of $GtCO_2e$	$GtCO_2e$
Transportation	17%	12
Energy	38%	28
Agriculture	14%	10
Nature	11%	8
Industry	20%	14
Total	100%	72

Our emissions estimate for 2050 is 72 gigatons. This number is aggregated from various climate reports (BNEF, IPCC, IEA, EIA, EPA, WRI, and CAT) that offer a wide range of "business as usual" scenarios with varying adoption and policy assumptions.

The key results of the Speed & Scale Plan are aggressive and intend to draw down and net out global emissions by 2050. But given the variability and uncertainty in 2050 emissions estimates, we chose to define our plan's impact on 2019 emissions (59 gigatons), a figure well understood and accepted.

How the Speed & Scale Plan Adds Up

Our plan has six objectives. The first five correspond to sectors (Transportation, Energy, Agriculture, Nature, and Industry) and what needs to be done to cut emissions. Though the prescribed actions are demanding, they are insufficient to reach zero. To do so, we add a sixth objective to address our remaining emissions.

Transportation → Electrify Transportation

Based on the UN report, the Transportation sector accounts for ~14 percent of total GHG emissions, or 8 gigatons. To provide further detail, we rely on IEA's 2018 Transport report. For aviation, we rely on Our World in Data's report on transport emissions, which uses data sourced from IEA and ICCT.

To reduce emissions from this sector we must *electrify transportation*. There are six key results for this objective. The first three are early indicators of progress in this sector: KR 1.1 (Price), KR 1.2 (Cars), and KR 1.3 (Buses and Trucks).

Resources

Transportation

Sector Breakdown	Current GtCO$_2$e	Reduction GtCO$_2$e
Road Transport	**6.0**	**1.0**
Passenger	3.6	0.2
→ Passenger Vehicles	3.2	0.2
→ Light Commerical Vehicles	0.1	0.0
→ Buses & Minibuses	0.2	0.0
→ Two/Three Wheelers	0.1	0.0
Freight (Heavy & Medium Trucks)	2.4	0.8
Aviation	**0.9**	**0.6**
Passenger	0.7	0.5
→ International	0.4	0.3
→ Domestic	0.2	0.1
Freight	0.2	0.1
Maritime	**0.9**	**0.3**
Rail	**0.1**	**0.0**
Other (Pipe, etc.)	**0.4**	**0.1**
Total Transportation Sector	**8.3**	**2.0**

The gigaton reductions come from progress on KR 1.4 (Miles), KR 1.5 (Planes), and KR 1.6 (Maritime):

KR 1.4—Miles: 50% of the miles driven (2-wheelers, 3-wheelers, cars, buses, and trucks) on the world's roads are electric by 2040, and 95% by 2050 → Will lead to a 5.0 gigaton reduction.

KR 1.5—Planes: 20% of miles flown use low-carbon fuel by 2025; 40% of miles flown are carbon-neutral by 2040 → Will lead to a 0.3 gigaton reduction.

KR 1.6—Maritime: Shift all new construction to "zero ready" ships by 2030 → Will lead to a 0.6 gigaton reduction.

For reductions we assume that almost all passenger road transport is electrified by 2050, a 95 percent reduction. This will require an acceleration in the turnover of vehicles. We are more conservative for harder-to-abate areas such as freight, aviation, and maritime. Both freight and maritime will eventually decarbonize but at a much slower pace than road transport—and likely beyond our 2050 goal, due to a lack of scalable options. We therefore assume a 65 percent reduction. Aviation will be the hardest subsector to decarbonize. While we hope for innovative solutions to carbon-neutral fuels, our plan assumes only a 20 percent reduction.

Together, these key results lead to a reduction of almost 6 gigatons.

Energy → Decarbonize the Grid

Based on the UN report, the Energy sector accounts for ~41 percent of total GHG emissions or 24 gigatons. We rely on IEA's 2018 CO_2 Emissions from Fuel Combustion report for the detailed breakdown.

Energy

Sector Breakdown	Current GtCO₂e	Reduction GtCO₂e
Power (Electricity & Heat Producers)	**14.0**	**1.9**
Coal	10.1	0.5
Oil	0.6	0.2
Natural Gas	3.1	1.1
Other	0.2	0.1
Other Energy Industries	**1.6**	**0.4**
Buildings (Residential + Commercial & Public Service)	**2.9**	**0.9**
Coal	0.4	0.0
Oil	0.8	0.3
Natural Gas	1.6	0.6
Other	0.1	0.0
Other & Fugitive Emissions	**5.9**	**0.3**
Total Energy Sector	**24.4**	**3.5**

Resources

To reduce emissions from this sector we must *decarbonize the grid*.
There are six key results for this objective. Four of them are specific
metrics that track our progress in reducing emissions from our grid
and shifting away from fossil fuels for heating and cooking: KR 2.2 for
Solar and Wind, KR 2.3 for Storage, KR 2.4 for Coal and Gas, and KR
2.7 for Clean Economy.

The gigaton reductions come from KR 2.1 on Zero Emissions, KR
2.5 on Methane Emissions, and KR 2.6 on Heating and Cooking:

KR 2.1—Zero Emissions: 50% of electricity worldwide comes
from zero-emissions sources by 2025 and 90% by 2035 (up
from 38% in 2020) → Will lead to a 16.5 gigaton reduction.

KR 2.5—Methane Emissions: Eliminate leaks, vents, and
flares from coal, oil, and gas sites by 2025 → Will lead to a
3 gigaton reduction.

KR 2.6—Heating and Cooking: Cut the use of gas and
oil for heating and cooking in half by 2040 → Will lead to
a 1.5 gigaton reduction.

Our model assumes that almost all coal usage will be shut down by
2050, a 95 percent reduction in emissions. Natural gas will be much
harder to replace with renewable energy due to its abundance,
reliability, and low cost; therefore, we assume a best-case scenario by
2050 is a 65 percent reduction in emissions. Last, we looked at oil.
Though most oil usage is captured within the Transport sector,
there remains slightly more than 1 gigaton left within the power
and buildings subsectors, where we assume a 70 percent reduction
in emissions from KR 2.1 and KR 2.6. We've accounted for this
additional gigaton in KR 2.1. For methane leaks, with the technologies
we have, we assume an 80 percent reduction.

Together these key results lead to a reduction of 21 gigatons.

Agriculture → Fix Food

Based on the UN report, the Agriculture sector accounts for ~15 percent of total GHG emissions or 9 gigatons. To provide a breakdown of this sector we rely on WRI's Sustainability Report.

Agriculture

Sector Breakdown	Current GtCO$_2$e	Reduction GtCO$_2$e
Agricultural Production	6.9	2.8
Ruminant Enteric Fermentation	2.3	1.4
Energy (On-Farm)	1.5	0.1
Rice (Methane)	1.1	0.5
Soil Fertilization	0.9	0.4
Manure Management	0.6	0.2
Ruminant Waste on Pastures	0.5	0.2
Farm Soils	0.0	-2.0
Energy (Ag Energy Sources)	0.4	0.0
Waste	1.6	0.9
Total Agriculture Sector	8.9	1.7

To reduce emissions from this sector, we must *fix food*. This includes changes to the agriculture system and shifts in consumption. Five key results drive this objective and all directly tie to gigaton reductions:

KR 3.1—Farm Soils: Improve soil health through practices that increase carbon content in top soils to a minimum of 3% → Will lead to absorption of 2 gigatons.

KR 3.2—Fertilizers: Stop the overuse of nitrogen-based fertilizers to cut N20 emissions in half by 2050 → Will lead to a 0.5 gigaton reduction.

KR 3.3—Consumption: Promote lower-emissions proteins, cutting annual consumption of beef and dairy 25% by 2030 and 50% by 2050 → Will lead to a 3 gigaton reduction.

KR 3.4—Rice: Reduce methane and nitrous oxide from rice farming by half by 2050 → Will lead to a 0.5 gigaton reduction.

KR 3.5—Food Waste: Lower the food waste ratio from 33% of all food produced to 10% → Will lead to a 1 gigaton reduction.

Our model assumes that people will not stop eating meat and dairy, but we do encourage a shift to lower-emissions proteins, leading to a 60 percent reduction in emissions. It is critical that agricultural practices are improved. By increasing the carbon content in soils, we account for an additional 2 gigatons of carbon absorption. Some research on the potentials of soil carbon absorption is quite optimistic. We are more conservative. We can cut emissions from fertilizers by 50 percent through precision application and switching to "green fertilizers." One third of all food is wasted. Curtailing that waste will require changes at farms, in storage, in transit, and when food is prepared at home or in a restaurant. Our plan cuts waste by more than half to 10 percent by 2050, leading to a 1 gigaton reduction. Other reports, including one from FAO, have a higher gigaton reduction potential for eliminating food waste. But their calculations include emissions reductions from the energy grid transition and land-use change, which we include in our other objectives.

Together these key results lead to a reduction of 7 gigatons, 5 from emissions avoidance and 2 from absorption.

Nature → Protect Nature

Based on the UN report, land-use changes account for ~10 percent of total GHG emissions or 6 gigatons. This sector "covers emissions and removals of greenhouse gases resulting from direct human-induced land use, land-use change and forestry activities."

Nature

Sector Breakdown	Current GtCO$_2$e	Reduction GtCO$_2$e
Land-use Change, "LUC"	5.9	-5.9
Total Nature Sector	5.9	-5.9

To reduce emissions from this sector we must *protect nature*. There are two key results that drive the gigaton reductions and absorption:

KR 4.1—Forests: Eliminate all deforestation by 2030 → Will eliminate 6 gigatons of emissions.

KR 4.2—Oceans: Eliminate deep-sea bottom trawling and protect at least 30% of oceans by 2030 and 50% by 2050 → Will reduce 1 gigaton of emissions.

By protecting more lands (50 percent) and our oceans (50 percent) and ending deforestation, we aim to eliminate the emissions from land-use change and return nature into a carbon sink.

Together these key results lead to a reduction of 7 gigatons.

Industry → Clean Up Industry

Based on the UN report, the Industry sector makes up ~20 percent of total GHG emissions or 12 gigatons. To break this section down further, we rely on the UNEP Emissions Gap Report in 2019.

Industry

Sector Breakdown	Current GtCO$_2$e	Reduction GtCO$_2$e
Iron & Steel	3.8	0.9
Cement	3.0	1.2
Other Materials	5.0	1.7
Chemicals (Plastics & Rubber)	1.4	0.4
Other Minerals	1.1	0.4
Wood Products	0.9	0.3
Aluminium	0.7	0.3
Other Metals	0.5	0.2
Glass	0.4	0.1
Total Industry Sector	**11.8**	**3.8**

Resources

To reduce emissions from this sector, we must *clean up industry*.
There are three key results for this objective that drive the gigaton
reductions from this sector's emissions:

KR 5.1—Steel: Reduce total carbon intensity of steel
production 50% by 2030 and 90% by 2040 → Will reduce
3 gigatons.

KR 5.2—Cement: Reduce total carbon intensity of
cement production 25% by 2030 and 90% by 2040 →
Will reduce 2 gigaton.

KR 5.3—Other Industries: Reduce emissions from
other industrial sources (i.e., plastics, chemicals, paper,
aluminum, glass, apparel) 80% by 2050 → Will reduce
3 gigatons.

These emissions will be difficult to decarbonize; many new innovations are needed for deployment at scale. If successful, we assume the
impact of achieving these key results will reduce industrial emissions
by 2050 by two thirds.

Together these key results lead to a reduction of 8 gigatons.

Remaining Emissions → Remove Carbon

The Speed & Scale Plan is hopeful but also realistic. While driving to
eliminate all of our emissions we must acknowledge there will be
hard-to-abate sectors. Developing countries will be relying on fossil
fuels in the short term to assist in their growth.

Remove Carbon

Sector Breakdown	Current GtCO$_2$e	Reduction GtCO$_2$e
Nature-Based Removal	0.0	-5.0
Engineered Removal	0.0	-5.0
Total Remove Carbon	0.0	-10.0

To close these gaps we must *remove carbon*. There are two key results that drive our gigaton reductions:

KR 6.1—Nature-Based Removal: Remove 1 gigaton per year by 2025, 3 gigatons by 2030, and 5 gigatons by 2040 → Will lead to 5 gigatons of removal.

KR 6.2 – Engineered Removal: Remove at least 1 gigaton per year by 2030, 3 by 2040, and 5 by 2050 → Will lead to 5 gigatons of removal.

Our model assumes that the world will prioritize eliminating emissions and creating more efficient uses of energy. But that will not be enough. To reach our 10 gigatons, our planet needs a portfolio of carbon removal solutions—some nature-based, some engineered, some a hybrid of the two.

Summary of Total Reductions

Speed & Scale Plan—Reductions

Objective	Reduction (GtCO$_2$e)
Electrify Transportation	6
Decarbonize the Grid	21
Fix Food	7
Protect Nature	7
Clean Up Industry	8
Remove Carbon	10
Total	59

Policy Needed in the United States

Policy Needed in the United States

Speed & Scale includes objectives and key results (OKRs) for critical policies needed for countries and their governments to accelerate and achieve net zero by 2050.

KR 7.1—Commitments: Each country enacts a national commitment to net-zero emissions by 2050 and gets at least halfway there by 2030.

KR 7.1.1—Power: Set an electricity sector requirement to cut emissions 50% by 2025, 80% by 2030, 90% by 2035, and 100% by 2040.

KR 7.1.2—Transportation: Decarbonize all new cars, buses, and light trucks by 2035; freight ships by 2030; semi trucks by 2045; and make 40% of flights carbon neutral by 2040.

KR 7.1.3—Buildings: Enforce zero-emissions buildings standards for new residential by 2025 and commercial by 2030 and prohibit sales of nonelectric equipment by 2030.

KR 7.1.4—Industry: Phase out fossil fuel use for industrial processes by 2050, and at least halfway by 2040.

KR 7.1.5—Carbon Labeling: Require emissions footprint labels on all goods.

KR 7.1.6—Leaks: Control flaring, prohibit venting, and mandate prompt capping of methane leaks.

KR 7.2—Subsidies: End direct and indirect subsidies to fossil fuel companies and for harmful agricultural practices.

KR 7.3—Price on Carbon: Set national prices on greenhouse gases at a minimum of $55 per ton, rising 5% annually.

KR 7.4—Global Bans: Prohibit HFCs as refrigerants and ban single-use plastics for all nonmedical purposes.

KR 7.5—Government R&D: Double (at minimum) public investment into research and development, times 5 in the United States.

Without policies that meet these marks, it is likely a country will overshoot a net-zero target.

In developing these OKRs, we employed many tools. Prominent among them is the Energy Policy Simulator developed by Energy Innovation.

The simulator models both what each policy can do and how it interacts with others. The model shows how each policy can reduce carbon compared to business as usual.

The entire model is open source, can be run on a web browser, produces instant results, and is thoroughly documented. It has been peer reviewed by researchers at three national labs and a half dozen universities.

Here's what the model surfaces:

There are technology tailwinds: The costs of many zero-carbon technologies, including highly efficient electric appliances and equipment, have fallen dramatically in the last ten years and will continue to fall further—making this transition possible and affordable.

The transition to clean energy will not happen on its own: It is clear that this transition will not happen on the timeline outlined by the Intergovernmental Panel on Climate Change without additional policy. Only well-designed policies can drive this technological transformation at the required pace.

It takes a collection of policies: Each sector needs its own policy package. There is no silver bullet in carbon abatement; we need to transform all parts of our economy.

The Energy Policy Simulator has been used in eight countries and helps policy leaders identify the most impactful climate and energy policies and how they can contribute to their climate targets.

Try the simulator yourself at https://energypolicy.solutions

Inputting *Speed & Scale* OKRs into Energy Policy Simulator

To put our OKRs to the test, we simulated the effect of implementing policies in the United States that match each of our key results. Here's what we found:

The solid black line represents business as usual in the U.S., where we emit around 6 gigatons each year. Each wedge represents the impact of

Emissions in the United States

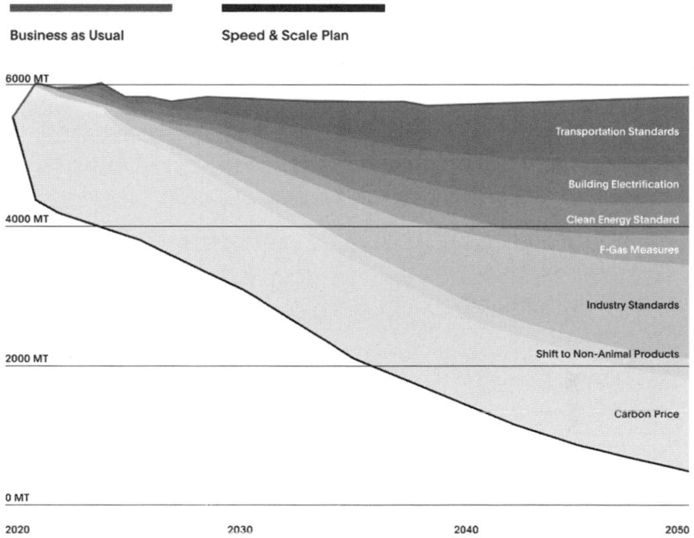
Business as Usual
Speed & Scale Plan
6000 MT
4000 MT
2000 MT
0 MT
2020
2030
2040
2050
Transportation Standards
Building Electrification
Clean Energy Standard
F-Gas Measures
Industry Standards
Shift to Non-Animal Products
Carbon Price

a set of policies on our emissions. These ambitious but necessary policies reduce U.S. emissions to 0.5 gigatons. Our expectation is that the remaining emissions are offset through nature-based or engineered carbon removal.

Key Emission Reduction Policies

Power

Clean Electricity Standard

Implement a Clean Electricity Standard cutting emissions at least 50% by 2025, 80% by 2030, 90% by 2035, and 100% by 2040.

Support it with policies building out the transmission system and deploying flexible resources such as grid battery storage and demand response.

Modeling Assumption: This scenario sets a standard of 100% zero-carbon electricity by 2040, which includes renewables, nuclear, and a small amount of natural gas equipped with CCS. It also includes a doubling of the transmission system above BAU, 510 GW of storage, and 450 GW of demand response by 2050.

Resources

Transportation

Vehicle Standards	Implement zero-emissions vehicle standards requiring 100% ZEV sales by 2035 for light-duty vehicles, buses, and light trucks; by 2045 for heavy-duty trucks; and by 2030 for ships.
	Support it with policies that accelerate fleet turnover, such as a subsidy for purchasers or a scrappage incentive.
	Modeling Assumption: This scenario models 100% ZEV sales for light-duty vehicles and buses by 2035, for heavy-duty trucks by 2045, and for ships by 2030.
Promote Sustainable Aviation	Set standards requiring a minimum use of carbon-neutral fuel in aviation.
	Modeling Assumption: This scenario includes 40% carbon-neutral fuel use in aviation by 2040.

Buildings

| Building Component Standards | Establish building codes and appliance standards that require all new building equipment to be electric by 2030. Complement these standards with incentives for building retrofits. |
| | Modeling Assumption: This scenario models standards that require all newly sold building components to be electric by 2030. |

Industry

| Industrial Fuel Shifting | Issue standards and incentives to shift industrial fuel consumption to 100% zero carbon sources by 2050, including production of zero carbon fuels, such as hydrogen. |
| | Modeling Assumption: This scenario shifts 100% of industrial fossil fuel use to a mix of electricity and hydrogen by 2050, depending on the potential for electrification in each industry category tracked in the model. All hydrogen is produced by electrolysis, the process of splitting water into hydrogen and oxygen using electricity. |

Stop Methane Leaks	**Establish standards requiring elimination of methane leaks and venting, and controlling flaring emissions.** Modeling Assumption: This scenario implements the full methane mitigation potential identified by the International Energy Agency by 2030.
Prohibit HFCs	**Adopt and enforce restrictions on the consumption and production of HFCs in line with the requirements under the Kigali Amendment to the Montreal Protocol.** Modeling Assumption: This scenario models compliance with the Kigali Amendment to the Montreal Protocol.

Agriculture

Reduce Demand for Meat and Dairy Products	**Require the food industry to print carbon footprints of their products on packaging, equipping consumers with the information needed to help them pick lower emissions foods.** Modeling Assumption: Nutrition labeling has shifted people's diets. Not enough research or studies have been done to show the effect of carbon labeling. Our hope is that it will lead to a meaningful change in consumer behavior. This scenario models a 50% reduction in consumption of animal products.

Cross Sector

Carbon Price	**Implement an economy-wide carbon price starting at $55 per ton and rising 5% annually.** Modeling Assumption: This scenario models a carbon price starting at $55 per ton in 2021, which rises 5% annually. This price applies to all greenhouse gases.

For Further Reading

Resource 3

For Further Reading

Understanding the Climate Crisis

Thank You for Being Late, Thomas Friedman
(New York: Farrar, Straus and Giroux, 2016)

Hot, Flat, and Crowded, Thomas Friedman
(New York: Farrar, Straus and Giroux, 2008)

An Inconvenient Sequel, Al Gore
(Emmaus, PA: Rodale, 2017)

An Inconvenient Truth, Al Gore
(Emmaus, PA: Rodale, 2006)

Under a White Sky, Elizabeth Kolbert
(New York: Crown, 2021)

The Physics of Climate Change, Lawrence Krauss
(New York: Post Hill Press, 2021)

How to Prepare for Climate Change, David Pogue
(New York: Simon & Schuster, 2021)

Climate Change, Joseph Romm
(New York: Oxford University Press, 2018)

Energy & Civilization, Vaclav Smil
(Cambridge, MA: MIT Press, 2018)

Plans for Net-Zero Emissions

How to Avoid a Climate Disaster, Bill Gates
(New York: Knopf, 2021)

The 100% Solution, Salomon Goldstein-Rose
(Melville House, 2020)

The Big Fix, Hal Harvey and Justin Gillis
(New York: Simon & Schuster, forthcoming, 2022)

Drawdown, Paul Hawken et al.
(New York: Penguin Books, 2017)

Nature-Based Solutions

Sunlight and Seaweed, Tim Flannery
(Washington, DC: Swann House, 2017)

Growing a Revolution, David Montgomery
(New York: W. W. Norton, 2017)

The Nature of Nature, Enric Sala
(Washington, DC: National Geographic, 2020)

Lo-TEK: Design by Radical Indigenism, Julia Watson
(Cologne, Germany: Taschen, 2020)

Half-Earth, Edward O. Wilson
(New York: Liveright, 2016)

Policy and Movements

Designing Climate Solutions, Hal Harvey
(Washington, DC: Island Press, 2018)

All We Can Save, Ayana Elizabeth Johnson et al.
(London: One World, 2020)

The New Climate War, Michael Mann
(New York: PublicAffairs, 2021)

Falter, Bill McKibben
(New York: Henry Holt, 2019)

Winning the Green New Deal, Varshini Prakash et al.
(New York: Simon & Schuster, 2020)

Short Circuiting Policy, Leah Stokes
(New York: Oxford University Press, 2020)

Leadership

Invent & Wander, Jeff Bezos with Walter Isaacson
(Cambridge, MA: Harvard Business Review Press, 2021)

BE 2.0, Jim Collins et al.
(New York: Penguin/Portfolio, 2020)

Good to Great, Jim Collins et al.
(New York: Penguin/Portfolio, 2001)

Freedom's Forge, Arthur Herman
(New York: Random House, 2012)

No One Is Too Small to Make a Difference, Greta Thunberg
(New York: Penguin, 2018)

Made in America, Sam Walton
(New York: Doubleday, 1992)

Web Resources

Speed & Scale—Tracking the OKRs, speedandscale.com

AAAS, whatweknow.aaas.org

Bloomberg New Energy Finance, bnef.com

Breakthrough Energy innovation, breakthroughenergy.org

Carbon Dioxide Removal Primer, cdrprimer.org

CarbonPlan, carbonplan.org

Carbon Tracker, carbontracker.org

COP26 U.N. Climate Change Conference, unfccc.int

International Energy Agency Report, iea.org/reports/net-zero-by-2050

IPCC Report from the Intergovernmental Panel
on Climate Change, ipcc.ch

Measure What Matters—Resource on OKRs, whatmatters.com

NASA, climate.nasa.gov/evidence/

National Geographic Climate Coverage, natgeo.com/climate

Our World in Data, ourworldindata.org

Paris Agreement, unfccc.int/sites/default/files/
english_paris_agreement.pdf

TED Countdown—Videos and Event, countdown.ted.com

Advocacy Groups

350.org

Agora Energiewende (Germany)

C40 Cities

Center for Biological Diversity

Climate Power

Climate Reality Project

Resources

Coalition for Rainforest Nations

Conservation International

EarthJustice

Energy Foundation

Environmental Defense Fund

European Climate Foundation

Institute of Public & Environmental Affairs (China)

National Resources Defense Council

Nature Conservancy

Rainforest Action Network

Rainforest Alliance

Renewable Energy Institute (Japan)

RMI

Sierra Club

Sunrise Movement

U.S. Climate Action Network

World Resources Institute

World Wildlife Fund

Climate Foundations

Bezos Earth Fund

The Campaign for Nature

Children's Investment Fund Foundation

Hewlett Foundation

IKEA Foundation

MacArthur Foundation

McKnight Foundation

Michael Bloomberg

Packard Foundation

Quadrature

Sequoia Foundation

Climate-Focused Investors

Breakthrough Energy Ventures

Climate and Nature Fund (Unilever)

Climate Innovation Fund (Microsoft)

Climate Pledge Fund (Amazon.com)

Congruent Ventures

DBL (Double Bottom Line) Partners

Earthshot Ventures

Elemental Excelerator

The Engine (built by MIT)

Generation Investment Management

G2 Venture Partners

Green Climate Fund

Greenhouse Capital Partners

Khosla Ventures

Kleiner Perkins

Imperative Science Ventures

Incite

Lower Carbon Capital

OGCI Climate Investments

Pale Blue Dot

Prime Impact Fund

Prelude Ventures

S2G Ventures

Sequoia Capital

Union Square Ventures

Y Combinator

See more at Climate 50, climate50.com

Disclosures & Investments

Resource 4   # Disclosures & Investments

Businesses mentioned in the book that Kleiner
Perkins, Breakthrough Energy Ventures, or John
Doerr have invested in:

Alphabet / Google

Amazon

Beyond Meat

Bloom Energy

Chargepoint

Charm Industrial

Commonwealth Fusion

Cypress Semiconductor

Enphase

Farmer's Business Network

Fisker

G2 Venture Partners

Generation Investment

Nest (acquired by Google)

OPower (acquired by Oracle)

Proterra

Quantumscape

Redwood Materials

Solidia

Stripe

Tradesy

Watershed

Notes

Notes

Prologue

xii Steve Jobs invited us to launch: Apple. "Apple launches iPhone SDK." 6 March 2008, www.speedandscale.com/ifund.

xii I said this in 2007: "Salvation (and Profit) in Greentech." Ted, uploaded by TEDxTalks, 1 March 2007, www.ted.com/talks/john_ doerr_salvation_and_profit_in_greentech/transcript.

xv bright orange skies of the wildfires: Alberts, Elizabeth. "'Off the Chart': CO_2 from California Fires Dwarf State's Fossil Fuel Emissions." Mongabay.com, 18 September 2020, news.mongabay.com/2020/09/ off-the-chart-co2-from-california-fires-dwarf-states-fossil-fuel-emissions.

Introduction: What's the Plan?

xviii Some of It Is absorbed: "Climate and Earth's Energy Budget." NASA Earth Observatory, earthobservatory.nasa.gov/features/ EnergyBalance/page6.php. Accessed 14 June 2021.

xix In the preindustrial era: European Environment Agency. "Atmospheric Greenhouse Gas Concentrations." European Environment Agency, 4 October 2020, www.cca.europa.eu/data-and-maps/indica tors/atmospheric-greenhouse-gas-concentrations-7/assessment.

xix more than 500 parts per million: "NOAA Global Monitoring Laboratory—The NOAA Annual Greenhouse Gas Index (AGGI)." NOAA Annual Greenhouse Gas Index (AGGI), 2021, gml.noaa.gov/aggi/aggi. html.

xx the weight of 10,000 fully loaded: Conlen, Matt. "Visualizing the Quantities of Climate Change." Global Climate Change: Vital Signs of the Planet, 12 March 2020, climate.nasa.gov/news/2933/visualizing-the-quantities-of-climate-change.

xx In terms of emissions: CO_2e is typically weighed by the gigaton, or one billion tons: "Greenhouse Gas Equivalencies Calculator." U.S. Environmental Protection Agency, 26 May 2021, www.epa.gov/energy/ greenhouse-gas-equivalencies-calculator.

xx risen by about 1 degree Celsius: "World of Change: Global Temperatures." NASA Earth Observatory, earthobservatory.nasa.gov/ world-of-change/global-temperatures. Accessed 13 June 2021.

xx more than half of those emissions: Stainforth, Thorfinn. "More Than Half of All CO_2 Emissions Since 1751 Emitted in the Last 30 Years." Institute for European Environmental Policy, 29 April 2020, ieep.eu/news/more-than-half-of-all-co2-emissions-since-1751-emitted-in-the-last-30-years.

xxii 4 degrees Celsius of warming: Roston, Eric. "Economists Warn That a Hotter World Will Be Poorer and More Unequal." *Bloomberg Green*, 7 July 2020, www.bloomberg.com/news/articles/2020-07-07/global-gdp-could-fall-20-as-climate-change-heats-up.

xxii "You say you got a real solution": Beatles. "Revolution 1." Recorded 1968. By John Lennon and Paul McCartney. Apple Records, 1968.

xxii TED talk on climate change: "Salvation (and Profit) in Greentech." Doerr, John. TEDxTalk, 1 March 2007, www.ted.com/talks/john_doerr_salvation_and_profit_in_greentech/transcript.

xxiii 3 degrees Celsius or more by 2100: "Temperatures." Climate Action Tracker, 4 May 2021, climateactiontracker.org/global/temperatures.

xxv 59 gigatons of CO_2e per year: UNEP and UNEP DTU Partnership. "UNEP Report—The Emissions Gap Report 2020." *Management of Environmental Quality: An International Journal*, 2020, https://www.unep.org/emissions-gap-report-2020.

xxx in particular the top five: UNEP and UNEP DTU Partnership. "UNEP Report—The Emissions Gap Report 2020."

xxx at least fourteen countries: The Energy & Climate Intelligence Unit and Oxford Net Zero. "Taking Stock: A Global Assessment of Net Zero Targets." The Energy & Climate Intelligence Unit, 2021, ca1-eci.edcdn.com/reports/ECIU-Oxford_Taking_Stock.pdf.

xxxi carbon pollution came roaring back: Tollefson, Jeff. "COVID Curbed Carbon Emissions in 2020—but Not by Much." *Nature* 589, no. 7842, 2021, 343, doi:10.1038/d41586-021-00090-3.

Chapter 1: Electrify Transportation

3 freely shared its patents: Tesla. "All Our Patent Are Belong to You." Tesla, 27 July 2019, www.tesla.com/blog/all-our-patent-are-belong-you.

3 Tesla was selling one of every five: "EV Sales." BloombergNEF, www.bnef.com/interactive-datasets/2d5d59acd9000014?data-hub=11. Accessed 13 June 2021.

3 it sold half a million: "Q4 and FY2020 Update." Tesla, 2020, tesla-cdn.thron.com/static/1LRLZK_2020_Q4_Quarterly_Update_

Deck_-_Searchable_LVA2GL.pdf?xseo=&response-content-disposition=
inline%3Bfilename%3D%22TSLA-Q4-2020-Update.pdf%22.

3 value of around $600 billion: TSLA Stock Price, Tesla Inc. Stock
Quote (U.S.: Nasdaq). MarketWatch, 20 June 2021, www.marketwatch.
com/investing/stock/tsla.

3 reasons ranging from price: Degen, Matt. "2012 Fisker Karma
Review." Kelly Blue Book, 23 December 2019, www.kbb.com/fisker/
karma.

3 A pair of sedan fires triggered a recall: Lavrinc, Damon. "At
Least 16 Fisker Karmas Drown, Catch Fire at New Jersey Port."
Wired, 30 October 2012, www.wired.com/2012/10/fisker-fire-new
-jersey.

3 More than three hundred cars: "Fisker Says $30 Million in
Luxury Cars Destroyed by Sandy in NJ Port." Reuters, 7 November
2012, www.reuters.com/article/us-fisker-sandy/fisker-says-30-million-
in-luxury-cars-destroyed-by-sandy-in-nj-port-
idUSBRE8A603820121107.

3 nearly 10 million EVs: Frangoul, Anmar. "Global Electric Vehicle
Numbers Set to Hit 145 Million by End of the Decade, IEA Says."
CNBC, 29 April 2021, www.cnbc.com/2021/04/29/global-electric-
vehicle-numbers-set-to-hit-145-million-by-2030-iea-.html.

3 miles driven with combustion vehicles: "New Energy Outlook
2020." BloombergNEF, 20 April 2021, about.bnef.com/new-energy-
outlook.

3 new car's life span: Budd, Ken. "How Today's Cars Are Built to
Last." AARP, 1 November 2018, www.aarp.org/auto/trends-lifestyle/
info-2018/how-long-do-cars-last.html.

4 cause 350,000 premature deaths: Harvard University et al.
"Fossil Fuel Air Pollution Responsible for 1 in 5 Deaths Worldwide."
C-CHANGE, Harvard T. H. Chan School of Public Health, 9 February
2021, www.hsph.harvard.edu/c-change/news/fossil-fuel-air-pollution-
responsible-for-1-in-5-deaths-worldwide.

4 linked to heart disease and lung cancer: Integrated Science
Assessment (ISA) for Particulate Matter (Final Report, December
2019). U.S. Environmental Protection Agency, Washington, DC,
EPA/600/R-19/188, 2019.

7 most people won't pay: "Who Is Willing to Pay More for Renewable
Energy?" Yale Program on Climate Change Communication, 16 July
2019, climatecommunication.yale.edu/publications/who-is-willing-to-
pay-more-for-renewable-energy; Walton, Robert. "Americans Could
Pay More for Clean Energy. But Will They Really?" Utility Dive, 9

March 2015, www.utilitydive.com/news/americans-could-pay-more-for-clean-energy-but-will-they-really/372381.

7 The green premium varies widely across sectors:

Electricity: "Electric Power Monthly—U.S. Energy Information Administration (EIA)." U.S. Energy Information Administration, www.eia.gov/electricity/monthly/epm_table_grapher.php. Accessed 13 June 2021; Matasci, Sara. "Understanding Your Sunrun Solar Lease, PPA and Solar Contract Agreement." Solar News, 15 July 2020, https://news.energysage.com/sunrun-solar-lease-ppa-solar-contract-agreement/.

Passenger EVs: "Google." Google Search—2021 Chevy Bolt MSRP, www.google.com. Accessed 23 June 2021; "Google." Google Search—2021 Toyota Camry MSRP, www.google.com. Accessed 23 June 2021.

Long-haul trucking/ shipping transportation fuel: "Alternative Fuel Price Report." U.S. Department of Energy, January 2021, https://afdc.energy.gov/fuels/prices.html.

Cement: "IBISWorld—Industry Market Research, Reports, and Statistics." IBISWorld, www.ibisworld.com/us/bed/price-of-cement/190. Accessed 22 June 2021; "Jet Fuel Price Monitor." IATA, www.iata.org/en/publications/economics/fuel-monitor. Accessed 14 June 2021.

Aviation: "Jet Fuel Price Monitor." IATA, www.iata.org/en/publications/economics/fuel-monitor. Accessed 14 June 2021; Robinson, Daisy. "Sustainable Aviation Fuel (Part 1): Pathways to Production." BloombergNEF, 29 March 2021, www.bnef.com/insights/25925?query=eyJxdWVyeSI6IlNBRiIsInBhZ2UiOjEsIm9yZGVyIjoicmVzZXRhbmNlIn0%3D.

Round trip (economy) SFO to Hawaii: "Google." Travel, www.google.com/travel/unsupported?ucpp=CiVodHRwczovL3d3dy5nb29nbGUuY29tL3RyYXZlbC9mbGlnaHRz. Accessed 4 May 2021.

Ground beef hamburger meat: "Average Retail Food and Energy Prices, U.S. and Midwest Region: Mid-Atlantic Information Office: U.S. Bureau of Labor Statistics." U.S. Bureau of Labor Statistics, www.bls.gov/regions/mid-atlantic/data/averageretailfoodandenergyprices_usandmidwest_table.htm. Accessed 20 June 2021.

8 the green premium is a rough measure: See the table on page 7, and also Breakthrough Energy. "The Green Premium." Breakthrough Energy, 2020, www.breakthroughenergy.org/our-challenge/the-green-premium.

8 75 percent EV market share: "Trends and Developments in Electric Vehicle Markets—Global EV Outlook 2021—Analysis." International

Energy Agency, 2021, www.iea.org/reports/global-ev-outlook-2021/
trends-and-developments-in-electric-vehicle-markets.

8 China has passed 5 percent: "Transportation: In China's Biggest
Cities, 1 in 5 Cars Sold Is Electric." E&E News, 11 May 2021, www.
eenews.net/energywire/2021/05/11/stories/1063732167.

9 Volkswagen is investing: Rauwald, Christoph. "VW Boosts Tech
Spending Within $177 Billion Investment Plan." *Bloomberg Green*, 13
November 2020, www.bloomberg.com/news/articles/2020-11-13/
vw-boosts-tech-spending-in-177-billion-budget-amid-virus-hit.

9 represent 10 percent: "Electric Vehicle Outlook." BloombergNEF,
www.bnef.com/interactive-datasets/
2d5d59acd900003d?data-hub=11&tab=Buses. Accessed 13 June 2021.

9 30 percent of the sector's global greenhouse gases: "Transport
Sector CO_2 Emissions by Mode in the Sustainable Development Scenario, 2000–2030—Charts—Data & Statistics." IEA, www.iea.org/data-and
-statistics/charts/transport-sector-co2-emissions-by-mode-in-the
-sustainable-development-scenario-2000-2030. Accessed 13 June 2021.

0 total passenger car miles: "Electric Vehicle Outlook."

9 exposing hundreds of millions: Gallucci, Maria. "At Last, the
Shipping Industry Begins Cleaning Up Its Dirty Fuels." Yale E360, Yale
Environment 260, 28 June 2018, e360.yale.edu/features/at-last-the-
shipping-industry-begins-cleaning-up-its-dirty-fuels.

9 fifteen-year life span: "Review of Maritime Transport 2011,
Chapter 2." United Nations Conference on Trade and Development,
2011, unctad.org/system/files/official-document/rmt2011ch2_en.pdf.

9 catch those deadly small particles: Gallucci, Maria. "At Last, the
Shipping Industry Begins Cleaning Up Its Dirty Fuels."

10 "I cannot conceive of one": Strohl, Daniel. "Fact Check: Did a GM
President Really Tell Congress 'What's Good for GM Is Good for
America?'" Hemmings, 5 September 2019, www.hemmings.com/
stories/2019/09/05/fact-check-did-a-gm-president-really-tell-congress-
whats-good-for-gm-is-good-for-america.

13 governor Gavin Newsom ordered a ban for 2035: "Twelve U.S.
States Urge Biden to Back Phasing Out Gas-Powered Vehicle Sales by
2035." Reuters, 21 April 2021, www.reuters.com/business/twelve-us-
states-urge-biden-back-phasing-out-gas-powered-vehicle-sales-by-2035-
2021-04-21.

14 But Wang had an ace up his sleeve: Huang, Echo. "How Much
Financial Help Does China Give EV Maker BYD?" Quartz, 27 March
2019, qz.com/1579568/how-much-financial-help-does-china-give-ev-
maker-byd.

Notes

15 Assured of public sector funding: Vincent, Danny. "The Uncertain Future for China's Electric Car Makers." BBC News, 27 March 2020, www.bbc.com/news/business-51711019.

22 $73,000 and $173,000 over a diesel bus: Quarles, Neil, et al. "Costs and Benefits of Electrifying and Automating Bus Transit Fleets." Multidisciplinary Digital Publishing Institute, 2020, www. caee.utexas.edu/prof/kockelman/public_html/TRB18AeBus.pdf.

22 removes the emissions equivalent of: Gilpin, Lyndsey. "These City Bus Routes Are Going Electric—and Saving Money." Inside Climate News, 23 October 2017, insideclimatenews.org/ news/23102017/these-city-bus-routes-are-going-all-electric.

22 operating in forty-three states: "Revolutionizing Commercial Vehicle Electrification." Proterra, April 2021, www.proterra.com/ wp-content/uploads/2021/04/PTRA-ACTC-Analyst-Day-Presentation-4.8.21-FINAL-1.pdf.

22 25 percent in China: "Long-Term Electric Vehicle Outlook 2021." BloombergNEF, 9 June 2021, www.bnef.com/insights/26533/view.

26 Wright's Law could be applied: Bui, Quan, et al. "Statistical Basis for Predicting Technological Progress." Santa Fe Institute, 5 July 2012, www.santafe.edu/research/results/working-papers/statistical-basis-for-predicting-technological-pro.

26 each doubling of production: "Evolution of Li-Ion Battery Price, 1995–2019—Charts—Data & Statistics." IEA, 30 June 2020, www.iea. org/data-and-statistics/charts/evolution-of-li-ion-battery-price-1995-2019. Accessed 13 June 2021.

26 cost only eight thousand dollars: Gold, Russell, and Ben Foldy. "The Battery Is Ready to Power the World." *Wall Street Journal*, 5 February 2021, www.wsj.com/articles/the-battery-is-ready-to-power-the-world-11612551578.

26 electric version of its F-150 pickup: Boudette, Neal. "Ford's Electric F-150 Pickup Aims to Be the Model T of E.V.s." *New York Times*, 19 May 2021, www.nytimes.com/2021/05/19/business/ford-electric-vehicle-f-150.html.

26 "The future of the auto industry is electric": Watson, Kathryn. "Biden Drives Electric Vehicle and Touts It as the 'Future of the Auto Industry.'" CBS News, 18 May 2021, www.cbsnews.com/news/biden-ford-electric-car-plant-michigan-watch-live-stream-today-05-18-2021.

26 "a giant battery on wheels": "The Ford Electric F-150 Lightning's Astonishing Price." *Atlantic*, 19 May 2021, www.theatlantic.com/technology/archive/2021/05/f-150-lightning-fords-first-electric-truck/618932.

26 In India, the most popular car: "Car Prices in India—Latest Models & Features 23 Jun 2021." BankBazaar, www.bankbazaar.com/ car-loan/car-prices-in-india.html. Accessed 22 June 2021; Mehra, Jaiveer. "Best Selling Cars in November 2020: Maruti Swift Remains Top Seller." Autocar India, 5 December 2020, www.autocarindia.com/ car-news/best-selling-cars-in-november-2020-maruti-swift-remains-top-seller-419341.

27 twenty-seven miles per day: "2020 Global Automotive Consumer Study." Deloitte, 2020, www2.deloitte.com/content/dam/Deloitte/us/ Documents/manufacturing/us-2020-global-automotive-consumer-study-global-focus-countries.pdf.

Chapter 2: Decarbonize the Grid

29 "I'd put my money on the sun": Newton, James D. *Uncommon Friends: Life with Thomas Edison, Henry Ford, Harvey Firestone, Alexis Carrel, & Charles Lindbergh.* New York: Mariner Books, 1989.

30 Scheer's Law specified a pay rate: Schwartz, Evan. "The German Experiment." *MIT Technology Review*, 2 April 2020, www.technology review.com/2010/06/22/26637/the-german-experiment; "Feed-in Tariffs in Germany." Wikipedia, 21 March 2021, en.wikipedia.org/ wiki/Feed-in_tariffs_in_Germany.

30 most German citizens backed the plan: Schwartz, Evan. "The German Experiment." *MIT Technology Review*, 22 June 2010, www. technologyreview.com/2010/06/22/26637/the-german-experiment.

30 A Bavarian livestock farmer: *Nova.* PBS, 24 April 2007, www.pbs. org/wgbh/nova/video/saved-by-the-sun.

30 three hundred thousand sorely needed jobs: Schwartz, Evan. "The German Experiment."

32 70 percent of the global panel market: Buchholz, Katharina. "China Dominates All Steps of Solar Panel Production." Statista Infographics, 21 April 2021, www.statista.com/chart/24687/solar-panel-global-market-shares-by-production-steps.

32 the panels plummeted: Sun, Xiaojing. "Solar Technology Got Cheaper and Better in the 2010s. Now What?" Wood Mackenzie, 18 December 2019, www.woodmac.com/news/opinion/solar-technology-got-cheaper-and-better-in-the-2010s.-now-what.

33 42 percent of Germany's electricity: "Renewables Meet 46.3% of Germany's 2020 Power Consumption, up 3.8 Pts." Reuters, 14 December 2020, www.reuters.com/article/germany-power-renewables-idUKKBN28O1AH.

33 Renewables became the leading energy source: Randowitz, Bernd. "Germany's Renewable Power Share Surges to 56% amid Covid-19 Impact." Recharge, July 2020, www.rechargenews.com/transition/germany-s-renewable-power-share-surges-to-56-amid-covid-19-impact/2-1-837212.

35 twenty-eight states have: "U.S. Nuclear Industry—U.S. Energy Information Administration (EIA)." U.S. Energy Information Administration, 6 April 2021, www.eia.gov/energyexplained/nuclear/us-nuclear-industry.php.

35 Hydroelectricity already accounts for 16 percent: "World Energy Outlook 2020—Analysis." IEA, October 2020, www.iea.org/reports/world-energy-outlook-2020.0.

35 Wind and solar have global shares: "Renewable Energy Market Update 2021," World Energy Outlook 2020—Analysis, International Energy Agency, https://www.iea.org/reports/renewable-energy-market-update-2021/renewable-electricity; "New Global Solar PV Installations to Increase 27% to Record 181 GW This Year," IHS Markit, 29 March 2021, https://www.reuters.com/business/energy/new-global-solar-pv-installations-increase-27-record-181-gw-this-year-ihs-markit-2021-03-29.

35 New installations of these renewables: Brandily, Tifenn, and Amar Vasdev. "2H 2020 LCOE Update." BloombergNEF, 10 December 2020, www.bnef.com/login?r=%2Finsights%2F24999%2Fview.

36 new development of coal, oil, and gas: "Net Zero by 2050—Analysis." International Energy Agency, May 2021, www.iea.org/reports/net-zero-by-2050.

37 "a total transformation": "Net Zero by 2050—Analysis."

37 "stop international financing": Piper, Elizabeth, and Markus Wacket. "In Climate Push, G7 Agrees to Stop International Funding for Coal." Reuters, 21 May 2021, www.reuters.com/business/energy/g7-countries-agree-stop-funding-coal-fired-power-2021-05-21.

37 push renewable energy sources: "Net Zero by 2050—Analysis."

37 Our Methane Emissions KR (2.5): "Methane Emissions from Oil and Gas—Analysis." International Energy Agency, www.iea.org/reports/methane-emissions-from-oil-and-gas. Accessed 18 June 2021.

37 Modern electric heat pumps: McKenna, Claire, et al. "It's Time to Incentivize Residential Heat Pumps." RMI, 22 July 2020, rmi.org/its-time-to-incentivize-residential-heat-pumps.

38 the sun delivers as much energy: "Solar Energy Basics." National Renewable Energy Laboratory, 2021, www.nrel.gov/research/re-solar.html.

38 solar installations today are outpacing: "Renewable Energy Market Update 2021." IEA, 2021, www.iea.org/reports/renewable-energy-market-update-2021/renewable-electricity.

43 "net metering" policies: "Net Metering." Solar Energy Industries Association, May 2017, www.seia.org/initiatives/net-metering.

44 100 gigawatts of installed solar capacity: "U.S. Solar Market Insight." Solar Energy Industries Association, 2021, www.seia.org/us-solar-market-insight. Updated 16 March 2021.

44 India set a target: "India Exceeding Paris Targets; to Achieve 450 GW Renewable Energy by 2030: PM Modi at G20 Summit." *Business Today*, 22 November 2020, www.businesstoday.in/current/economy-politics/india-exceeding-paris-targets-to-achieve-450-gw-renewable-energy-by-2030-pm-modi-at-g20-summit/story/422691.html.

45 Wind is growing faster: Russi, Sofia. "Global Wind Report 2021." Global Wind Energy Council, 30 April 2021, gwec.net/global-wind-report-2021.

45 China's Gansu Wind Farm: Besta, Shankar. "Profiling Ten of the Biggest Onshore Wind Farms in the World." NS Energy, 9 December 2019, www.nsenergybusiness.com/features/worlds-biggest-onshore-wind-farms.

45 onshore wind power faces several constraints: Gross, Samantha. "Renewables, Land Use, and Local Opposition in the United States." Brookings Institution, January 2020, www.brookings.edu/wp-content/uploads/2020/01/FP_20200113_renewables_land_use_local_opposition_gross.pdf.

46 global prices collapsed: "Natural Gas Prices—Historical Chart." MacroTrends, 2021, www.macrotrends.net/2478/natural-gas-prices-historical-chart.

49 one of the first makers of wind turbines: Vestas focused on wind power in 1987. "Vestas History." Vestas, 2021, www.vestas.com/en/about/profile#!from-1987-1998.

49 90 percent of the energy generated by Ørsted: "Our Green Business Transformation: What We Did and Lessons Learned." Ørsted, April 2021, https://orsted.com/en/about-us/whitepapers/green-transformation-lessons-learned.

49 the world's most sustainable company: Scott, Mike. "Top Company Profile: Denmark's Ørsted Is 2020's Most Sustainable Corporation." Corporate Knights, 21 January 2020, www.corporateknights.com/reports/2020-global-100/top-company-profile-orsted-sustainability-15795648.

Notes

52 "the most emissions ever measured": "Satellite Data Reveals Extreme Methane Emissions from Permian Oil & Gas Operations; Shows Highest Emissions Ever Measured from a Major U.S. Oil and Gas Basin." Environmental Defense Fund, 22 April 2020, www.edf.org/media/satellite-data-reveals-extreme-methane-emissions-permian-oil-gas-operations-shows-highest.

54 cutting human-caused methane emissions: Chung, Tiy. "Global Assessment: Urgent Steps Must Be Taken to Reduce Methane Emissions This Decade." United Nations Environment Programme (UNEP), 6 May 2021, www.unep.org/news-and-stories/press-release/global-assessment-urgent-steps-must-be-taken-reduce-methane.

54 methane leaks occur: Plant, Genevieve. "Large Fugitive Methane Emissions from Urban Centers Along the U.S. East Coast." *AGU Journals*, 28 July 2019, agupubs.onlinelibrary.wiley.com/doi/full/10.1029/2019GL082635; Lebel, Eric D., et al. "Quantifying Methane Emissions from Natural Gas Water Heaters." ACS Publications, 6 April 2020, pubs.acs.org/doi/10.1021/acs.est.9b07189; "Major U.S. Cities Are Leaking Methane at Twice the Rate Previously." *Science | AAAS*, 19 July 2019, www.sciencemag.org/news/2019/07/major-us-cities-are-leaking-methane-twice-rate-previously-believed.

56 The technology exists today: "Gas Leak Detection & Repair." MBS Engineering, 2021, www.mbs.engineering/gas-leak-detection-repair.html; "Perform Valve Leak Repair During Pipeline Replacement." U.S. Environmental Protection Agency, 31 August 2016, www.epa.gov/sites/production/files/2016-06/documents/perform leakrepairduringpipelinereplacement.pdf.

56 leaks from U.S. fracking sites: Lipton, Eric, and Hiroko Tabuchi. "Driven by Trump Policy Changes, Fracking Booms on Public Lands." *New York Times*, 27 October 2018, www.nytimes.com/2018/10/27/climate/trump-fracking-drilling-oil-gas.html; Davenport, Coral. "Trump Eliminates Major Methane Rule, Evenas Leaks Are Worsening," updated 18 April 2021, https://www.nytimes.com/2020/08/13/climate/trump-methane.html.

56 The practice of flaring: "Natural Gas Flaring and Venting: State and Federal Regulatory Overview, Trends and Impacts." Office of Fossil Energy (FE) of the U.S. Department of Energy, June 2019, www.energy.gov/sites/prod/files/2019/08/f65/Ntural%20Gas%20Flaring%20and%20Venting%20Report.pdf.

57 Roughly half of American homes and restaurants: Jacobs, Nicole. "New Poll: Natural Gas Still the Top Choice for Cooking." Energy in Depth, 16 February 2021, www.energyindepth.org/new-poll-natural-gas-still-the-top-choice-for-cooking.

58 Another approach is "net metering": National Renewable Energy Laboratory, 2020, www.nrel.gov/state-local-tribal/basics-net-metering.html.

58 27,000 terawatt-hours of electricity: "Net Zero by 2050—Analysis."

59 LED lighting, for one example, uses 75 percent less: Popovich, Nadja. "America's Light Bulb Revolution." *New York Times*, 8 March 2019, www.nytimes.com/interactive/2019/03/08/climate/light-bulb-efficiency.html.

59 More efficiently designed pipes and ducts: Lovins, Amory B. "How Big Is the Energy Efficiency Resource?" IOP Science, IOP Publishing Ltd, 18 September 2018, iopscience.iop.org/article/10.1088/1748-9326/aad965/pdf.

59 retrofit of New York City's Empire State Building: Carmichael, Cara, and Eric Harrington. "Project Case Study: Empire State Building." Rocky Mountain Institute, 2009, rmi.org/wp-content/uploads/2017/04/Buildings_Retrofit_EmpireStateBuilding_CaseStudy_2009.pdf.

59 buildings use nearly 75 percent: "Quadrennial Technology Review," Chapter 5: Increasing Efficiency of Building Systems and Technologies." United States Department of Energy, September 2015, www.energy.gov/sites/prod/files/2017/03/f34/qtr-2015-chapter5.pdf.

59 install an electric heat pump: "How Much Does an Electric Furnace Cost to Install?" Modernize Home Services, 2021, modernize.com/hvac/heating-repair-installation/furnace/electric.

59 the Energy Star program: "ENERGY STAR Impacts." ENERGY STAR, 2019, www.energystar.gov/about/origins_mission/impacts.

60 United States ranked an uninspiring tenth: Castro-Alvarez, Fernando, et al. "The 2018 International Energy Efficiency Scorecard." ©American Council for an Energy-Efficient Economy, June 2018, www.aceee.org/sites/default/files/publications/researchreports/i1801.pdf.

60 Had the rest of the United States kept pace: Komanoff, Charles, et al. "California Stars Lighting the Way to a Clean Energy Future." Natural Resources Defense Council, May 2019, www.nrdc.org/sites/default/files/california-stars-clean-energy-future-report.pdf.

Chapter 3: Fix Food

64 2,500 gigatons of carbon: Ontl, Todd A., and Lisa A. Schulte. "Soil Carbon Storage." Knowledge Project, Nature Education, 2012, www.nature.com/scitable/knowledge/library/soil-carbon-storage-84223790/.

Notes

64 fully one third has been depleted: "Global Plans of Action Endorsed to Halt the Escalating Degradation of Soils." Food and Agriculture Organization of the United States, 24 July 2014, www.fao.org/news/story/en/item/239341/icode.

66 Fertilizers alone account for 2 gigatons: Tian, Hanqin, et al. "A Comprehensive Quantification of Global Nitrous Oxide Sources and Sinks." *Nature,* 7 October 2020, www.nature.com/articles/s41586-020-2780-0.

66 15 percent of the entire emissions: UNEP and UNEP DTU Partnership. "UNEP Report—The Emissions Gap Report 2020." *Management of Environmental Quality: An International Journal,* 2020, https://www.unep.org/emissions-gap-report-2020.

66 up to 60 percent more calories: Ranganathan, Janet, et al. "How to Sustainably Feed 10 Billion People by 2050, in 21 Charts." World Resources Institute, 5 December 2018, www.wri.org/insights/how-sustainably-feed-10-billion-people-2050-21-charts.

68 absorb 2 gigatons of CO_2: Zomer, Robert. "Global Sequestration Potential of Increased Organic Carbon in Cropland Soils." *Scientific Reports,* 14 November 2017, www.nature.com/articles/s41598-017-15794-8?error=cookies_not_supported&code=4f2be93e-fd6c-4958-814b-d7ea0649ee8e.

68 one third of all food produced today is wasted: "Worldwide Food Waste." UN Environment Programme, 2010, www.unep.org/thinkeatsave/get-informed/worldwide-food-waste.

68 generates nearly 2 gigatons of CO_2-equivalent emissions: Ott, Giffen. "We're a Climate Fund—Why Start with Waste?" FullCycle, www.fullcycle.com/insights/were-a-climate-fund-why-start-with-waste. Accessed 13 June 2021.

68 Soil is created over time: Funderburg, Eddie. "What Does Organic Matter Do in Soil?" North Noble Research Institute, 31 July 2001, www.noble.org/news/publications/ag-news-and-views/2001/august/what-does-organic-matter-do-in-soil.

68 Healthy, undisrupted soil: Kautz, Timo. "Research on Subsoil Biopores and Their Functions in Organically Managed Soils: A Review," *Renewable Agriculture and Food Systems,* Cambridge University Press, 15 January 2014, www.cambridge.org/core/journals/rcnewable-agriculture-and-food-systems/article/research-on-subsoil-biopores-and-their-functions-in-organicallymanaged-soils-a-review/A72F0E0E7B86FE904A5EC5EE37F6D6C9.

69 Limited to less than 7 percent: Plumer, Brad. "No-Till Farming Is on the Rise. That's Actually a Big Deal." *Washington Post,* 9 November

2013, www.washingtonpost.com/news/wonk/wp/2013/11/09/no-till-farming-is-on-the-rise-thats-actually-a-big-deal; "USDA ERS—No-Till and Strip-Till Are Widely Adopted but Often Used in Rotation with Other Tillage Practices." Economic Research Service, U.S. Department of Agriculture, www.ers.usda.gov/amber-waves/2019/march/no-till-and-strip-till-are-widely-adopted-but-often-used-in-rotation-with-other-tillage-practices. Accessed 13 June 2021.

69 expanded to 21 percent in the United States: Creech, Elizabeth. "Saving Money, Time and Soil: The Economics of No-Till Farming." U.S. Department of Agriculture, 30 November 2017, www.usda.gov/media/blog/2017/11/30/saving-money-time-and-soil-economics-no-till-farming.

69 croplands across South America: Gianessi, Leonard. "Importance of Herbicides for No-Till Agriculture in South America." CropLife International, 16 November 2014, croplife.org/case-study/importance-of-herbicides-for-no-till-agriculture-in-south-america.

69 less labor intensive: Smil, Vaclav. *Energy and Civilization: A History.* Boston: The MIT Press, 2018.

71 If 25 percent of the world's farmlands: Poeplau, Christopher, and Axel Don. "Carbon Sequestration in Agricultural Soils via Cultivation of Cover Crops—A Meta-Analysis." *Agriculture, Ecosystems & Environment* 200, 2015, 33–41, doi:10.1016/j.agee.2014.10.024.

71 20 million acres lay fallow after floods: Ahmed, Amal. "Last Year's Historic Floods Ruined 20 Million Acres of Farmland." *Popular Science*, 26 April 2021, www.popsci.com/story/environment/2019-record-floods-midwest.

71 nitrous oxide is: UNEP and UNEP DTU Partnership. "UNEP Report—The Emissions Gap Report 2020." *Management of Environmental Quality: An International Journal*, 2020, https://www.unep.org/emissions-gap-report-2020.

71 nitrous oxide emissions can be reduced: Waite, Richard, and Alex Rudee. "6 Ways the US Can Curb Climate Change and Grow More Food." World Resources Institute, 20 August 2020, www.wri.org/insights/6-ways-us-can-curb-climate-change-and-grow-more-food.

72 Creating synthetic fertilizer: Boerner, Leigh Krietsch. "Industrial Ammonia Production Emits More CO_2 than Any Other Chemical-Making Reaction. Chemists Want to Change That." *Chemical & Engineering News*, 15 June 2019, cen.acs.org/environment/green-chemistry/Industrial-ammonia-production-emits-CO2/97/i24.

72 using less fertilizer: Tullo, Alexander H. "Is Ammonia the Fuel of the Future?" *Chemical & Engineering News*, 8 March 2021, cen.acs.org/business/petrochemicals/ammonia-fuel-future/99/i8.

Notes

72 the United States eats more beef: "Agricultural Output—Meat Consumption—OECD Data." OECD.org, 2020, data.oecd.org/agroutput/meat-consumption.htm.

72 Typical Americans annually consume: Durisin, Megan, and Shruti Singh. "Americans Will Eat a Record Amount of Meat in 2018." *Bloomberg*, 2 February 2018, www.bloomberg.com/news/articles/2018-01-02/have-a-meaty-new-year-americans-will-eat-record-amount-in-2018.

72 a bonanza for the fast-food industry: Wood, Laura. "Fast Food Industry Analysis and Forecast 2020-2027." Business Wire, 16 July 2020, www.businesswire.com/news/home/20200716005498/en/Fast-Food-Industry-Analysis-and-Forecast-2020-2027---ResearchAndMarkets.com.

72 7 gigatons of CO_2e per year: "Key Facts and Findings." Food and Agriculture Organization of the United States, 2020, www.fao.org/news/story/en/item/197623/icode.

72 cattle is king, at 4.6 gigatons: "Tackling Climate Change Through Livestock." Food and Agriculture Organization of the United Nations, 2013, http://www.fao.org/3/i3437e/i3437e.pdf.

74 the cow's digestive process: "Which Is a Bigger Methane Source: Cow Belching or Cow Flatulence?" Climate Change: Vital Signs of the Planet, 2021, climate.nasa.gov/faq/33/which-is-a-bigger-methane-source-cow-belching-or-cow-flatulence.

74 80 pounds of daily manure: "Animal Manure Management." U.S. Department of Agriculture, December 1995, www.nrcs.usda.gov/wps/portal/nrcs/detail/null/?cid=nrcs143_014211.

74 75 percent of farmland: "How Much of the World's Land Would We Need in Order to Feed the Global Population with the Average Diet of a Given Country?" Our World in Data, 3 October 2017, ourworldindata.org/agricultural-land-by-global-diets.

74 these animals supply only: "How Much of the World's Land Would We Need in Order to Feed the Global Population with the Average Diet of a Given Country?"

74 small amounts of seaweed: Nelson, Diane. "Feeding Cattle Seaweed Reduces Their Greenhouse Gas Emissions 82 Percent." University of California, Davis, 17 March 2021, www.ucdavis.edu/news/feeding-cattle-seaweed-reduces-their-greenhouse-gas-emissions-82-percent.

75 Nutrition Facts labels: Shangguan, Siyi, et al. "A Meta-Analysis of Food Labeling Effects on Consumer Diet Behaviors and Industry

Practices." *American Journal of Preventive Medicine* 56, no. 2, 2019, 300–314, doi:10.1016/j.amepre.2018.09.024.

75 "underestimate the emissions": Camilleri, Adrian, et al. "Consumers Underestimate the Emissions Associated with Food but Are Aided by Labels." *Nature Climate Change* 9, 17 December 2018, www.nature.com/articles/s41558-018-0354-z.

75 "environmental price tags": Donnellan, Douglas. "Climate Labels on Food to Become a Reality in Denmark." Food Tank, 11 April 2019, foodtank.com/news/2019/04/climate-labels-on-food-to-become-a-reality-in-denmark.

75 "Cool Food Meal": "RELEASE: New 'Cool Food Meals' Badge Hits Restaurant Menus Nationwide, Helping Consumers Act on Climate Change." World Resources Institute, 14 October 2020, www.wri.org/news/release-new-cool-food-meals-badge-hits-restaurant-menus-nationwide-helping-consumers-act.

76 "two-thirds vegan diet": "How Much Would Giving Up Meat Help the Environment?" *Economist*, 18 November 2019, www.economist.com/graphic-detail/2019/11/15/how-much-would-giving-up-meat-help-the-environment; Klm, Brent F., et al. "Country-Specific Dietary Shifts to Mitigate Climate and Water Crises." ScienceDirect, 1 May 2020, www.sciencedirect.com/science/article/pii/S0959378018306101.

82 plant-based burger customers are neither vegan nor vegetarian: O'Connor, Anahad. "Fake Meat vs. Real Meat." *New York Times*, 2 December 2020, www.nytimes.com/2019/12/03/well/eat/fake-meat-vs-real-meat.html.

82 the category grew 45 percent: Mount, Daniel. "Retail Sales Data: Plant-Based Meat, Eggs, Dairy." Good Food Institute, 9 June 2021, gfi.org/marketresearch/#:%7E:text.

82 no signs of flattening: Poinski, Megani. "Plant-Based Food Sales Outpace Growth in Other Categories during Pandemic." Food Dive, 27 May 2020, www.fooddive.com/news/plant-based-food-sales-outpace-growth-in-other-categories-during-pandemic/578653.

82 price parity with beef by 2024: Lucas, Amelia. "Beyond Meat Unveils New Version of Its Meat-Free Burgers for Grocery Stores." CNBC, 27 April 2021, www.cnbc.com/2021/04/27/beyond-meat-unveils-new-version-of-its-meat-free-burgers-in-stores.html.

82 "entirely remove the animal": Card, Jon. "Lab-Grown Food: 'The Goal Is to Remove the Animal from Meat Production.'" *Guardian*, 9 August 2018, www.theguardian.com/small-business-network/2017/jul/24/lab-grown-food-indiebio-artificial-intelligence-walmart-vegetarian.

Notes

82 Fifteen percent of total milk sales: Mount, Daniel. "U.S. Retail Market Data for Plant-Based Industry."

83 cheese, the third-highest-emitting food: Ritchie, Hannah. "You Want to Reduce the Carbon Footprint of Your Food? Focus on What You Eat, Not Whether Your Food Is Local." Our World in Data, 24 January 2020, ourworldindata.org/food-choice-vs-eating-local.

83 single cow emits about 250 pounds of methane per year: University of Adelaide. "Potential for Reduced Methane from Cows." ScienceDaily, 8 July 2019, www.sciencedaily.com/releases/2019/07/190708112514.htm.

83 Rice—a dietary cornerstone: "System of Rice Intensification." Project Drawdown, 7 August 2020, drawdown.org/solutions/system-of-rice-intensification.

83 12 percent of the globe's methane emissions: Proville, Jeremy, and K. Kritee. "Global Risk Assessment of High Nitrous Oxide Emissions from Rice Production." Environmental Defense Fund, 2018, www.edf.org/sites/default/files/documents/EDF_White_Paper_Global_Risk_Analysis.pdf.

83 increase in nitrous oxide emissions: "Overview of Greenhouse Gases." U.S. Environmental Protection Agency, 20 April 2021, www.epa.gov/ghgemissions/overview-greenhouse-gases#nitrous-oxide.

83 Shallow flooding, together with nitrogen and organic matter management: "Nitrous Oxide Emissions from Rice Farms Are a Cause for Concern for Global Climate." Environmental Defense Fund, 10 September 2018, www.edf.org/media/nitrous-oxide-emissions-rice-farms-are-cause-concern-global-climate.

84 Mars, Inc., reached the 99 percent mark: Dawson, Fiona. "Mars Food Works to Deliver Better Food Today." Mars, 2020, www.mars.com/news-and-stories/articles/how-mars-food-works-to-deliver-better-food-today-for-a-better-world-tomorrow.

86 the global cattle population: "Cattle Population Worldwide 2012–2021." Statista, 20 April 2021, www.statista.com/statistics/263979/global-cattle-population-since-1990.

86 Milk prices have declined: Nepveux, Michael. "USDA Report: U.S. Dairy Farm Numbers Continue to Decline." American Farm Bureau Federation, 26 February 2021, fb.org/market-intel/usda-report-u.s.-dairy-farm-numbers-continue-to-decline.

86 subsidies for the dairy industry: Calder, Alice. "Agricultural Subsidies: Everyone's Doing It." Hinrich Foundation, 15 October 2020, www.hinrichfoundation.com/research/article/protectionism/agricultural-subsidies/#:%7E:text.

87 A staggering 33 percent: "Food Loss and Food Waste." Food and Agriculture Organization of the United Nations, 2021, http://www.fao.org/food-loss-and-food-waste/flw-data.

87 more than 800 million people: "World Hunger Is Still Not Going Down After Three Years and Obesity Is Still Growing—UN Report." World Health Organization, 15 July 2019, www.who.int/news/item/15-07-2019-world-hunger-is-still-not-going-down-after-three-years-and-obesity-is-still-growing-un-report.

87 consumers throw out 35 percent: Center for Food Safety and Applied Nutrition. "Food Loss and Waste." U.S. Food and Drug Administration, 23 February 2021, www.fda.gov/food/consumers/food-loss-and-waste.

87 Annual waste amounts to $240 billion: Yu, Yang, and Edward C. Jaenicke. "Estimating Food Waste as Household Production Inefficiency." *American Journal of Agricultural Economics* 102, no. 2, 2020, 525–47, doi:10.1002/ajae.12036; Bandoim, Lana. "The Shocking Amount of Food U.S. Households Waste Every Year." *Forbes*, 27 January 2020, www.forbes.com/sites/lanabandoim/2020/01/26/the-shocking-amount-of-food-us-households-waste-every-year.

87 more than 2,700 French supermarkets: "Is France's Groundbreaking Food-Waste Law Working?" PBS *NewsHour*, 31 August 2019, www.pbs.org/newshour/show/is-frances-groundbreaking-food-waste-law-working.

89 U.S. farmers have trended in this direction: "United States Summary and State Data." U.S. Department of Agriculture, April 2019, www.nass.usda.gov/Publications/AgCensus/2017/Full_Report/Volume_1,_Chapter_1_US/usv1.pdf.

89 reduced the amount of land and water used: Capper, J. L. "The Environmental Impact of Beef Production in the United States: 1977 Compared with 2007." *Journal of Animal Science* 89, no. 12, 2011, 4249–61, doi:10.2527/jas.2010-3784.

89 50 percent more calories: Ranganathan, Janet. "How to Sustainably Feed 10 Billion People by 2050, in 21 Charts." World Resources Institute, www.wri.org/insights/how-sustainably-feed-10-billion-people-2050-21-charts. Accessed 18 June 2021.

Chapter 4: Protect Nature

94 no way to hit the pause button: Schädel, Christina. "Guest Post: The Irreversible Emissions of a Permafrost 'Tipping Point.'" Carbon Brief, 12 February 2020, www.carbonbrief.org/guest-post-the-irreversible-emissions-of-a-permafrost-tipping-point.

94 280 parts per million of carbon dioxide: Prentice, L. C. "The Carbon Cycle and Atmospheric Carbon Dioxide." IPCC, www.ipcc.ch/site/assets/uploads/2018/02/TAR-03.pdf.

94 up 50 percent since the mid-1700s: Betts, Richard. "Met Office: Atmospheric CO2 Now Hitting 50% Higher than Pre-Industrial Levels." Carbon Brief, 16 March 2021, www.carbonbrief.org/met-office-atmospheric-co2-now-hitting-50-higher-than-pre-industrial-levels.

94 "The Half-Earth proposal": Wilson, Edward O. *Half-Earth.* New York: Liveright, 2017.

96 stretch goal from the 15 percent protected in 2020: Mark, Jason. "A Conversation with E. O. Wilson." *Sierra*, 13 May 2021, www.sierraclub.org/sierra/conversation-eo-wilson.

97 a football field of forest: Roddy, Mike. "We Lost a Football Pitch of Primary Rainforest Every 6 Seconds in 2019." *Global Forest Watch* (Blog), 2 June 2020, www.globalforestwatch.org/blog/data-and-research/global-tree-cover-loss-data-2019/.

97 6 gigatons of annual CO$_2$: Gibbs, David, et al. "By the Numbers: The Value of Tropical Forests in the Climate Change Equation." World Resources Institute, 4 October 2018, www.wri.org/insights/numbers-value-tropical-forests-climate-change-equation; Mooney, Chris, et al. "Global Forest Losses Accelerated Despite the Pandemic, Threatening World's Climate Goals." *Washington Post*, 31 March 2021, www.washingtonpost.com/climate-environment/2021/03/31/climate-change-deforestation.

97 The Amazon rainforest alone: Helmholtz Centre for Environmental Research. "The Forests of the Amazon Are an Important Carbon Sink." ScienceDaily, 18 November 2019, www.sciencedaily.com/releases/2019/11/191118100834.htm.

97 tropical deforestation would place third: "By the Numbers: The Value of Tropical Forests in the Climate Change Equation." World Resources Institute, 4 October 2018, www.wri.org/insights/numbers-value-tropical-forests-climate-change-equation.

101 These avoided emissions, known as offsets: Cullenward, Danny, and David Victor. *Making Climate Policy Work.* Polity, 2020.

103 considerable progress on deforestation: Ritchie, Hannah. "Deforestation and Forest Loss." Our World in Data, 2020, ourworldindata.org/deforestation.

104 Kraft Foods, a company with more than $30 billion in revenues: "Kraft's Annual Report 2001." Kraft, 2001, www.annual

reports.com/HostedData/AnnualReportArchive/m/NASDAQ_
mdlz_2001.pdf.

104 reduced its carbon footprint by 15 percent: Kraft Foods, "Kraft
Foods Maps Its Total Environmental Footprint." PR Newswire, 14
December 2011, www.prnewswire.com/news-releases/kraft-foods-
maps-its-total-environmental-footprint-135585188.html.

105 carbon emissions dipped 25 percent, from 4 gigatons to 3:
"Carbon Emissions from Forests down by 25% Between 2001–2015."
Food and Agriculture Organization of the United Nations, 20 March
2015, www.fao.org/news/story/en/item/281182/icode.

105 development of a powerful metric: "Return on Sustainability
Investment (ROSITM)." New York University Stern School of Business,
2021, www.stern.nyu.edu/experience-stern/about/departments-
centers-initiatives/centers-of-research/center-sustainable-business/
research/return-sustainability-investment-rosi.

106 "Parties should take action": "Paris Agreement." United Nations
Framework Convention on Climate Change, 12 December 2015, unfccc.
int/sites/default/files/english_paris_agreement.pdf.

106 deforestation funding outpaces: "Where We Focus: Global."
Climate and Land Use Alliance, 16 November 2018, www.climateand
landusealliance.org/initiatives/global.

106 80 percent of the world's biodiversity: "Indigenous Peoples."
World Bank, 2020, www.worldbank.org/en/topic/indigenouspeoples.

106 1.2 billion acres of forest: "Indigenous Peoples' Forest Tenure."
Project Drawdown, 30 June 2020, www.drawdown.org/solutions/
indigenous-peoples-forest-tenure.

106 When forests are managed: Blackman, Allen. "Titled Amazon
Indigenous Communities Cut Forest Carbon Emissions." ScienceDirect,
1 November 2018, www.sciencedirect.com/science/article/abs/pii/
S0921800917309746.

107 tenure-secure lands generated: Veit, Peter, and Katie Reytar.
"By the Numbers: Indigenous and Community Land Rights." World
Resources Institute, 20 March 2017, www.wri.org/insights/numbers-
indigenous-and-community-land-rights.

107 most cost-effective mechanisms: "New Study Finds 55% of
Carbon in Amazon Is in Indigenous Territories and Protected Lands,
Much of It at Risk." Environmental Defense Fund, www.edf.org/
media/new-study-finds-55-carbon-amazon-indigenous-territories-and-
protected-lands-much-it-risk. Accessed 18 June 2021.

107 The world's oceans provide: "How Much Oxygen Comes from the Ocean?" National Oceanic and Atmospheric Administration, 26 February 2021, oceanservice.noaa.gov/facts/ocean-oxygen.html.

107 oceans have mostly functioned as a receptacle: Sabine, Chris. "Ocean-Atmosphere CO_2 Exchange Dataset, Science on a Sphere." National Oceanic and Atmospheric Administration, 2020, sos.noaa. gov/datasets/ocean-atmosphere-co2-exchange.

107 the seas nearest our coastlines: Thomas, Ryan. *Marine Biology: An Ecological Approach.* Waltham Abbey, U.K.: ED-TECH Press, 2019.

107 prevent 1 gigaton of carbon emissions: "The Ocean as a Solution to Climate Change." World Resources Institute: Ocean Panel Secretariat, 2019, live-oceanpanel.pantheonsite.io/sites/default/ files/2019-10/19_4PAGER_HLP_web.pdf.

107 the deep seas: Diaz, Cristobal. "Open Ocean." National Oceanic and Atmospheric Administration, 26 February 2021, oceana.aorg/ marine-life/marine-science-and-ecosystems/open-ocean.

107 contain thousands of times more carbon: "The Carbon Cycle." NASA Earth Observatory, earthobservatory.nasa.gov/features/ CarbonCycle. Accessed 22 June 2021.

107 Deep-sea mining and fishing: Sala, Enric, et al. "Protecting the Global Ocean for Biodiversity, Food and Climate." *Nature* 592, no. 7854, 2021, 397–402, doi:10.1038/s41586-021-03371-z.

108 enormous nets release 1.5 gigatons: Sala, Enric, et al. "Protecting the Global Ocean for Biodiversity, Food and Climate."

108 Australia's Great Barrier Reef: Cave, Damien, and Justin Gillis. "Large Sections of Australia's Great Reef Are Now Dead, Scientists Find." *New York Times*, 22 August 2020, www.nytimes.com/2017/03/15/ science/great-barrier-reef-coral-climate-change-dieoff.html.

108 "Instead of spending": Sala, Enric. "Let's Turn the High Seas into the World's Largest Nature Reserve." TED Talks, 28 June 2018, https://www.ted.com/talks/enric_sala_let_s_turn_the_high_seas_into_ the_world_s_largest_nature_reserve.

109 "The jury is in": Bland, Alastair. "Could a Ban on Fishing in International Waters Become a Reality?" NPR, 14 September 2018, www.npr.org/sections/thesalt/2018/09/14/647441547/could-a-ban-on- fishing-in-international-waters-become-a-reality.

109 Russia, China, Taiwan, Japan, Korea, and Spain: "The Economics of Fishing the High Seas." *Science Advances* 4, no. 6, 6 June 2018, advances.sciencemag.org/content/4/6/eaat2504.

110 international ban on bottom trawling: Bland, Alastair. "Could a Ban on Fishing in International Waters Become A Reality?"

111 Sea kelp absorbs: Hurlimann, Sylvia. "How Kelp Naturally Combats Global Climate Change." Science in the News, 4 July 2019, sitn.hms.harvard.edu/flash/2019/how-kelp-naturally-combats-global-climate-change. https://sitn.hms.harvard.edu/flash/2019/how-kelp-naturally-combats-global-climate-change/.

111 cited by Project Drawdown as a "coming attraction": Hawken, Paul. *Drawdown: The Most Comprehensive Plan Ever Proposed to Reverse Global Warming.* New York: Penguin Books, 2017.

111 "can be considered permanently sequestered": Bryce, Emma. "Can the Forests of the World's Oceans Contribute to Alleviating the Climate Crisis?" GreenBiz, 16 July 2020, www.greenbiz.com/article/can-forests-worlds-oceans-contribute-alleviating-climate-crisis.

113 second-largest store of carbon in the world: "Peatland Protection and Rewetting." Project Drawdown, 1 March 2020, www.drawdown.org/solutions/peatland-protection-and-rewetting.

113 Drained peatlands, for example: Günther, Anke. "Prompt Rewetting of Drained Peatlands Reduces Climate Warming despite Methane Emissions." Nature Communications, 2 April 2020, www.nature.com/articles/s41467-020-15499-z?error=cookies_not_supported&code=3a9e399b ff81-4cb7-a65a-2cdc90c77af1.

114 8 to 9 million plant or animal species: Zimmer, Carl. "How Many Species? A Study Says 8.7 Million, but It's Tricky." *New York Times*, 29 August 2011, www.nytimes.com/2011/08/30/science/30species.html.

114 more than 1 million species: "UN Report: Nature's Dangerous Decline 'Unprecedented'; Species Extinction Rates 'Accelerating.'" United Nations Sustainable Development Group, 6 May 2019, www.un.org/sustainabledevelopment/blog/2019/05/nature-decline-unprecedented-report.

115 "prevent a mass extinction crisis": "50 Countries Announce Bold Commitment to Protect at Least 30% of the World's Land and Ocean by 2030." Campaign for Nature, 10 June 2021, www.campaignfornature.org.

Chapter 5: Clean Up Industry

117 "the vision-bearer of banning plastics": "King Kibe Meets the Guy behind #BANPLASTICKE, James Wakibia." YouTube, 13 September 2017, www.youtube.com/watch?v=aOMSp-IssHU.

Notes

117 "Photography has helped me": "Meet James Wakibia, the Campaigner Behind Kenya's Plastic Bag Ban." United Nations Environment Programme, 4 May 2018, www.unep.org/news-and-stories/story/meet-james-wakibia-campaigner-behind-kenyas-plastic-bag-ban.

118 world's strictest ban on single-use plastics: Reality Check Team. "Has Kenya's Plastic Bag Ban Worked?" BBC News, 28 August 2019, www.bbc.com/news/world-africa-49421885.

119 80 percent of the population: Reality Check Team. "Has Kenya's Plastic Bag Ban Worked?"

119 Wakibia was hailed: "Meet James Wakibia, the Campaigner behind Kenya's Plastic Bag Ban." United Nations Environment Programme, 4 May 2018, www.unep.org/news-and-stories/story/meet-james-wakibia-campaigner-behind-kenyas-plastic-bag-ban.

119 "I just want to say one word: plastics": Nichols, Mike. *The Graduate*. Los Angeles: Embassy Pictures, 1967.

119 half of all plastic in human history: Parker, Laura. "The World's Plastic Pollution Crisis Explained." *National Geographic*, 7 June 2019, www.nationalgeographic.com/environment/article/plastic-pollution.

121 Our Steel KR (5.1) targets the single: "Emissions Gap Report 2019." United Nations Environment Programme, 2019, www.unep.org/resources/emissions-gap-report-2019.

121 The Cement KR (5.2) applies to: "Emissions Gap Report 2019."

122 Nearly 75 percent of all aluminum: Leahy, Meredith. "Aluminum Recycling in the Circular Economy." Rubicon, 11 September 2019, www.rubicon.com/blog/aluminum-recycling.

122 7 million tons of scrap per year: Joyce, Christopher. "Where Will Your Plastic Trash Go Now That China Doesn't Want It?" NPR, 13 March 2019, https://www.npr.org/sections/goatsandsoda/209/03/13/702501726/where-will-your-plastic-trash-go-now-that-china-doesnt-want-it.

122 blighted China's rivers: Joyce, Christopher. "Where Will Your Plastic Trash Go Now That China Doesn't Want It?"

122 lobbied for recycling symbols: Sullivan, Laura. "How Big Oil Misled the Public into Believing Plastic Would Be Recycled." NPR, 11 September 2020, www.npr.org/2020/09/11/897692090/how-big-oil-misled-the-public-into-believing-plastic-would-be-recycled.

122 The ubiquitous 1–7 code: Hocevar, John. "Circular Claims Fall Flat: Comprehensive U.S. Survey of Plastics Recyclability." Greenpeace

Inc., 18 February 2020, www.greenpeace.org/usa/research/report-circular-claims-fall-flat.

122 U.S. consumers took to chucking: Katz, Cheryl. "Piling Up: How China's Ban on Importing Waste Has Stalled Global Recycling." Yale Environment 360, 7 March 2019, e360.yale.edu/features/piling-up-how-chinas-ban-on-importing-waste-has-stalled-global-recycling.

123 Coca-Cola experimented with bottles: Herring, Chris. "Coke's New Bottle Is Part Plant." *Wall Street Journal*, 24 January 2010, www.wsj.com/articles/SB10001424052748703672104574654212774510476.

123 Other bioplastics may be even worse: Cho, Renee. "The Truth About Bioplastics." Columbia Climate School, 13 December 2017, news.climate.columbia.edu/2017/12/13/the-truth-about-bioplastics.

124 it struggles to decompose: Oakes, Kelly. "Why Biodegradables Won't Solve the Plastic Crisis." BBC Future, 5 November 2019, www.bbc.com/future/article/20191030-why-biodegradables-wont-solve-the-plastic-crisis.

124 PLA for single-use food servings: Oakes, Kelly. "Why Biodegradables Won't Solve the Plastic Crisis." BBC Future, 5 November 2019, www.bbc.com/future/article/20191030-why-biodegradables-wont-solve-the-plastic-crisis.

124 Twelve percent is incinerated: Geyer, Roland, et al. "Production, Use, and Fate of All Plastics Ever Made." *Science Advances* 3, no. 7, 2017, p. e1700782, doi:10.1126/sciadv.1700782.

124 It kills up to a million seabirds: "Plastic Pollution Affects Sea Life Throughout the Ocean." Pew Charitable Trusts, 24 September 2018, www.pewtrusts.org/en/research-and-analysis/articles/2018/09/24/plastic-pollution-affects-sea-life-throughout-the-ocean; "New UN Report Finds Marine Debris Harming More Than 800 Species, Costing Countries Millions." 5 December 2016, https://news.un.org/en/story/2016/12/547032-new-un-report-finds-marine-debris-harming-more-00-species-costing-countries.

125 European Union has banned single-use straws: Leung, Hillary. "E.U. Sets Standard with Ban on Single-Use Plastics by 2021." *Time*, 28 March 2019, time.com/5560105/european-union-plastic-ban.

125 127 countries have some kind of regulation: Excell, Carole. "127 Countries Now Regulate Plastic Bags. Why Aren't We Seeing Less Pollution?" World Resources Institute, 11 March 2019, www.wri.org/insights/127-countries-now-regulate-plastic-bags-why-arent-we-seeing-less-pollution.

Notes

125 advent of "fast fashion": Thomas, Dana. "The High Price of Fast Fashion." *Wall Street Journal*, 29 August 2019, www.wsj.com/articles/the-high-price-of-fast-fashion-11567096637.

125 recent *Vogue* report: Webb, Bella. "Fashion and Carbon Emissions: Crunch Time." Vogue Business, 26 August 2020, www.voguebusiness.com/sustainability/fashion-and-carbon-emissions-crunch-time.

126 "We treat the environment": Schwartz, Evan. "Anchoring OKRs to Your Mission." What Matters, 26 June 2020, www.whatmatters.com/articles/okrs-mission-statement-allbirds-sustainability.

126 shared its carbon footprint calculator: Verry, Peter. "Allbirds Is Making Its Carbon Footprint Calculator Open-Source Ahead of Earth Day." Footwear News, 18 April 2021, footwearnews.com/2021/business/sustainability/allbirds-carbon-footprint-calculator-open-source-earth-day-1203132233; "Carbon Footprint Calculator & Tools." Allbirds, 2021, www.allbirds.com/pages/carbon-footprint-calculator.

127 Heat for industrial processes: Bellevrat, Elie, and Kira West. "Clean and Efficient Heat for Industry." International Energy Agency, 23 January 2018, www.iea.org/commentaries/clean-and-efficient-heat-for-industry.

127 electrify the manufacturing processes: Roelofsen, Occo, et al. "Plugging in: What Electrification Can Do for Industry." McKinsey & Company, 28 May 2020, www.mckinsey.com/industries/electric-power-and-natural-gas/our-insights/plugging-in-what-electrification-can-do-for-industry#.

128 cost of producing clean hydrogen: "1H 2021 Hydrogen Levelized Cost Update." BloombergNEF, www.bnef.com/insights/26011. Accessed 14 June 2021.

130 For every ton of concrete that's produced: "Available and Emerging Technologies for Reducing Greenhouse Gas Emissions from the Portland Cement Industry." U.S. Environmental Protection Agency, October 2010, www.epa.gov/sites/production/files/2015-12/documents/cement.pdf.

130 Institutional Investors Group on Climate Change: "Investors Call on Cement Companies to Address Business-Critical Contribution to Climate Change." Institutional Investors Group on Climate Change, 22 July 2019, www.iigcc.org/news/investors-call-on-cement-companies-to-address-business-critical-contribution-to-climate-change.

130 "We don't take this challenge lightly": Frangoul, Anmar. "'We Have to Improve Our Operations to Be More Sustainable,' LafargeHolcim CEO Says." CNBC, 31 July 2020, www.cnbc.com/2020/07/31/lafargeholcim-ceo-stresses-importance-of-sustainability.html.

130 the company committed to net-zero emissions: "LafargeHolcim Signs Net Zero Pledge with Science-Based Targets." BusinessWire, 21 September 2020, www.businesswire.com/news/home/20200921005750/en/LafargeHolcim-Signs-Net-Zero-Pledge-with-Science-Based-Targets.

134 use an electric current to melt recycled scrap steel: "Steel Production." American Iron and Steel Institute, 2 November 2020, www.steel.org/steel-technology/steel-production; Hites, Becky. "The Growth of EAF Steelmaking." Recycling Today, 30 April 2020, www.recyclingtoday.com/article/the-growth-of-eaf-steelmaking.

135 1.8 billion tons of global production: "Steel Statistical Yearbook 2020 Concise Version." WorldSteel Association, www.worldsteel.org/en/dam/jcr:5001dac8-0083-46f3-aadd-35aa357acbcc/Steel%2520Statistical%2520Yearbook%25202020%2520%2528concise%2520version%2529.pdf. Accessed 21 June 2021.

135 the first time that hydrogen was used to heal steel: "First in the World to Heat Steel Using Hydrogen." Ovako, 2021, www.ovako.com/en/newsevents/stories/first-in-the-world-to-heat-steel-using-hydrogen.

135 called it "ridiculous" to skip LNG hydrogen: Collins, Leigh. " 'Ridiculous to Suggest Green Hydrogen Alone Can Meet World's H2 Needs.' " Recharge, 27 April 2020, www.rechargenews.com/transition/-ridiculous-to-suggest-green-hydrogen-alone-can-meet-world-s-h2-needs-/2-1-797831.

135 "We are embarking": "Speech by Prime Minister Stefan Löfven at Inauguration of New HYBRIT Pilot Plant." Government Offices of Sweden, 31 August 2020, www.government.se/speeches/2020/08/speech-by-prime-minister-stefan-lofven-at-inauguration-of-new-hybrit-pilot-plant.

135 "We must seize this chance": "HYBRIT: SSAB, LKAB and Vattenfall to Start up the World's First Pilot Plant for Fossil-Free Steel." SSAB, 21 August 2020, www.ssab.com/news/2020/08/hybrit-ssab-lkab-and-vattenfall-to-start-up-the-worlds-first-pilot-plant-for-fossilfree-steel.

Chapter 6: Remove Carbon

139 what exactly is carbon dioxide removal: Wilcox, J., et al. "CDR Primer." CDR, 2021, cdrprimer.org/read/concepts.

Notes

140 "the steepest climb of all": Cembalest, Michael. "Eye on the Market: 11th Annual Energy Paper." J.P. Morgan Assset Management, 2021, am.jpmorgan.com/us/en/asset-management/institutional/insights/market-insights/eye-on-the-market/annual-energy-outlook.

140 In both natural and engineered carbon removal: Wilcox, J., et al. "CDR Primer." CDR, 2021, cdrprimer.org/read/chapter-1.

143 The process would soak up: Sönnichsen, N. "Distribution of Primary Energy Consumption in 2019, by Country." Statista, 2021, www.statista.com/statistics/274200/countries-with-the-largest-share-of-primary-energy-consumption.

143 direct air capture is $600 per ton: Lebling, Katie. "Direct Air Capture: Resource Considerations and Costs for Carbon Removal." World Resources Institute, 6 January 2021, www.wri.org/insights/direct-air-capture-resource-considerations-and-costs-carbon-removal.

143 "fraught with moral, practical and political difficulties": Masson-Delmotte, Valérie. "Global Warming of 1.5°C." Intergovernmental Panel on Climate Change, 2018, www.ipcc.ch/site/assets/uploads/sites/2/2019/06/SR15_Full_Report_Low_Res.pdf.

143 programs that enable companies: Wilcox, J., et al. "CDR Primer." CDR, 2021, cdrprimer.org/read/glossary.

144 prone to overestimates and even fraud: Badgley, Grayson, et al. "Systematic Over-Crediting in California's Forest Carbon Offsets Program." BioRxiv, doi.org/10.1101/2021.04.28.441870.

145 to remove the emissions of Americans alone: Gates, Bill. *How to Avoid a Climate Disaster: The Solutions We Have and the Breakthroughs We Need*. New York: Knopf, 2021.

145 With efforts under way from China to Ethiopia: Welz, Adam. "Are Huge Tree Planting Projects More Hype than Solution?" Yale E360, 8 April 2021, e360.yale.edu/features/are-huge-tree-planting-projects-more-hype-than-solution. https://e360.yale.edu/features/are-huge-tree-planting-projects-more-hype-than-solution.

145 eighteen gigantic fanlike units: Gertner, Jon. "The Tiny Swiss Company That Thinks It Can Help Stop Climate Change." *New York Times*, 14 February 2019, www.nytimes.com/2019/02/12/magazine/climeworks-business-climate-change.html.

146 4,000 tons of CO_2 a year: Doyle, Alister. "Scared by Global Warming? In Iceland, One Solution Is Petrifying." Reuters, 4 February 2021, https://www.reuters.com/article/us-climate-change-technology-emissions-f/scared-by-global-warming-in-iceland-one-solution-is-petrifying-idUSKBN2A415R.

146 capacity to remove 1 million tons of CO$_2$ per year: Carbon Engineering Ltd. "Carbon Engineering Breaks Ground at Direct Air Capture Innovation Centre." Oceanfront Squamish, 11 June 2021, oceanfrontsquamish.com/stories/carbon-engineering-breaking-ground-on-their-innovation-centre.

148 "We're really founding a new industry": Gertner, Jon. "The Tiny Swiss Company That Thinks It Can Help Stop Climate Change." *New York Times*, 14 February 2019, www.nytimes.com/2019/02/12/magazine/climeworks-business-climate-change.html.

148 two thousand businesses purchasing carbon removal: "Stripe Commits $8M to Six New Carbon Removal Companies." Stripe, 26 May 2021, stripe.com/newsroom/news/spring-21-carbon-removal-purchases.

151 to become carbon *negative* by 2030: Smith, Brad. "Microsoft Will Be Carbon Negative by 2030." *Official Microsoft Blog*, 16 January 2020, blogs.microsoft.com/blog/2020/01/16/microsoft-will-be-carbon-negative-by-2030.

151 Microsoft's first crop of removal projects: "Microsoft Carbon Removal: Lessons from an Early Corporate Purchase." Microsoft, 2021, query.prod.cms.rt.microsoft.com/cms/api/am/binary/RE4MDlc.

Chapter 7: Win Politics and Policy

157 testified at a U.S. Senate hearing: "Investing in Green Technology as a Strategy for Economic Recovery." U.S. Senate Committee on Environment and Public Works, 2009, www.epw.senate.gov/public/index.cfm/2009/1/full-committee-briefing-entitled-investing-in-green-technology-as-a-strategy-for-economic-recovery.

157 the Earth Summit: Editors of Encyclopaedia Britannica, "United Nations Conference on Environment and Development | History & Facts." Britannica.com, 27 May 2021, www.britannica.com/event/United-Nations-Conference-on-Environment-and-Development.

157 They put their distinguished heads together: Palmer, Geoffrey. "The Earth Summit: What Went Wrong at Rio?" *Washington University Law Review* 70, no. 4, 1992, openscholarship.wustl.edu/cgi/viewcontent.cgi?article=1867&context=law_lawreview; UNCED Secretary General Maurice Strong, https://openscholarship.wustl.edu/cgi/viewcontent.cgi?article=1867&context=law_lawreview.

158 Bush threatened to boycott the convention: Palmer, Geoffrey. "The Earth Summit: What Went Wrong at Rio?"

158 "Rich countries and poor countries": Plumer, Brad. "The 1992 Earth Summit Failed. Will This Year's Edition Be Different?" *Washing-*

ton Post, 7 June 2012, www.washingtonpost.com/blogs/ezra-klein/post/
the-1992-earth-summit-failed-will-this-years-edition-be-
different/2012/06/07/gJQAARikLV_blog.html.

158 the Senate voted to block ratification: Dewar, Helen, and Kevin
Sullivan. "Senate Republicans Call Kyoto Pact Dead." *Washington Post*,
1997, www.washingtonpost.com/wp-srv/inatl/longterm/climate/
stories/clim121197b.htm.

158 "significantly reduce the risks": "Paris Agreement." United
Nations Framework Convention on Climate Change (UNFCCC),
December 2015, cop23.unfccc.int/sites/default/files/english_paris_
agreement.pdf.

159 "If you did everything": Lustgarten, Abraham. "John Kerry,
Biden's Climate Czar, Talks About Saving the Planet." *ProPublica*, 18
December 2020, www.propublica.org/article/john-kerry-biden-
climate-czar.

164 California's green building codes: "Achieving Energy
Efficiency," California Energy Commission, https://www.energy.ca.
gov/about/core-responsibility-fact-sheets/achieving-energy-efficiency.
Accessed 22 June 2021.

164 average annual household electricity bill: "California's Energy
Efficiency Success Story: Saving Billions of Dollars and Curbing Tons
of Pollution." Natural Resources Defense Council, July 2013, www.
nrdc.org/sites/default/files/ca-success-story-FS.pdf.

165 "fugitive emissions," mostly methane: "Methane Emissions
from Oil and Gas—Analysis." International Energy Agency. www.iea.
org/reports/methane-emissions-from-oil-and-gas. Accessed 21 June
2021.

165 Fossil fuels receive $296 billion: Coady, David, et al. "Global
Fossil Fuel Subsidies Remain Large: An Update Based on Country-
Level Estimates." International Monetary Fund, 2 May 2019, www.imf.
org/en/Publications/WP/Issues/2019/05/02/Global-Fossil-Fuel-
Subsidies-Remain-Large-An-Update-Based-on-Country-Level-
Estimates-4650.

165 health-care costs from air pollution: Coady, David, et al. "Global
Fossil Fuel Subsidies Remain Large: An Update Based on Country-
Level Estimates." International Monetary Fund, 2 May 2019, www.imf.
org/en/Publications/WP/Issues/2019/05/02/Global-Fossil-Fuel-
Subsidies-Remain-Large-An-Update-Based-on-Country-Level-
Estimates-4650.

165 the United States alone spends $81 billion: DiChristopher, Tom.
"US Spends $81 Billion a Year to Protect Global Oil Supplies, Report
Estimates." CNBC, 21 September 2018, www.cnbc.com/2018/09/21/

us-spends-81-billion-a-year-to-protect-oil-supplies-report-estimates. html.

169 Five entities account for nearly two thirds: UNEP and UNEP DTU Partnership. "UNEP Report—The Emissions Gap Report 2020." *Management of Environmental Quality: An International Journal*, 2020, https://www.unep.org/emissions-gap-report-2020.

169 energy industry accounts for 80 percent of Russia's greenhouse gas pollution: "Summary of GHG Emissions for Russian Federation." United Nations Framework Convention on Climate Change, 2018, di.unfccc.int/ghg_profiles/annexOne/RUS/RUS_ghg_profile.pdf.

169 Europe, where emissions from transportation rose in both 2018 and 2019: "Average Car Emissions Kept Increasing in 2019, Final Data Show." European Environment Agency, 1 June 2021, www.eea.europa.eu/highlights/average-car-emissions-kept-increasing.

170 policies of the top 5 emitters:
7.1: Frangoul, Anmar. "President Xi Tells UN That China Will Be 'Carbon Neutral' within Four Decades." CNBC, 23 September 2020, www.cnbc.com/2020/09/23/china-claims-it-will-be-carbon-neutral-by-the-year-2060.html; "FACT SHEET: President Biden Sets 2030 Greenhouse Gas Pollution Reduction Target Aimed at Creating Good-Paying Union Jobs and Securing U.S. Leadership on Clean Energy Technologies." White House, 22 April 2021, www.whitehouse.gov/briefing-room/statements-releases/2021/04/22/fact-sheet-president-biden-sets-2030-greenhouse-gas-pollution-reduction-target-aimed-at-creating-good-paying-union-jobs-and-securing-u-s-leadership-on-clean-energy-technologies; "2050 Long-Term Strategy." European Commission, 23 November 2016, ec.europa.eu/clima/policies/strategies/2050_en.

7.1.1: "China's Xi Targets Steeper Cut in Carbon Intensity by 2030." Reuters, 12 December 2020, www.reuters.com/world/china/chinas-xi-targets-steeper-cut-carbon-intensity-by-2030-2020-12-12; Shields, Laura. "State Renewable Portfolio Standards and Goals." National Conference of State Legislatures, 7 April 2021, www.ncsl.org/research/energy/renewable-portfolio-standards.aspx; "2030 Climate & Energy Framework." European Commission, 16 February 2017, ec.europa.eu/clima/policies/strategies/2030_en; "India Targeting 40% of Power Generation from Non-Fossil Fuel by 2030: PM Modi." Economic Times, 2 October 2018, economictimes.indiatimes.com/industry/energy/power/india-targeting-40-of-power-generation-from-non-fossil-fuel-by-2030-pm-modi/articleshow/66043374.cms?from=mdr.

7.1.2: "Electric Vehicles." Guide to Chinese Climate Policy, 2021, chineseclimatepolicy.energypolicy.columbia.edu/en/electric-vehicles; Tabeta, Shunsuke. "China Plans to Phase Out Conventional Gas-

Notes

Burning Cars by 2035." Nikkei Asia, 27 October 2020, asia.nikkei. com/Business/Automobiles/China-plans-to-phase-out-conventional-gas-burning-cars-by-2035; "Overview—Electric Vehicles: Tax Benefits & Purchase Incentives in the European Union." ACEA—European Automobile Manufacturers' Association, 9 July 2020, www.acea.auto/ fact/overview-electric-vehicles-tax-benefits-purchase-incentives-in-the-european-union; "Faster Adoption and Manufacturing of Hybrid and EV (FAME) II." International Energy Agency, 30 June 2020, www.iea.org/policies/7450-faster-adoption-and-manufacturing-of-hybrid-and-ev-fame-ii; Kireeva, Anna. "Russia Cancels Import Tax for Electric Cars in Hopes of Enticing Drivers." Bellona.org, 16 April 2020, bellona.org/news/transport/2020-04-russia-cancels-import-tax-for-electric-cars-in-hopes-of-enticing-drivers.

7.1.3: "A New Industrial Strategy for Europe." European Commission, 10 March 2020, ec.europa.eu/info/sites/default/files/communication-eu-industrial-strategy-march-2020_en.pdf.

7.1.4: "Zero Net Energy." California State Portal, 2021, www.cpuc.ca. gov/zne; Energy Efficiency Division. "High Performance Buildings." Mass.gov, 2021, www.mass.gov/high-performance-buildings; "Nzeb." European Commission, 17 October 2016, ec.europa.eu/energy/content/ nzeb-24_en.

7.1.5: University of Copenhagen Faculty of Science. "Carbon Labeling Reduces Our CO_2 Footprint—Even for Those Who Try to Remain Uninformed." ScienceDaily, 29 March 2021, www.sciencedaily.com/ releases/2021/03/210329122841.htm.

7.1.6: Adler, Kevin. "US Considers Stepping up Methane Emissions Reductions." IHS Markit, 7 April 2021, ihsmarkit.com/research-analysis/us-considers-stepping-up-methane-emissions-reductions. html; "Press Corner." European Commission, 14 October 2020, ec. europa.eu/commission/presscorner/detail/en/QANDA_20_1834.

7.2: Coady, David, et al. "Global Fossil Fuel Subsidies Remain Large: An Update Based on Country-Level Estimates." IMF Working Papers 19, no. 89, 2019, 1, doi:10.5089/9781484393178.001.

7.3: Buckley, Chris. "China's New Carbon Market, the World's Largest: What to Know." *New York Times*, 26 July 2021, www.nytimes.com/ 2021/07/16/business/energy-environment/china-carbon-market .html.

7.4: "EU Legislation to Control F-Gases." Climate Action—European Commission, 16 February 2017, ec.europa.eu/clima/policies/f-gas/ legislation_en.

7.5: "R&D and Technology Innovation—World Energy Investment 2020." World Energy Investment, 2020, www.iea.org/reports/

world-energy-investment-2020/rd-and-technology-innovation; "India 2020: Energy Policy Review." International Energy Agency, 2020, iea. blob.core.windows.net/assets/2571ae38-c895-430e-8b62-bc19019c6807/India_2020_Energy_Policy_Review.pdf.

172 China would aim to reach carbon net zero by 2060: "The Secret Origins of China's 40-Year Plan to End Carbon Emissions." *Bloomberg Green*, 22 November 2020, www.bloomberg.com/news/features/2020-11-2/china-s-2060-climate-pledge-inside-xi-jinping-s-secret-plan-to-end-emissions.

172 Finding jobs for more than 2 million miners: Feng, Hao. "2.3 Million Chinese Coal Miners Will Need New Jobs by 2020." China Dialogue, 7 August 2017, chinadialogue.net/en/energy/9967-2-3-million-chinese-coal-miners-will-need-new-jobs-by-2-2.

172 relies on it for 60 percent of its electricity: "International—U.S. Energy Information Administration (EIA)." China, www.eia.gov/international/analysis/country/CHN. Accessed 18 June 2021.

173 Chinese enterprises were financing: McSweeney, Eoin. "Chinese Coal Projects Threaten to Wreck Plans for a Renewable Future in Sub-Saharan Africa." CNN, 9 December 2020, edition.cnn.com/2020/12/09/business/africa-coal-energy-goldman-prize-dst-hnk-intl/index.html.

173 Xie Zhenhua is the most powerful voice: "The Secret Origins of China's 40-Year Plan to End Carbon Emissions."

174 killed an estimated 49,000 people: "CORRECTED: Smog Causes an Estimated 49,000 Deaths in Beijing, Shanghai in 2020—Tracker." Reuters, 9 July 2020, www.reuters.com/article/china-pollution/corrected-smog-causes-an-estimated-49000-deaths-in-beijing-shanghai-in-2020-tracker-idUSL4N2EG1T5.

174 more than 400 gigatons: Statista. "Global Cumulative CO_2 Emissions by Country 1750–2019." Statista, 29 March 2021, www.statista.com/statistics/1007454/cumulative-co2-emissions-worldwide-by-country.

174 George W. Bush took his cues: Goldenberg, Suzanne. "The Worst of Times: Bush's Environmental Legacy Examined." *Guardian*, 16 January 2009, www.theguardian.com/politics/2009/jan/16/greenpolitics-georgebush.

175 federal spending for energy research and development: Clark, Corrie E. "Renewable Energy R&D Funding History: A Comparison with Funding for Nuclear Energy, Fossil Energy, Energy Efficiency, and Electric Systems R&D." Congressional Research Service Report, 2018, fas.org/sgp/crs/misc/RS22858.pdf.

Notes

175 Americans spend in a week on gasoline: "Use of Gasoline—U.S. Energy Information Administration (EIA)." U.S. Energy Information Association, 26 May 2021, www.eia.gov/energyexplained/gasoline/use-of-gasoline.php.

175 less than we spend annually on potato chips: "Salty Snacks: U.S. Market Trends and Opportunities: Market Research Report." Packaged Facts, 21 June 2018, www.packagedfacts.com/Salty-Snacks-Trends-Opportunities-11724010.

175 match what it now allocates to the National Institutes of Health: "National Institutes of Health (NIH) Funding: FY1995-FY2021." Congressional Research Service, 2021, fas.org/sgp/crs/misc/R43341.pdf.

176 EU set its own 2050 net-zero target: Frangoul, Anmar. "EU Leaders Agree on 55% Emissions Reduction Target, but Activist Groups Warn It Is Not Enough." CNBC, 11 December 2020, www.cnbc.com/2020/12/11/eu-leaders-agree-on-55percent-greenhouse-gas-emissions-reduction-target.html.

176 cut emissions in line: "EU." Climate Action Tracker, 2020, climateactiontracker.org/countries/eu.

176 Berlin pledged carbon neutrality by 2045: Jordans, Frank. "Germany Maps Path to Reaching 'Net Zero' Emissions by 2045." AP News, 12 May 2021, apnews.com/article/europe-germany-climate-business-environment-and-nature-6437e64891d8117a9c0bff7cabb200eb.

176 Germany is almost certain to fall short: Amelang, Sören. "Europe's 55% Emissions Cut by 2030: Proposed Target Means Even Faster Coal Exit." Energy Post, 5 October 2020, energypost.eu/europes-55-emissions-cut-by-2030-proposed-target-means-even-faster-coal-exit.

177 poverty rate of more than 60 percent: Manish, Sai. "Coronavirus Impact: Over 100 Million Indians Could Fall Below Poverty Line." *Business Standard*, 2020, www.business-standard.com/article/economy-policy/coronavirus-impact-over-100-million-indians-could-fall-below-poverty-line-120041700906_1.html.

177 Modi announced a national moonshot: "India Exceeding Paris Targets; to Achieve 450 GW Renewable Energy by 2030: PM Modi at G20 Summit." *Business Today*, 22 November 2020, www.businesstoday.in/current/economy-politics/india-exceeding-paris-targets-to-achieve-450-gw-renewable-energy-by-2030-pm-modi-at-g20-summit/story/422691.html.

177 Historically, the United States has emitted 25 percent: Ritchie, Hannah. "Who Has Contributed Most to Global CO_2 Emissions?" Our

World in Data, 1 October 2019, ourworldindata.org/contributed-most-global-co2.

178 Meeting India's renewable goals: Jaiswal, Anjali. "Climate Action: All Eyes on India." Natural Resources Defense Council, 12 December 2020, www.nrdc.org/experts/anjali-jaiswal/climate-action-all-eyes-india.

179 doubting the science of climate change: "Russia's Putin Says Climate Change in Arctic Good for Economy." CBC, 30 March 2017, www.cbc.ca/news/science/russia-putin-climate-change-beneficial-economy-1.4048430.

179 Russian lands are warming: Agence France-Presse. "Russia Is 'Warming 2.5 Times Quicker' Than the Rest of the World." The World, 25 December 2015, www.pri.org/stories/2015-12-25/russia-warming-25-times-quicker-rest-world.

179 Arctic permafrost stores 1,400 gigatons of carbon: Struzik, Ed. "How Thawing Permafrost Is Beginning to Transform the Arctic." Yale Environment 360, 21 January 2020, e360.yale.edu/features/how-melting-permafrost-is-beginning-to-transform-the-arctic.

179 Russia's Energy Strategy 2035: Alekseev, Alexander N., et al. "A Critical Review of Russia's Energy Strategy in the Period Until 2035." *International Journal of Energy Economics and Policy* 9, no. 6, 2019, 95–102, doi:10.32479/ijeep.8263.

179 Russia's own 2050 projections: Ross, Katie. "Russia's Proposed Climate Plan Means Higher Emissions Through 2050." World Resources Institute, 13 April 2020, www.wri.org/insights/russias-proposed-climate-plan-means-higher-emissions-through-2050.

183 California showed that you could cut emissions: "California Leads Fight to Curb Climate Change." Environmental Defense Fund, 2021, www.edf.org/climate/california-leads-fight-curb-climate-change.

184 the climate bill was strangled: Weiss, Daniel. "Anatomy of a Senate Climate Bill Death." Center for American Progress, 12 October 2010, www.americanprogress.org/issues/green/news/2010/10/12/8569/anatomy-of-a-senate-climate-bill-death.

184 reduced California's greenhouse gas emissions: Song, Lisa. "Cap and Trade Is Supposed to Solve Climate Change, but Oil and Gas Company Emissions Are Up." *ProPublica*, 15 November 2019, www.propublica.org/article/cap-and-trade-is-supposed-to-solve-climate-change-but-oil-and-gas-company-emissions-are-up.

185 "We were seeing lots of interest": Descant, Skip. "In a Maryland County, the Yellow School Bus Is Going Green." GovTech, 17 June 2021, www.govtech.com/fs/in-a-maryland-county-the-yellow-school-bus-is-going-green.

186 Black neighborhoods got paved over: Beyer, Scott. "How the U.S. Government Destroyed Black Neighborhoods." Catalyst, 2 April 2020, catalyst.independent.org/2020/04/02/how-the-u-s-government-destroyed-black-neighborhoods.

188 disinformation campaigns by the likes of ExxonMobil and the Koch family: "Exxon's Climate Denial History: A Timeline." Greenpeace USA, 16 April 2020, www.greenpeace.org/usa/ending-the-climate-crisis/exxon-and-the-oil-industry-knew-about-climate-change/exxons-climate-denial-history-a-timeline; Mayer, Jane. " 'Kochland' Examines the Koch Brothers' Early, Crucial Role in Climate-Change Denial." *New Yorker*, 13 August 2019, www.newyorker.com/news/daily-comment/kochland-examines-how-the-koch-brothers-made-their-fortune-and-the-influence-it-bought.

188 *The Washington Post* found: Westervelt, Amy. "How the Fossil Fuel Industry Got the Media to Think Climate Change Was Debatable." *Washington Post*, 10 January 2019, www.washingtonpost.com/outlook/2019/01/10/how-fossil-fuel-industry-got-media-think climate-change-was-debatable.

189 nearly half the American public: Newport, Frank. "Americans' Global Warming Concerns Continue to Drop." Gallup, 11 March 2010, news.gallup.com/poll/126560/americans-global-warming-concerns-continue-drop.aspx.

189 nearly two thirds of Republicans: Funk, Cary, and Meg Hefferon. "U.S. Public Views on Climate and Energy." Pew Research Center Science & Society, 25 November 2019, www.pewresearch.org/science/2019/11/25/u-s-public-views-on-climate-and-energy.

189 create millions of well-paying jobs: "Net Zero by 2050—Analysis." International Energy Agency, May 2021, www.iea.org/reports/net-zero-by-2050.

Chapter 8: Turn Movements into Action

191 "But I don't want your hope": Workman, James. " 'Our House Is on Fire.' 16-Year-Old Greta Thunberg Wants Action." World Economic Forum, 25 January 2019, www.weforum.org/agenda/2019/01/our-house-is-on-fire-16-year-old-greta-thunberg-speaks-truth-to-power.

191 4 million people across the world: Sengupta, Somini. "Protesting Climate Change, Young People Take to Streets in a Global Strike." *New York Times*, 20 September 2019, www.nytimes.com/2019/09/20/climate/global-climate-strike.html.

191 "You have stolen my dreams": "Transcript: Greta Thunberg's Speech at the U.N. Climate Action Summit." NPR, 23 September 2019, https://www.npr.org/2019/09/23/763452863/transcript-greta-thunbergs-speech-at-the-u-n-climate-action-summit.

192 it passed a law: Department for Business, Energy & Industrial Strategy, and Chris Skidmore. "UK Becomes First Major Economy to Pass Net Zero Emissions Law." GOV.UK, 27 June 2019, www.gov.uk/government/news/uk-becomes-first-major-economy-to-pass-net-zero-emissions-law.

192 "We can't just continue": Alter, Charlotte, et al. "Greta Thunberg: TIME's Person of the Year 2019." *Time*, 11 December 2019, time.com/person-of-the-year-2019-greta-thunberg.

194 there is *political power*: Prakash, Varshini, and Guido Girgenti, eds. *Winning the Green New Deal: Why We Must, How We Can.* New York: Simon & Schuster, 2020.

194 National Labor Relations Act: Glass, Andrew. "FDR Signs National Labor Relations Act, July 5, 1935." *Politico*, 5 July 2018, www.politico.com/story/2018/07/05/fdr-signs-national-labor-relations-act-july-5-1935-693625.

194 according to a Harvard University study: Nicholasen, Michelle. "Why Nonviolent Resistance Beats Violent Force in Effecting Social, Political Change." *Harvard Gazette*, 4 February 2019, news.harvard.edu/gazette/story/2019/02/why-nonviolent-resistance-beats-violent-force-in-effecting-social-political-change.

196 only 3 percent of voters: Saad, Lydia. "Gallup Election 2020 Coverage." Gallup, 29 October 2020, news.gallup.com/opinion/gallup/321650/gallup-election-2020-coverage.aspx.

196 European Union's twenty-eight member countries: "Europeans and the EU Budget." Standard Eurobarometer 89, 2018, publications.europa.eu/resource/cellar/9cacfd6b-9b7d-11e8-a408-01aa75ed71a1.0002.01/DOC_1.

196 the issue leapt to number two: "Autumn 2019 Standard Eurobarometer: Immigration and Climate Change Remain Main Concerns at EU Level." European Commission, 20 December 2019, https://ec.europa.eu/commission/presscorner/detail/en/IP_19_6839.

196 pressed demands for cleaner air: Rooij, Benjamin van. "The People vs. Pollution: Understanding Citizen Action against Pollution in China." Taylor & Francis, 27 January 2010, www.tandfonline.com/doi/full/10.1080/10670560903335777.

196 National Clean Air Quality Action Plan: "China: National Air Quality Action Plan (2013)." Air Quality Life Index, 10 July 2020, aqli.

epic.uchicago.edu/policy-impacts/china-national-air-quality-action-plan-2014.

196 reduced smog in its big cities: Greenstone, Michael. "Four Years After Declaring War on Pollution, China Is Winning." *New York Times*, 12 March 2018, www.nytimes.com/2018/03/12/upshot/china-pollution-environment-longer-lives.html.

196 "Climate Change in the Chinese Mind": "Climate Change in the Chinese Mind Survey Report 2017." Energy Foundation China, 2017, www.efchina.org/Attachments/Report/report-comms-20171108/Climate_Change_in_the_Chinese_Mind_2017.pdf.

196 top concerns in 2019: Crawford, Alan. "Here's How Climate Change Is Viewed Around the World." *Bloomberg*, 25 June 2019, www.bloomberg.com/news/features/2019-06-26/here-s-how-climate-change-is-viewed-around-the-world.

197 she focused on small things, like recycling: First-Arai, Leanna. "Varshini Prakash Has a Blueprint for Change." *Sierra*, 4 November 2019, www.sierraclub.org/sierra/2019-4-july-august/act/varshini-prakash-has-blueprint-for-change.

198 "We didn't have time to waste": Prakash, Varshini. "Varshini Prakash on Redefining What's Possible." *Sierra*, 22 December 2020, www.sierraclub.org/sierra/2021-1-january-february/feature/varshini-prakash-redefining-whats-possible.

199 for the Green New Deal: Friedman, Lisa. "What Is the Green New Deal? A Climate Proposal, Explained." *New York Times*, 21 February 2021, www.nytimes.com/2019/02/21/climate/green-new-deal-questions-answers.html.

199 "we know you guys have your own plan": Krieg, Gregory. "The Sunrise Movement Is an Early Winner in the Biden Transition. Now Comes the Hard Part." CNN, 2 January 2021, edition.cnn.com/2021/01/02/politics/biden-administration-sunrise-movement-climate/index.html.

199 100 percent clean electricity: "2020 Presidential Candidates on Energy and Environmental Issues." Ballotpedia, 2021, ballotpedia.org/2020_presidential_candidates_on_energy_and_environmental_issues.

200 "keep one foot inside the halls of power": Krieg, Gregory. "The Sunrise Movement Is an Early Winner in the Biden Transition. Now Comes the Hard Part."

200 Sierra Club was preparing for: Hattam, Jennifer. "The Club Comes Together." *Sierra*, 2005, vault.sierraclub.org/sierra/200507/bulletin.asp.

200 "We were about to find ourselves": Bloomberg, Michael, and Carl Pope. *Climate of Hope*. New York: St. Martin's Press, 2017.

204 shut down all *existing* U.S. coal plants: "Bruce Nilles." Energy Innovation: Policy and Technology, 7 January 2021, energyinnovation. org/team-member/bruce-nilles.

206 just one hundred companies: Riley, Tess. "Just 100 Companies Responsible for 71% of Global Emissions, Study Says." *Guardian*, 10 July 2017, www.theguardian.com/sustainable-business/2017/jul/10/100-fossil-fuel-companies-investors-responsible-71-global-emissions-cdp-study-climate-change.

206 American Business Act on Climate: "American Business Act on Climate Pledge." White House, 2016, obamawhitehouse.archives.gov/climate-change/pledge.

206 Google has matched: Hölzle, Urs. "Google Achieves Four Consecutive Years of 100% Renewable Energy." Google Cloud Blog, cloud.google.com/blog/topics/sustainability/google-achieves-four-consecutive-years-of-100-percent-renewable-energy. Accessed 21 June 2021.

206 Apple has been carbon neutral: Jackson, Lisa. "Environmental Progress Report." Apple, 2020, www.apple.com/environment/pdf/Apple_Environmental_Progress_Report_2021.pdf.

206 "net-zero emissions" pledges: "Net zero emissions." Glossary, Intergovernmental Panel on Climate Change, 2021, www.ipcc.ch/sr15/chapter/glossary.

206 balance any residual emissions: "Foundations for Science Based Net-Zero Target Setting in the Corporate Sector." Science Based Targets, September 2020, sciencebasedtargets.org/resources/legacy/2020/09/foundations-for-net-zero-full-paper.pdf.

207 Amazon built a team of sustainability experts: Day, Matt. "Amazon Tries to Make the Climate Its Prime Directive." *Bloomberg Green*, 21 September 2020, www.bloomberg.com/news/features/2020-09-21/amazon-made-a-climate-promise-without-a-plan-to-cut-emissions.

207 a plan for Amazon to reach net zero by 2040: Palmer, Annie. "Jeff Bezos Unveils Sweeping Plan to Tackle Climate Change." CNBC, 19 September 2019, www.cnbc.com/2019/09/19/jeff-bezos-speaks-about-amazon-sustainability-in-washington-dc.html.

209 movement called the Climate Pledge: "The Climate Pledge." Amazon Sustainability, 2021, sustainability.aboutamazon.com/about/the-climate-pledge.

Notes

209 When Colgate-Palmolive signed: "Colgate-Palmolive." Climate Pledge, 2021, www.theclimatepledge.com/us/en/Signatories/colgate-palmolive.

209 When PepsiCo signed: "PepsiCo Announces Bold New Climate Ambition." PepsiCo, 14 January 2021, www.pepsico.com/news/story/pepsico-announces-bold-new-climate-ambition.

210 "Statement on the Purpose of a Corporation": "Business Roundtable Redefines the Purpose of a Corporation to Promote 'An Economy That Serves All Americans.'" Business Roundtable, 19 August 2019, www.businessroundtable.org/business-roundtable-redefines-the-purpose-of-a-corporation-to-promote-an-economy-that-serves-all-americans.

211 Sam Walton founded Walmart in 1962: Walton, Sam, and John Huey. *Sam Walton: Made in America*. New York: Bantam Books, 1993.

216 world's largest investment manager with $8.7: "About Us." BlackRock, 2021, www.blackrock.com/sg/en/about-us.

216 2021 open letter to the heads: Fink, Larry. "Larry Fink's 2021 Letter to CEOs." BlackRock, 2021, www.blackrock.com/corporate/investor-relations/larry-fink-ceo-letter.

219 total return has dropped by 20 percent: Engine No. 1, LLC. "Letter to the ExxonMobil Board of Directors." Reenergize Exxon, 7 December 2020, reenergizexom.com/materials/letter-to-the-board-of-directors.

219 a campaign called "Reenergize Exxon": Engine No. 1, LLC. "Letter to the ExxonMobil Board of Directors."

219 Engine No. 1 led a shareholder revolt: Merced, Michael. "How Exxon Lost a Board Battle with a Small Hedge Fund." *New York Times*, 28 May 2021, www.nytimes.com/2021/05/28/business/energy-environment/exxon-engine-board.html.

219 "a landmark moment": Krauss, Clifford, and Peter Eavis. "Climate Change Activists Notch Victory in ExxonMobil Board Elections." *New York Times*, 26 May 2021, www.nytimes.com/2021/05/26/business/exxon-mobil-climate-change.html.

219 "a social tipping point": Sengupta, Somini. "Big Setbacks Propel Oil Giants Toward a 'Tipping Point.'" *New York Times*, 29 May 2021, www.nytimes.com/2021/05/29/climate/fossil-fuel-courts-exxon-shell-chevron.html.

223 Each additional year of secondary school: Herz, Barbara, and Gene Sperling. "What Works in Girls' Education: Evidence and Policies from the Developing World by Barbara Herz." 30 June 2004. Paperback. Council on Foreign Relations, 2004.

223 Better-educated women marry later: Sperling, Gene, et al. "What Works in Girls' Education: Evidence for the World's Best Investment." Brookings Institution Press, 2015.

223 By 2025, that number is projected to grow: "Malala Fund Publishes Report on Climate Change and Girls' Education." Malala Fund, 2021, malala.org/newsroom/archive/malala-fund-publishes-report-on-climate-change-and-girls-education.

223 130 million girls are being denied: Evans, David K., and Fei Yuan. "What We Learn about Girls' Education from Interventions That Do Not Focus on Girls." Policy Research Working Papers, 2019, doi:10.1596/1813-9450-8944.

226 all girls—and boys—worldwide: Cohen, Joel E. "Universal Basic and Secondary Education." American Academy of Arts and Sciences, 2006, www.amacad.org/sites/default/files/publication/downloads/ubase_universal.pdf.

226 "Getting millions of girls into school": Sperling, Gene, et al. "What Works in Girls' Education: Evidence for the World's Best Investment." Brookings Institution Press, 2015.

226 "voluntary reproductive health resources": "Health and Education." Project Drawdown, 12 February 2020, drawdown.org/solutions/health-and-education/technical-summary.

226 one of five premature deaths worldwide: Chaisson, Clara. "Fossil Fuel Air Pollution Kills One in Five People." NRDC, www.nrdc.org/stories/fossil-fuel-air-pollution-kills-one-five-people. Accessed 20 June 2021.

226 toxic air killed more than 1.6 million people: Pandey, Anamika, et al. "Health and Economic Impact of Air Pollution in the States of India: The Global Burden of Disease Study 2019." *Lancet Planetary Health* 5, no. 1, 2021, e25–38, doi:10.1016/s2542-5196(20)30298-9.

226 The Black community in particular: Mikati et al. "Disparities in Distribution of Particulate Matter Emission Sources by Race and Poverty Status." *American Journal of Public Health* 108, 2018, 480–85, http://ajph.aphapublications.org/doi/pdf/10.2105/AJPH.2017.304297.

227 The economic opportunity of a transition: "Unlocking the Inclusive Growth Story of the 21st Century." New Climate Economy, 2018, newclimateeconomy.report/2018/key-findings.

227 65 million new jobs: "Unlocking the Inclusive Growth Story of the 21st Century." New Climate Economy, 2018, newclimateeconomy.report/2018/key-findings.

229 a crop of recorded talks: "Countdown." TED, 2021, www.ted.com/series/countdown.

229 "How to Be a Good Ancestor": Krznaric, Roman. "How to Be a Good Ancestor." TED Countdown, 10 October 2020, www.ted.com/talks/roman_krznaric_how_to_be_a_good_ancestor.

229 The justices quoted Roman's TED Talk: Supreme Court of Pakistan. *D. G. Khan Cement Company Ltd. Versus Government of Punjab through its Chief Secretary, Lahore, etc.* 2021. Climate Change Litigation Databases, http://climatecasechart.com/climate-change-litigation/non-us-case/d-g-khan-cement-company-v-government-of-punjab/.

229 Amanda Gorman, may have captured this feeling: "24 Hours of Reality: 'Earthrise' by Amanda Gorman." YouTube, 4 December 2018, www.youtube.com/watch?v=xwOvBv8RLmo.

Chapter 9: Innovate!

231 Advanced Research Projects Agency: Lyon, Matthew, and Katie Hafner. *Where Wizards Stay Up Late: The Origins of the Internet.* New York: Simon & Schuster, 1999, 20.

231 ARPANET, the 1960s precursor to the internet: "Paving the Way to the Modern Internet." Defense Advanced Research Projects Agency, 2021, www.darpa.mil/about-us/timeline/modern-internet.

231 groundwork for the Global Positioning System: "Where the Future Becomes Now." Defense Advanced Research Projects Agency, 2021, www.darpa.mil/about-us/darpa-history-and-timeline.

232 jump-started a global tech sector: Henry-Nickie, Makada, et al. "Trends in the Information Technology Sector." Brookings Institution, 29 March 2019, www.brookings.edu/research/trends-in-the-information-technology-sector.

232 led to ARPA-E: "ARPA-E History." ARPA-E, 2021, arpa-e.energy.gov/about/arpa-e-history.

232 U.S. spending on energy R&D was less: Clark, Corrie E. "Renewable Energy R&D Funding History: A Comparison with Funding for Nuclear Energy, Fossil Energy, Energy Efficiency, and Electric Systems R&D." Congressional Research Service Report, 2018, fas.org/sgp/crs/misc/RS22858.pdf.

232 A small chunk, $400 million, went to ARPA-E: "ARPA-E: Accelerating U.S. Energy Innovation." ARPA-E, 2021, arpa-e.energy.gov/technologies/publications/arpa-e-accelerating-us-energy-innovation.

235 TED talk on climate and energy: Gates, Bill. "Innovating to Zero!" TED, 18 February 2010, www.ted.com/talks/bill_gates_innovating_to_zero.

235 "breakthrough energy effort": Wattles, Jackie. "Bill Gates Launches Multi-Billion Dollar Clean Energy Fund." CNN Money, 30 November 2015, money.cnn.com/2015/11/29/news/economy/bill-gates-breakthrough-energy-coalition.

236 10,000 gigawatt hours (GWh): "2020 Battery Day Presentation Deck." Tesla, 22 September 2019, tesla-share.thron.com/content/?id=96ea71cf-8fda-4648-a62c-753af436c3b6&pkey=S1dbei4.

242 Volta's first battery: "BU-101: When Was the Battery Invented?" Battery University, 14 June 2019, batteryuniversity.com/learn/article/when_was_the_battery_invented.

242 Over the past twenty years, energy density: Field, Kyle. "BloombergNEF: Lithium-Ion Battery Cell Densities Have Almost Tripled Since 2010." CleanTechnica, 19 February 2020, cleantechnica.com/2020/02/19/bloombergnef-lithium-ion-battery-cell-densities-have-almost-tripled-since-2010.

244 ARPA-E awarded them $1.5 million: Heidel, Timothy, and Kate Chesley. "The All-Electron Battery." ARPA-E, 29 April 2010, arpa-e.energy.gov/technologies/projects/all-electron-battery.

245 QuantumScape and VW created a joint venture: "Volkswagen Partners with QuantumScape to Secure Access to Solid-State Battery Technology." Volkswagen Aktiengesellschaft, 21 June 2018, www.volkswagenag.com/en/news/2018/06/volkswagen-partners-with-quantumscape-.html.

245 VW committed another $200 million: Korosec, Kirsten. "Volkswagen-Backed QuantumScape to Go Public via SPAC to Bring Solid-State Batteries to EVs." *TechCrunch*, 3 September 2020, techcrunch.com/2020/09/03/vw-backed-quantumscape.

245 To electrify every new car: Xu, Chengjian, et al. "Future Material Demand for Automotive Lithium-Based Batteries." *Communications Materials* 1, no. 1, 2020, doi:10.1038/s43246-020-00095-x, https://www.nature.com/articles/s43246-020-00095-x.

245 Tesla's Gigafactory in Nevada: "Tesla Gigafactory." Tesla, 14 November 2014, www.tesla.com/gigafactory.

245 produce just 35 GWh of cells per year: Lambert, Fred. "Tesla Increases Hiring Effort at Gigafactory 1 to Reach Goal of 35 GWh of Battery Production." Electrek, 3 January 2018, electrek.co/2018/01/03/tesla-gigafactory-hiring-effort-battery-production.

Notes

245 a hundred like-size plants: Mack, Eric. "How Tesla and Elon Musk's 'Gigafactories' Could Save the World." *Forbes*, 30 October 2016, www.forbes.com/sites/ericmack/2016/10/30/how-tesla-and-elon-musk-could-save-the-world-with-gigafactories/?sh=67e44ead2de8.

245 "accelerate the transition to sustainable energy": "Welcome to the Gigafactory:| Before the Flood." YouTube, 27 October 2016, www.youtube.com/watch?v=iZm_NohNm6I&ab_channel=NationalGeographic.

245 Sixty percent of the world's supply: Frankel, Todd C., et al. "The Cobalt Pipeline." *Washington Post*, 30 September 2016, www.washingtonpost.com/graphics/business/batteries/congo-cobalt-mining-for-lithium-ion-battey.

245 But given the limited: Harvard John A. Paulson School of Engineering and Applied Sciences. "A Long-Lasting, Stable Solid-State Lithium Battery: Researchers Demonstrate a Solution to a 40-Year Problem." ScienceDaily, 12 May 2021, www.sciencedaily.com/releases/2021/05/210512115651.htm.

246 Sixty percent of Texans' homes: Webber, Michael E. "Opinion: What's Behind the Texas Power Outages?" MarketWatch, 16 February 2021, www.marketwatch.com/story/whats-behind-the-texas-power-outages-11613508031#.

246 1989 state energy code: "Texas: Building Energy Codes Program." U.S. Department of Energy, 2 August 2018, www.energycodes.gov/adoption/states/texas.

246 Over 150 people died: Steele, Tom. "Number of Texas Deaths Linked to Winter Storm Grows to 151, Including 23 in Dallas-Fort Worth Area." *Dallas News*, 30 April 2021, www.dallasnews.com/news/weather/2021/04/30/number-of-texas-deaths-linked-to-winter-storm-grows-to-151-including-23-in-dallas-fort-worth-area.

246 almost 10 gigawatts have been installed: "Energy Storage Projects." BloombergNEF, www.bnef.com/interactive-datasets/2d5d59acd900000c?data-hub=17. Accessed 14 June 2021.

247 "the largest battery in the world": "Bath County Pumped Storage Station." Dominion Energy, 2020, www.dominionenergy.com/projects-and-facilities/hydroelectric-power-facilities-and-projects/bath-county-pumped-storage-station.

247 Energy Vault, lifts, drops, and stacks: Energy Vault. energyvault.com.

247 Bloom Energy can use green hydrogen produced and stored on site to power their fuel cells: Baker, David R. "Bloom Energy Surges After Expanding into Hydrogen Production." *Bloomberg Green*,

15 July 2020, www.bloomberg.com/news/articles/2020-07-15/fuel-cell-maker-bloom-energy-now-wants-to-make-hydrogen-too.

247 three significant reactor accidents: "Safety of Nuclear Reactors." World Nuclear Association, March 2021, www.world-nuclear.org/information-library/safety-and-security/safety-of-plants/safety-of-nuclear-power-reactors.aspx.

248 the plant's six reactors shut down automatically: "Fukushima Daiichi Accident—World Nuclear Association." World Nuclear Association, www.world-nuclear.org/information-library/safety-and-security/safety-of-plants/fukushima-daiichi-accident.aspx. Accessed 20 June 2021.

248 planning to dump it into the sea: "The Reality of the Fukushima Radioactive Water Crisis." Greenpeace East Asia and Greenpeace Japan, October 2020, storage.googleapis.com/planet4-japan-stateless/2020/10/5768c541-the-reality-of-the-fukushima-radioactive-water-crisis_en_summary.pdf.

248 safety add-ons to Fukushima-style reactors: Garthwaite, Josie. "Would a New Nuclear Plant Fare Better than Fukushima?" *National Geographic*, 23 May 2011, www.nationalgeographic.com/science/article/110323-fukushima-japan-new-nuclear-plant-design.

248 More than fifty labs or startups: Bulletin of the Atomic Scientists. "Can North America's Advanced Nuclear Reactor Companies Help Save the Planet?" Pulitzer Center, 7 February 2017, pulitzercenter.org/stories/can-north-americas-advanced-nuclear-reactor-companies-help-save-planet.

248 startup called TerraPower: "TerraPower, CNNC Team Up on Travelling Wave Reactor." World Nuclear News, 25 September 2015, www.world-nuclear-news.org/NN-TerraPower-CNNC-team-up-on-travelling-wave-reactor-250915 1.html.

248 Bill told *60 Minutes*: "Bill Gates: How the World Can Avoid a Climate Disaster." *60 Minutes*, CBS News, 15 February 2021, www.cbsnews.com/news/bill-gates-climate-change-disaster-60-minutes-2021-02-14.

249 TerraPower demonstration plant: Gardner, Timothy, and Valerie Volcovici. "Bill Gates' Next Generation Nuclear Reactor to Be Built in Wyoming." Reuters, 2 June 2021, www.reuters.com/business/energy/utility-small-nuclear-reactor-firm-select-wyoming-next-us-site-2021-06-02.

250 absurdly high temperatures and pressure: Freudenrich, Patrick Kiger, and Craig Amp. "How Nuclear Fusion Reactors Work." HowStuffWorks, 26 January 2021, science.howstuffworks.com/fusion-reactor2.htm.

Notes

250 Commonwealth Fusion Systems: Commonwealth Fusion Systems, 2021, cfs.energy.

250 hydrogen from a gallon of seawater: "DOE Explains . . . Deuterium-Tritium Fusion Reactor Fuel." Office of Science, Department of Energy, 2021, www.energy.gov/science/doe-explainsdeuterium-tritium-fusion-reactor-fuel.

250 the solar cell in the 1950s: Gertner, Jon. *The Idea Factory: Bell Labs and the Great Age of American Innovation.* New York: Penguin Random House, 2020.

251 Depending on the process: "LCFS Pathway Certified Carbon Intensities: California Air Resources Board." CA.Gov, ww2.arb.ca.gov/resources/documents/lcfs-pathway-certified-carbon-intensities. Accessed 24 June 2021.

251 But as demand rises, so does the risk: "Economics of Biofuels." U.S. Environmental Protection Agency, 4 March 2021, www.epa.gov/environmental-economics/economics-biofuels.

252 two thirds of all energy produced from fossil fuels: "Estimated U.S. Consumption in 2020: 92.9 Quads." Lawrence Livermore National Laboratory, 2020, flowcharts.llnl.gov/content/assets/images/energy/us/Energy_US_2020.png.

252 The BMW i3 EV hatchback: "I3 and I3s Electric Sedan Features and Pricing." BMW USA, 2021, www.bmwusa.com/vehicles/bmwi/i3/sedan/pric ng-features.html.

252 Ford's popular F-150 pickup: Boudette, Neal. "Ford Bet on Aluminum Trucks, but Is Still Looking for Payoff." *New York Times,* 1 March 2018, www.nytimes.com/2018/03/01/business/ford-f150-aluminum-trucks.html.

253 LEDs accounted for 30 percent: "LED Adoption Report." Energy.Gov, www.energy.gov/eere/ssl/led-adoption-report. Accessed 24 June 2021.

253 latest iPhone was shipped without power adapters: "Environmental Progress Report." Apple, 2020, www.apple.com/environment/pdf/Apple_Environmental_Progress_Report_2021.pdf.

253 first known seawall: Gannon, Megan. "Oldest Known Seawall Discovered Along Submerged Mediterranean Villages." *Smithsonian,* 18 December 2019, www.smithsonianmag.com/history/oldest-known-seawall-discovered-along-submerged-mediterranean-villages-180973819.

254 Mount Tambora gave rise to: Oppenheimer, Clive. "Climatic, Environmental and Human Consequences of the Largest Known Historic Eruption: Tambora Volcano (Indonesia) 1815." *Progress in*

Physical Geography: Earth and Environment 27, no. 2, 2003, 230–59, doi:10.1191/0309133303pp379ra.

254 spread more than 800 miles from the eruption site: Stothers, R. B. "The Great Tambora Eruption in 1815 and Its Aftermath." *Science* 224, no. 4654, 1984, 1191–98, doi:10.1126/science.224.4654.1191.

254 "the year without a summer": Briffa, K. R., et al. "Influence of Volcanic Eruptions on Northern Hemisphere Summer Temperature over the Past 600 Years." *Nature* 393, no. 6684, 1998, 450–55, doi:10.1038/30943.

254 sulfur dioxide gave rise to acid rain: "Volcano Under the City: Deadly Volcanoes." *Nova*, 2021, www.pbs.org/wgbh/nova/volcanocity/dead-nf.html.

254 David Keith founded Harvard University's Solar Geoengineering Research Program: "David Keith." Harvard's Solar Geoengineering Research Program, 2021, geoengineering.environment.harvard.edu/people/david-keith.

254 cloudiness—all day, every day: Kolbert, Elizabeth. *Under a White Sky.* New York: Crown Publishers 2021.

254 "because the real world": Kolbert, Elizabeth. *Under a White Sky.*

255 In 2000, 371 cities worldwide: "The World's Cities in 2018." United Nations, 2018, www.un.org/en/events/citiesday/assets/pdf/the_worlds_cities_in_2018_data_booklet.pdf.

255 pours more cement in two years: Hawkins, Amy. "The Grey Wall of China: Inside the World's Concrete Superpower." *Guardian*, 28 February 2019, www.theguardian.com/cities/2019/feb/28/the-grey-wall-of-china-inside-the-worlds-concrete-superpower.

255 fifty "near-zero" urban carbon zones: Campbell, Iain, et al. "Near-Zero Carbon Zones in China." Rocky Mountain Institute, 2019, rmi.org/insight/near-zero-carbon-zones-in-china.

255 Palava City is projected to be home: Bagada, Kapil. "Palava: An Innovative Answer to India's Urbanisation Conundrum." Palava, 21 January 2019, www.palava.in/blogs/An-innovative-answer-to-Indias-Urbanisation-conundrum; Stone, Laurie. "Designing the City of the Future and the Pursuit of Happiness." RMI, 22 July 2020, rmi.org/designing-the-city-of-the-future-and-the-pursuit-of-happiness.

255 Lodha Group, India's largest real estate developer: Coan, Seth. "Designing the City of the Future and the Pursuit of Happiness." Rocky Mountain Institute, 16 September 2019, rmi.org/designing-the-city-of-the-future-and-the-pursuit-of-happiness.

Notes

256 Copenhagen has cut emissions: Sengupta, Somini, and Charlotte Fuente. "Copenhagen Wants to Show How Cities Can Fight Climate Change." *New York Times*, 25 March 2019, www.nytimes.com/2019/03/25/climate/copenhagen-climate-change.html.

256 60 percent of the city's commuters and students: Kirschbaum, Erik. "Copenhagen Has Taken Bicycle Commuting to a Whole New Level." *Los Angeles Times*, 8 August 2019, www.latimes.com/world-nation/story/2019-08-07/copenhagen-has-taken-bicycle-commuting-to-a-new-level.

256 lack of protected bike lanes: Monsere, Christopher, et al. "Lessons from the Green Lanes: Evaluating Protected Bike Lanes in the U.S." PDXScholar, June 2014, pdxscholar.library.pdx.edu/cgi/viewcontent.cgi?article=1143&context=cengin_fac.

256 Barcelona is famous for its car-free zones: O'Sullivan, Feargus. "Barcelona Will Supersize Its Car-Free 'Superblocks.'" *Bloomberg*, 11 November 2020, https://www.bloomberg.com/news/articles/2020-11-11/barcelona-s-new-car-free-superblock-will-be-big.

257 push 125,000 cars off the streets: Burgen, Stephen. "Barcelona to Open Southern Europe's Biggest Low-Emissions Zone." *Guardian*, 31 December 2019, www.theguardian.com/world/2019/dec/31/barcelona-to-open-southern-europes-biggest-low-emissions-zone.

257 the green plot ratio: Ong, Boon Lay. "Green Plot Ratio: An Ecological Measure for Architecture and Urban Planning." *Landscape and Urban Planning* 63, no. 4, 2003, 197–211, doi:10.1016/s0169-2046(02)00191-3.

257 sky terraces, communal planter boxes, and communal ground gardens: "Health and Medical Care." Urban Redevelopment Authority, 15 January 2020, www.ura.gov.sg/Corporate/Guidelines/Development-Control/Non-Residential/HMC/Greenery.

257 Ground-level greenery: Wong, Nyuk Hien, et al. "Greenery as a Mitigation and Adaptation Strategy to Urban Heat." *Nature Reviews Earth & Environment* 2, no. 3, 2021, 166–81, doi:10.1038/s43017-020-00129-5.

257 High Line industrial railway: The High Line, 11 June 2021, www.thehighline.org.

258 four hundred miles of protected bikeways: Shankman, Samantha. "10 Ways Michael Bloomberg Fundamentally Changed How New Yorkers Get Around." *Business Insider*, 7 August 2013, www.businessinsider.com/how-bloomberg-changed-nyc-transportation-2013-8?international=true&r=US&IR=T.

258 cars were banned from Fourteenth Street: Hu, Winnie, and Andrea Salcedo. "Cars All but Banned on One of Manhattan's Busiest Streets." *New York Times*, 3 October 2019, www.nytimes. com/2019/10/03/nyregion/car-ban-14th-street-manhattan.html.

258 15 percent reduction in CO_2 emissions: "Inventory of New York City Greenhouse Gas Emissions in 2016." City of New York, December 2017, www1.nyc.gov/assets/sustainability/downloads/pdf/publications/ GHG%20Inventory%20Report%20Emission%20Year%202016.pdf.

258 80 percent by 2050: "New York City's Roadmap to 80 × 50." New York City Mayor's Office of Sustainability, www1.nyc.gov/assets/ sustainability/downloads/pdf/publications/New%20York%20City's%20 Roadmap%20to%2080%20x%2050.pdf. Accessed 23 June 2021.

258 "If you can make it there, you can make it anywhere": Sinatra, Frank. "(Theme from) New York New York." *Trilogy: Past Present Future*. Capitol, June 21, 1977.

Chapter 10: Invest!

261 "In other words, John Doerr may once again": Eilperin, Juliet. "Why the Clean Tech Boom Went Bust." *Wired*, 20 January 2012, www. wired.com/2012/01/ff_solyndra.

261 "disastrous detour into renewable energy": Marinova, Polina. "How the Kleiner Perkins Empire Fell." Fortune, 23 April 2019, fortune.com/longform/kleiner-perkins-vc-fall.

263 "the misfits, the rebels": "The Iconic Think Different Apple Commercial Narrated by Steve Jobs." Farnam Street, 5 February 2021, fs.blog/2016/03/steve-jobs-crazy-ones.

263 Beyond Meat skyrocketed in value: Shanker, Deena, et al. "Beyond Meat's Value Soars to $3.8 Billion in Year's Top U.S. IPO." *Bloomberg*, 1 May 2019, https://www.bloomberg.com/news/ articles/2019-05-01/beyond-meat-ipo-raises-241-million-as-veggie-foods-grow-fast.

265 eliminate financial subsidies for fossil fuels: Taylor, Michael. "Evolution in the Global Energy Transformation to 2050." International Renewable Energy Agency, 2020, www.irena.org/-/ media/Files/IRENA/Agency/Publication/2020/Apr/IRENA_Energy_ subsidies_2020.pdf.

265 what it now allocates to the National Institutes of Health: "National Institutes of Health (NIH) Funding: FY1995-FY2021." Congressional Research Service, updated 12 May 2020, fas.org/sgp/ crs/misc/R43341.pdf.

Notes

265 foundations that control nearly $1.5 trillion globally: Johnson, Paula D. "Global Philanthropy Report: Global Foundation Sector." Harvard University's John F. Kennedy School of Government, April 2018, cpl.hks.harvard.edu/files/cpl/files/global_philanthropy_report_final_april_2018.pdf.

266 to the tune of $447 billion: Taylor, Michael. "Evolution in the Global Energy Transformation to 2050." International Renewable Energy Agency, 2020, www.irena.org/-/media/Files/IRENA/Agency/Publication/2020/Apr/IRENA_Energy_subsidies_2020.pdf.

266 "No fossil fuel has borne": Smil, Vaclav. *Energy Myths and Realities*. Washington, D.C., AEI Press, 2010.

266 more than $3 trillion a year: Taylor, Michael. "Energy Subsidies: Evolution in the Global Energy Transformation to 2050." Irena, 2020, www.irena.org/-/media/Files/IRENA/Agency/Publication/2020/Apr/IRENA_Energy_subsidies_2020.pdf.

266 employment in the U.S. solar industry grew: "10th Annual National Solar Jobs Census 2019." Solar Foundation, February 2020, www.thesolarfoundation.org/wp-content/uploads/2020/03/SolarJobsCensus2019.pdf.

267 lent or guaranteed more than $35 billion: "Financing Options for Energy Infrastructure." Loan Programs Office, Department of Energy, May 2020, www.energy.gov/sites/default/files/2020/05/f74/DOE-LPO-Brochure-May2020.pdf.

267 a $465 million DOE loan: "TESLA." 2021, Loan Programs Office, Department of Energy, www.energy.gov/lpo/tesla.

267 Chinese government allocated an average of $77 billion: Koty, Alexander Chipman. "China's Carbon Neutrality Pledge: Opportunities for Foreign Investment." China Briefing News, 6 May 2021, www.china-briefing.com/news/chinas-carbon-neutrality-pledge-new-opportunities-for-foreign-investment-in-renewable-energy.

268 China now owns 70 percent: Rapoza, Kenneth. "How China's Solar Industry Is Set Up to Be the New Green OPEC." *Forbes*, 14 March 2021, www.forbes.com/sites/kenrapoza/2021/03/14/how-chinas-solar-industry-is-set-up-to-be-the-new-green-opec/?sh=2cfec9f91446.

268 $52 billion in global venture capital: Analysis by Ryan Panchadsaram, data from Crunchbase.com.

269 less than $400 million across eighty climate deals: Devashree, Saha and Mark Muro. "Cleantech Venture Capital: Continued Declines and Narrow Geography Limit Prospects." Brookings Institution, 1 December 2017, www.brookings.edu/research/cleantech-venture-capital-continued-declines-and-narrow-geography-limit-prospects.

269 almost $7 billion had been plowed into four hundred deals: Devashree, Saha and Mark Muro. "Cleantech Venture Capital: Continued Declines and Narrow Geography Limit Prospects."

272 Investments in clean technologies: "Technology Radar, Climate-Tech Investing." BloombergNEF, 16 February 2021, www.bnef.com/login?r=%2Finsights%2F25571%2Fview.

275 Investor enthusiasm is surging: Special Purpose Acquisition Company Database: SPAC Research. www.spacresearch.com.

275 Twenty percent are energy or climate related: Guggenheim Sustainability SPAC Market Update. June 6, 2021.

275 Through "creative destruction": Alm, Richard, and W. Michael Cox. "Creative Destruction." Library of Economics and Liberty, 2019, www.econlib.org/library/Enc/CreativeDestruction.html.

279 clean energy project financing: "Energy Transition Investment." BloombergNEF, www.bnef.com/interactive-datasets/2d5d59acd9000005. Accessed 14 June 2021.

281 The company hit that target three years early: "Achieving Our 100% Renewable Energy Purchasing Goal and Going Beyond." Google, December 2016, static.googleusercontent.com/media/www.google.com/en//green/pdf/achieving-100-renewable-energy-purchasing goal.pdf.

282 largest sustainability bond in corporate history: Porat, Ruth. "Alphabet Issues Sustainability Bonds to Support Environmental and Social Initiatives." Google, 4 August 2020, blog.google/alphabet/alphabet-issues-sustainability-bonds-support-environmental-and-social-initiatives.

289 "The green economy is poised": Kenis, Anneleen, and Matthias Lievens. *The Limits of the Green Economy: From Re-Inventing Capitalism to Re-Politicising the Present* (Routledge Studies in Environmental Policy). Abingdon, Oxfordshire, U.K.: Routledge, 2017.

289 philanthropic giving totaled $730 billion in 2019: Roeyer, Hannah, et al. "Funding Trends: Climate Change Mitigation Philanthropy." ClimateWorks Foundation, 11 June 2021, www.climateworks.org/report/funding-trends-climate-change-mitigation-philanthropy.

291 the IKEA Foundation: "FAQ." IKEA Foundation, 6 January 2021, ikeafoundation.org/faq.

291 The list of grantees: Palmer, Annie. "Jeff Bezos Names First Recipients of His $10 Billion Earth Fund for Combating Climate Change." CNBC, 16 November 2020, www.cnbc.com/2020/11/16/jeff-bezos-names-first-recipients-of-his-10-billion-earth-fund.html.

Notes

298 90 percent of the state's electricity: Daigneau, Elizabeth. "From Worst to First: Can Hawaii Eliminate Fossil Fuels?" Governing, 30 June 2016, www.governing.com/archive/gov-hawaii-fossil-fuels-renewable-energy.html.

300 an estimated ten thousand people: "Hawaii Clean Energy Initiative 2008-2018." Hawai'i Clean Energy Initiative, Jan. 2018, energy.hawaii.gov/wp-content/uploads/2021/01/HCEI-10Years.pdf.

300 the state exceeded its 30 percent clean energy goal: "Hawaiian Electric Hits Nearly 35% Renewable Energy, Exceeding State Mandate." Hawaiian Electric, 15 February 2021, www.hawaiianelectric.com/hawaiian-electric-hits-nearly-35-percent-renewable-energy-exceeding-state-mandate.

Conclusion

304 a wave of new technologies: "Bell Labs." Engineering and Technology History Wiki, 1 August 2016, ethw.org/Bell_Labs.

304 historic mobilization of wartime manufacturing: Herman, Arthur. *Freedom's Forge: How American Business Produced Victory in World War II*. New York: Random House, 2012.

306 "fundamental rights to a human future": Connolly, Kate. "'Historic' German Ruling Says Climate Goals Not Tough Enough." *Guardian*, 29 April 2021, www.theguardian.com/world/2021/apr/29/historic-german-ruling-says-climate-goals-not-tough-enough.

Acknowledgments

312 Call me Al: "Paul Simon—You Can Call Me Al (Official Video)." YouTube, 16 June 2011, www.youtube.com/watch?v=uq-gYOrU8bA.

Index

Index

Index

Index

Index

Index

Image Credits

Introduction

Page xvii	Photo: President Franklin D. Roosevelt grabbed a cocktail napkin	Photo courtesy of Jay S. Walker's Library of Human Imagination, Ridgefield, Connecticut.
Page xix	Infographic: Carbon dioxide in the atmosphere has risen dramatically over the last 200 years	Adapted from Max Roser and Hannah Ritchie, "Atmospheric Concentrations," Our World in Data, accessed June 2021, ourworldindata.org/atmospheric-concentrations.
Page xxi	Infographic: Policy scenarios, emissions, and temperature range projections	Adapted from "Temperatures," Climate Action Tracker, 4 May 2021, climateactiontracker.org/global/temperatures.
Page xxiv	Infographic: How our greenhouse gas emissions add up	Adapted from UNEP and UNEP DTU Partnership, "UNEP Report—The Emissions Gap Report 2020," *Management of Environmental Quality: An International Journal*, 2020, https://www.unep.org/emissions-gap-report-2020.
Page xxviii	Photo: Slide from Exxon Research and Engineering Co. from 1978	Slide by James F. Black and Exxon Research and Engineering Co.

Chapter One: Electrify Transportation

Page 2	Infographic: Electric vehicles are growing in popularity	Adapted from "EV Sales," BloombergNEF, accessed 13 June 2021, www.bnef.com/interactive-datasets/2d5d59acd9000014?data-hub=11.
Pages 4–5	Photo: Cars on road	Photo by Michael Gancharuk/Shutterstock.com.
Page 8	Infographic: Miles driven by electric vehicles lags across categories	Adapted from Max Roser and Hannah Ritchie, "Technological Progress," Our World in Data, 11 May 2013, ourworldindata.org/technological-progress; Wikipedia contributors, "Transistor Count," Wikipedia, 1 June 2021, en.wikipedia.org/wiki/Transistor_count.
Page 15	Photo: BVD fleet/charging in China	Photo by Qilai Shen/*Bloomberg* via Getty Images.
Page 22	Photo: Proterra bus	Photo courtesy of Proterra.

Image Credits

Page 23	Infographic: Moore's Law demonstrates exponential growth	Adapted from Max Roser and Hannah Ritchie, "Technological Progress," Our World in Data, 11 May 2013, ourworld indata.org/technological-progress; Wikipedia contributors, "Transistor Count," Wikipedia, 1 June 2021, en.wiki pedia.org/wiki/Transistor_count.
Page 24	Infographic: Wright's Law in Action: Solar	Adapted from Max Roser, "Why Did Renewables Become so Cheap so Fast? And What Can We Do to Use This Global Opportunity for Green Growth?" Our World in Data, 1 December 2020, ourworld indata.org/cheap-renewables-growth.
Page 25	Infographic: Wright's Law in Action: Batteries	Adapted from "Evolution of Li-Ion Battery Price, 1995-2019—Charts—Data & Statistics," IEA, 30 June 2020, www.iea. org/data-and-statistics/charts/evolution-of-li-ion-battery-price-1995-2019.
Page 27	Photo: President Biden test-driving an all-electric F-150.	Photo by NICHOLAS KAMM/AFP via Getty Images.

Chapter Two: Decarbonize the Grid

Page 31	Infographic: As solar prices fell, demand soared	Adapted from "The Solar Pricing Struggle," Renewable Energy World, 23 August 2013, www.renewableenergyworld.com/solar/the-solar-pricing-struggle/#gref.
Page 36	Infographic: Renewables are winning as prices drop and installed capacity grows	Adapted from Max Roser, "Why Did Renewables Become so Cheap so Fast? And What Can We Do to Use This Global Opportunity for Green Growth?" Our World in Data, 1 December 2020, ourworld indata.org/cheap-renewables-growth.
Page 38	Infographic: Europe generates more economic output with fewer emissions	Adapted from "Statistical Review of World Energy," BP, 2020, www.bp.com/en/global/corporate/energy-economics/statistical-re view-of-world-energy.html; "GDP per Capita (Current US$)," The World Bank, 2021, data.worldbank.org/indicator/NY.GDP.PCAP.CD; "Population, Total," The World Bank, 2021, data.worldbank.org/indicator/SP.POP.TOTL.

Page 42	**Photo: Sunrun**	Photo by Mel Melcon/*Los Angeles Times* via Getty Images.
Page 46	**Photo: Vindeby Offshore Wind Farm**	Photo courtesy of Wind Denmark, formerly the Danish Wind Industry Association.
Pages 50–51	**Infographic: Size Matters: Ørsted's Wind Turbines Produce More Power as They Get Larger**	Adapted from "Ørsted.Com—Love Your Home," Ørsted, accessed 13 June 2021, Orsted.com.
Page 57	**Photo: Induction range**	Photo by iStock.com/LightFieldStudios.

Chapter Three: Fix Food

Pages 64–65	**Photo: Al Gore from Edible**	Photo by Hartmann Studios.
Page 66	**Photo: Dust Bowl**	Photo by PhotoQuest via Getty Images.
Page 69	**Infographic: Less tilling creates healthier roots and soil**	Adapted from Ontario Ministry of Agriculture, Food and Rural Affairs, "No-Till: Making it Work," Best Management Practices Series BMP11E, Government of Ontario, Canada, 2008, available online at: http://www.omafra.gov.on.ca/english/environment/bmp/no-till.htm (verified 14 January 2009). ©2008 Queen's Printer for Ontario. Adapted by Joel Gruver, Western Illinois University.
Page 70	**Infographic: Regenerative agriculture explained**	Adapted from "Can Regenerative Agriculture Replace Conventional Farming?" EIT Food, accessed 22 June 2021, www.eitfood.eu/blog/post/can-regenerative-agriculture-replace-conventional-farming.
Page 73	**Infographic: Emissions by kilogram of food**	Adapted from "You Want to Reduce the Carbon Footprint of Your Food? Focus on What You Eat, Not Whether Your Food Is Local," Our World in Data, 24 January 2020, ourworldindata.org/food-choice-vs-eating-local.

Page 77	Infographic: A climate-friendly diet: lots of fruits and vegetables, limited animal-based proteins	Adapted from "Which Countries Have Included Sustainability Within Their National Dietary Guidelines?" Dietary Guidelines, Plant-Based Living Initiative, accessed 22 June 2021, themouthful.org/article-sustainable-dietary-guidelines.
Page 81	Photo: Beyond Burger	Photo courtesy of Beyond Burger.
Pages 84–85	Photo: Rice cultivation	Photo by BIJU BORO/AFP via Getty Images.

Chapter Four: Protect Nature

Pages 92–93	Infographic: Carbon moves through the land, atmosphere, and oceans	Adapted from "The Carbon Cycle," NASA: Earth Observatory, 2020, earthobservatory.nasa.gov/features/CarbonCycle.
Pages 98–99	Photo: Forest loss	Photo by Universal Images Group via Getty Images.
Page 100	Infographic: Tropical deforestation drives global forest loss	Adapted from Hannah Ritchie, "Deforestation and Forest Loss," Our World in Data, 2020, ourworldindata.org/deforestation.
Page 103	Photo: RainForest Alliance	Photo courtesy of the RainForest Alliance.
Page 110	Photo: Deep sea bottom trawling	Photo by Jeff J Mitchell via Getty Images.
Page 112	Photo: Kelp farm	Photo by Gregory Rec/*Portland Press Herald* via Getty Images.
Page 113	Photo: Peat	Photo by Muhammad A.F/Anadolu Agency via Getty Images.

Chapter Five: Clean Up Industry

Page 118	Photo: James Wakibia	Photo courtesy of James Wakibia.
Page 122	Photo: PLA	Photo by Brian Brainerd/*The Denver Post* via Getty Images.
Page 123	Infographic: Instructive labels can help consumers make the right recycling decisions	Adapted from "How2Recycle—A Smarter Label System," How2Recycle, accessed 17 June 2021, how2recycle.info.

Page 124	Infographic: Plastic pollutes at every stage of its lifestyle	Adapted from Roland Geyer et al., "Production, Use, and Fate of All Plastics Ever Made," *Science Advances*, vol. 3, no. 7, 2017, p. e1700782. Crossref, doi:10.1126/sciadv.1700782.
Page 127	Infographic: Replacing fossil fuels in many industrial processes is possible	Adapted from Occo Roelofsen et al., "Plugging In: What Electrification Can Do for Industry," McKinsey & Company, 28 May 2020, www.mckinsey.com/indus tries/electric-power-ad-natural-gas/ our-insights/plugging-in-what-electrifica tion-can-do-for-industry.
Pages 136–37	Photo: Steel production	Photo by Sean Gallup via Getty Images.

Chapter Six: Remove Carbon

Page 142	Infographic: Carbon removal: the many ways to do it	Adapted from J. Wilcox et al., "CDR Primer," CDR, 2021, cdrprimer.org/read/chapter-1.
Pages 146–47	Photo: Climeworks	Photo courtesy of Climeworks.
Page 152	Photo: Watershed	Photo courtesy of Watershed.

Chapter Seven: Win Politics and Policy

Pages 160–61	Photo: Paris Agreement	Photo by Arnaud BOUISSOU/COP21/Anadolu Agency via Getty Images.
Page 169	Infographic: Over two thirds of emissions come from just five countries	Adapted from UNEP and UNEP DTU Partnership, "UNEP Report—The Emissions Gap Report 2020," *Management of Environmental Quality: An International Journal*, 2020, https://www.unep.org/emissions-gap-report-2020.
Page 173	Photo: China	Photo by Costfoto/Barcroft Media via Getty Images.
Page 178	Photo: India solar	Photo by Pramod Thakur/*Hindustan Times* via Getty Images.

Chapter Eight: Turn Movements into Action

Pages 192–93	Photo: Greta Thunberg	Photo by Sarah Silbiger via Getty Images.
Page 198	Photo: Sunrise Movement	Photo by Rachael Warriner/Shutterstock.com.
Page 202	Photo: Beyond Coal	Photo courtesy of the Sierra Club.
Page 204	Photo: Arnold Schwarzenegger meets with Michael Bloomberg	Photo by Susan Watts-Pool via Getty Images.
Page 214	Photo: Walmart Sustainability	Graphic courtesy of Walmart.

Chapter Nine: Innovate!

Page 241	Photo: Breakthrough Energy Ventures	Photo courtesy of Breakthrough Energy Ventures.
Pages 258–59	Photo: New York City's High Line	Photo by Alexander Spatari via Getty Images.

Chapter Ten: Invest!

Page 268	Infographic: The first decade of early stage cleantech investing: from boom to bust	Adapted from Benjamin Gaddy et al., "Venture Capital and Cleantech: The Wrong Model for Clean Energy Innovation," MIT Energy Initiative, July 2016, energy.mit.edu/wp-content/uploads/2016/07/MITEI-WP-2016-06.pdf.
Page 280	Infographic: Project financing for clean energy is on the rise	Adapted from "Energy Transition Investment," BloombergNEF, accessed 14 June 2021, www.bnef.com/interactive-datasets/2d5d59acd9000005.
Page 290	Infographic: Foundations are rising to the moment to fight climate change	Adapted from Climate Leadership Initiative, climatelead.org.